RF & Microwave Engineering
VOLUME IV

The Foundation of
High Frequency Electronics
(CD-ROM Download)

Matthew M. Radmanesh, Ph.D.

Why This Book?

Waves of all types, shapes and frequencies abound all around us and are very commonplace in our daily existence; radio waves, sound waves, sunlight, heat waves, lasers, microwaves, are just a few examples.

In this book, we are laying down the essential foundation as a minimum requirement for professional engineers to understand the subject of electromagnetism and its subset "RF & Microwaves."

Moreover, this book provides the technical fundamentals that should be utilized by all professional engineers for the design of all high frequency circuits with confidence and composure. This would prove to be the cornerstone of all future inventions and developments in this arena.

This book simplifies the subject of electricity and electronics, particularly the high frequency electronics and RF/Microwaves, making it accessible to the general reader by bringing the fundamentals to the forefront, thus ushering in a new era of incredible developments for Mankind!

CD-ROM Download provides a powerful interactive software for circuit analysis or design.

For more information please visit: **www.KRCbooks.com**

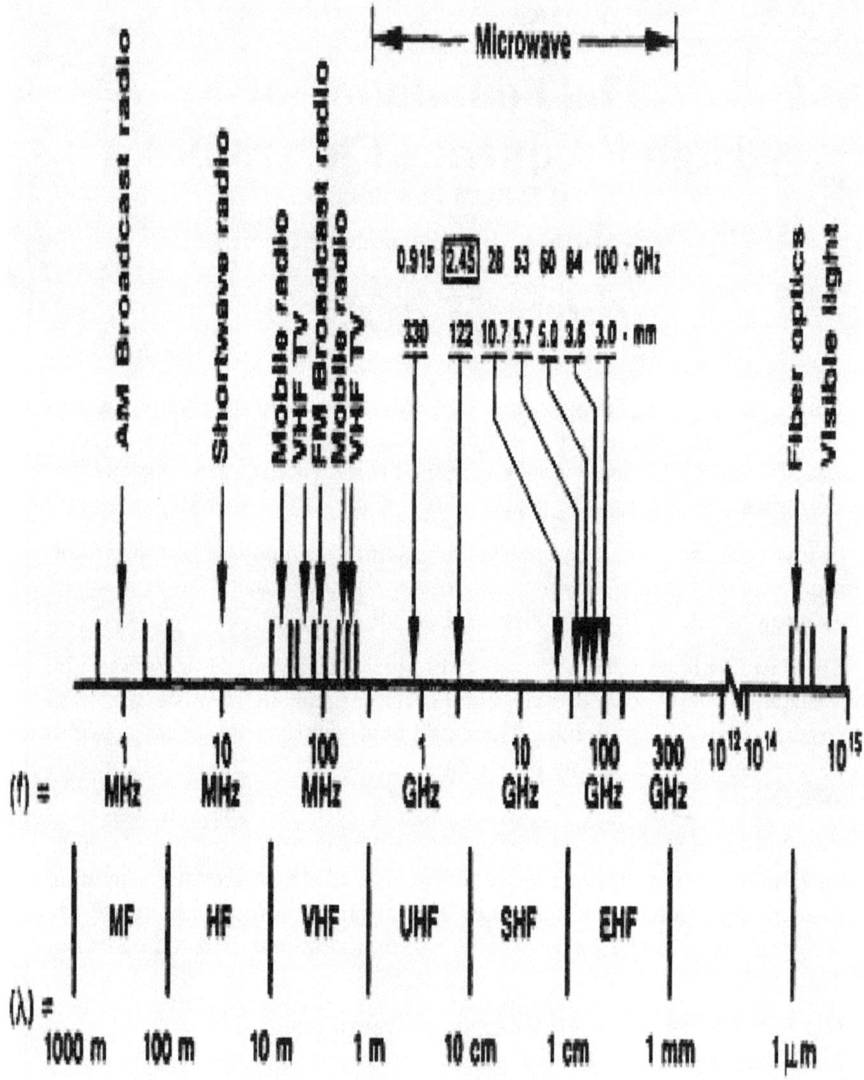

The Spectrum of Electromagnetic (EM) Radiations.

RF & Microwave
Engineering
VOLUME IV

The Foundation of High Frequency Electronics
(CD-ROM Download)

Matthew M. Radmanesh, Ph.D.

Professor of Electrical
& Computer Engineering,
California State University,
Northridge

ISBN: 9798639318573 (sc)

Library of Congress Control Number: 2007901367

Matthew M. Radmanesh, Ph.D.   v

*Dedicated to
the Forefathers of
RF & Microwaves!*

Contents

Chapter 6 SMITH CHART APPLICATIONS   215

****

PART II CIRCUIT DESIGN ESSENTIALS 271

Chapter 7   DESIGN OF MATCHING NETWORKS 273

Chapter 8   STABILITY IN ACTIVE NETWORKS 317

Chapter 9   GAIN CONCEPTS IN AMPLIFIERS 339

****

PART III APPENDICES   397

Preface

This book is the fourth installment of the RF & Microwave Engineering series, and if followed diligently from the first volume to the present will turn the "not-know" of any individual gradually into a state of "deep knowingness" about electricity, waves, and the universe at large.

The initial motivation was to bring the basics to the forefront and orient the reader in such a way that he or she can think with these fundamentals correctly. This eventually led to writing the first manuscript several years ago and then the final preparation of this book at present.

In preparing this book, the emphasis was shifted from rigorous and sophisticated mathematical solutions of Maxwell's equations and instead has been aptly placed on RF and microwave circuit analysis and design principles using simple concepts while emphasizing the basics all the way.

It is an interestingly uncommon book written primarily for the technical man. It is intended to serve several classes of our society
   a) The technically versed individuals,
   b) The microwave design engineers,
   c) The professional scientists,
   d) The RF/microwave students.

This book will surely serve also an important class of our society— the technical inventors who are looking for inspirations and new ideas to imbue them with enough understanding to finalize and materialize their thoughts into reality.

It is written for the technically versed individual who desires to learn about the microwave industry more in depth. It is intended to lift the aura of "black magic" surrounding the world of microwaves, to enlighten and demystify the subject of sciences in the minds of ordinary individuals.

This book is intended to be used as a basic course in RF and microwave electronics engineering for senior-level or graduate students and should serve as an excellent reference guide for the practicing RF and microwave engineers in the field as well.

A list of symbols used in each chapter and a series of problems are included at the end of each chapter to help the reader gain a fuller understanding of the presented materials.

The book ends with a glossary of technical terms and several important appendixes. These appendixes cover physical constants and other important data needed in the analysis or design process.

The Importance of Work

Within the confines of this book, one is given a chance for the first time to take an in-depth look and inspect first-hand the fundamentals of RF and Microwaves where its dominance and imposing characteristics in all aspects of our existence is truly remarkable.

The basics of this subject are stated in simple terms, whereas the clear explanations in this work express the powerful principles lucidly and dynamically, providing an unforgettable impression in the reader's mind.

In summary, this book has some unique strengths which make it different from prior literature and an attractive reference for the reader. For example, there is a shift of emphasis from rigorous mathematical solutions of Maxwell's equations, and instead has been aptly placed on simple yet fundamental concepts that underlie these equations. This shift of emphasis will promote a deeper understanding of the electronics, particularly at RF/Microwave frequencies.

New technical terms are precisely defined as they are first introduced, thereby keeping the subject matter in focus and preventing misunderstanding, and finally the abundant use of graphical illustrations and diagrams brings a great deal of clarity and conceptual understanding, enabling difficult concepts to be understood with ease.

A comprehensive glossary of technical terms provided at the end is a great aid in understanding RF and Microwaves and makes this book invaluable for anyone aspiring to master this field of study. The appendix section provides a list of all symbols used in the book, information on many physical constants, mathematical identities, and generally known laws and makes them sufficiently accessible for easy reference.

The Author's Goals
The author intends to bring forth a milestone achievement that can be summed up as:

To bring about a deeper awareness of how the waves particularly RF and microwaves affect us in many ways on a constant basis and how a basic knowledge of the subject can lead to a plethora of applications.

The technical reader will be invited to examine a series of basic materials that will enable him/her to understand this powerful far better than ever before. He/she will be exposed to materials of considerable significance, which surely would open up the gates of knowledge along with a wider horizon of understanding.

Any communications in the way of a healthy criticism and/or correction are welcome. Moreover, the author considers it one of the most rewarding things to have others grasp the materials in all of their simplicity and increase their own potential survival in this universe and help others to achieve their goals.

In the process, this helps make Man take control of his own destiny, without being shackled by the chains of higher authority or superstition.

Therefore, in order to improve the quality of this work, the author would like to have all comments or suggestions be sent directly to:
Dr. Matthew M. Radmanesh
18111 Nordhoff Street,

Department of Electrical and Computer Engineering,
California State University, Northridge, California 91330.
Or email to: **matt@csun.edu**
You can also check out these related websites for more information:
www.KRCbooks.com

Matthew M. Radmanesh, Ph.D.

Acknowledgements

Special thanks are due to the great eastern and western philosophers, whose philosophical works made many scientific discoveries concerning our physical universe possible, and allowed the physical sciences to soar to new heights where it can enrich Man's lives and enable him to achieve higher goals! These works initially inspired the author to write the original manuscript to simplify Electromagnetics, particularly RF and microwaves arena.

This book is filled with many discovered fundamentals that have been meticulously gathered and distilled through the use of the scientific methodology in order to obtain a workable set of design methodologies as well as operating rules for accurate analysis and rapid design work.

The author would like to thank many of his professional colleagues and mentors through the years, especially Dr. George Haddad, Dr. C. M. Chu, Dr. Emmet Leith, Dr. Chen-to Tai, Dr. Thomas Senior, Dr. Dean Peterson, the early mentors at the University of Michigan (Ann Arbor campus); Dr. M. Torfeh (Kettering University, Flint, MI), my great friend in Michigan; My great colleagues especially Dr. Amini, Dr. Roosta, Dr. Rengarajan, Dr. El Naga, Dr. Sedghisigarchi, Dr. Mirzaei, Prof. Vazan, and the late Dr. Gillespie at the ECE Department, California State University, Northridge (CSUN). Their friendship, support, and collegiality through the years are greatly appreciated.

My deep gratitude belongs to Jaime Rodriguez, an extremely loyal friend who has been instrumental in creating a peaceful environment where this book could be researched and assembled into a coherent piece of work.

Moreover, there are some very invaluable individuals who need to be thanked greatly, especially Danny Abergel whose work in providing great assistance in finance has contributed greatly to an affluent scene that is far above average. Also, contributions from

Farhad Hanasab and Jasmin Jimenez in securing my world from harmful intents and unholy men, are highly appreciated.

The author expresses deep appreciation toward his lovely wife, Jane, and his brilliant son, William, for making life encouraging and for providing much-needed support. They filled life with laughter and much happiness during this intense project.

Finally, the author's deepest gratitude belongs to his parents Mary and the late Dr. G. H. Radmanesh, for their true love and unconditional support. They were the first who instilled a sense of inquiry and curiosity toward life's higher principles and filled the author's mind with a thirst for true knowledge. They set the bar high for achievement in author's early years, thus this work would never have been possible without their initial goal setting as well as their love for mankind, and is therefore wholeheartedly dedicated to them for their contribution to the author's life!

Matthew M. Radmanesh, Ph.D.
Dept. of Electrical and Computer Engineering,
California State University, Northridge,
May 2020

What Sets This Book Apart

What sets this book apart is the fact that it is not just another microwave book describing scientific facts and phenomena. It would surely be redundant since that task has been done many times over with much more elegant prose and brighter narrators.

Here is a book where, for the first time, we have undertaken the task of breaking the subject of RF and microwaves into its many components.

Just like the light phenomenon, which was made to be a subset of electricity by James Clerk Maxwell, thus revolutionizing our world, so would this book by bringing about a new era of incredible design and applications in the microwave world!

This book is the road map of circuit design for high frequency signals where it, through the use of numerous examples, presents detailed and yet powerful analysis techniques that anyone can learn!

The Volume IV of the series, lays the technical foundation of high frequency electronics and provides the know-how to design any microwave circuit, whether passive or active. The list of circuit applications is endless. Prominent amongst them are filter circuits, oscillators, resonant circuits, amplifiers, mixers, detectors, so on and so forth.

It gradually dawns upon one that the knowledge contained within the confine of this book could be one's biggest asset in the design process of sophisticated RF and Microwave passive or active circuits, as these will be treated later in Volumes V and VI.

⊱ ᔥ ✶ ❀ ❍ ❀ ✶ ᔥ ⊰

Point of Caution

To understand this book fully and get the maximum possible benefit from reading it, the reader needs to familiarize himself with a certain amount of information, which will act as the prerequisite to this work.

There are three volumes that have preceded the current work, and thus can act as the necessary prerequisites that could help the reader greatly in his progress toward a greater understanding. The sequence of familiarity is as follows:

The Prerequisites

Volume I: deals with the engineering foundation of our universe. The reader needs to have a "**Low level of familiarity**."

Volume II: deals with the scientific foundation of electrical engineering. The reader needs to have a "**Medium level of familiarity**."

Volume III: deals with the principles of electricity and electromagnetics. The reader needs to have a "**High level of familiarity**."

Neglect in acquainting oneself with the above prerequisites in the prescribed proportions would prove to be a costly mistake as the presented materials in this work will be either greatly misunderstood, misapplied, or cast aside as useless and unworkable!

ᠣ~ 𐒐 ✳ ❀ ❌ ❀ ✳ ᠺ ~ᠣ

PART I

THE
FUNDAMENTALS

The Prerequisites

The sequence of familiarity is as follows:

Volume I: deals with the engineering foundation of our universe. The reader needs to have a "**Low level of familiarity**."

Volume II: deals with the scientific foundation of electrical engineering. The reader needs to have a "**Medium level of familiarity**."

Volume III: deals with the principles of electricity and electromagnetics. The reader needs to have a "**High level of familiarity**."

Neglect in acquainting oneself with the above prerequisites in the prescribed proportions would prove to be a costly mistake as the presented materials in this work will be either greatly misunderstood, misapplied, or cast aside as useless and unworkable!

CHAPTER 1

Basics of RF and Microwaves

1.1 INTRODUCTION

By "RF" we mean "Radio Frequency" signals propagating at frequencies in the range of 300 MHz to 1 GHz, whereas the waves at frequencies ranging from 1 GHz to 300 GHz are generally known as "microwaves."

This chapter lays the foundation for understanding higher frequency wave phenomena and compartments the task of active circuit design RF/MW frequencies into specific concept blocks. The concept blocks create a gradient approach to understanding and designing RF/MW circuits and represent specific realms of knowledge that need to be mastered in order to become an accomplished designer.

Before we proceed into analysis and description of these types of waves we need to consider why RF/Microwaves as a subject have become so important as to be placed at the forefront of our modern technology; and furthermore, we need to expand our minds to the many possibilities that these signals can provide for peaceful practices by exploring various commercial applications useful to mankind.

1.1.1 A Short History of RF & Microwaves

Circa 1864-1873, James Clark Maxwell integrated the entire man's extant knowledge on electricity and magnetism and introduced a series of four coherent and self-consistent equations that described the behavior of electric and magnetic fields on a classical level.

This was the beginning of microwave engineering as presented in a treatise by Maxwell at that time. He predicted, purely from a mathematical and theoretical standpoint, the existence of electromagnetic wave propagation and that light was also a form of electromagnetic energy, which were both completely new concepts at that time.

From 1885 to 1887, Oliver Heaviside simplified Maxwell's work in his published papers. From 1887 to 1891, a German physics professor by the name of Heinrich Hertz verified Maxwell's predictions experimentally and demonstrated the propagation of electromagnetic waves. He also investigated wave propagation phenomena along transmission lines and antennas and developed several useful structures. He could be called the first microwave engineer.

Marconi tried to commercialize Radio at a much lower frequency for long-distance communications, but he had a business interest in all of his work and developments. So this was not a purely scientific endeavor.

The possibility of electromagnetic wave propagation inside a hollow metal tube was never investigated by Hertz or Heaviside, since it was felt that two conductors were necessary for the transfer of electromagnetic waves or energy. In 1897, Lord Rayleigh mathematically showed that electromagnetic wave propagation was possible in a waveguide, both circular and rectangular. He showed that there is an infinite set of modes of the TE and TM type possible, each with its own cut-off frequency. These were all theoretical predictions with no experimental verifications.

From 1897 to 1936, waveguide was essentially forgotten until it was rediscovered by two men, George Southworth (AT&T) and W. L. Barron (MIT) who showed experimentally that waveguide could be used as a small bandwidth transmission media, capable of carrying high power signals.

With the invention of the transistor in the 1950s and the advent of microwave integrated circuits in the 1960s, the concept of a microwave system on a chip became a reality. There have been many other developments, mostly in terms of application mass, which has made RF and microwaves an enormously useful and popular subject.

Maxwell equations lay the foundation and laws of the science of electromagnetics, of which the field of RF and microwaves is a small subset. Due to the exact and all-encompassing nature of these laws in predicting electromagnetic phenomena along with the great body of analytical and experimental investigations performed since then, we can consider the field of RF and microwave engineering a "mature discipline" at this time.

1.1.2 Applications of Maxwell's Equations

Standard circuit theory can not be used at RF and particularly at microwave frequencies. This is because the dimensions of the device or components are comparable to the wavelength, which means that the phase of an electrical signal (e.g. a current or voltage) changes significantly over the physical length of the device or component. Thus the use of Maxwell's equations at these higher frequencies becomes imperative.

In contrast, the signal wavelengths at lower frequencies are so much larger than the device or component dimensions that there is negligible variation in phase across the dimensions of the circuit. Thus Maxwell's equations simplify into basic circuit theory that most circuit technicians are familiar with use in their analysis and design.

At the other extreme of the frequency range lies the optical field, where the wavelength is much smaller than the device or circuit dimensions. In this case, Maxwell's equations simplify into a subject commonly referred to as geometrical optics which treats light as a ray traveling on a straight line. These optical techniques may be applied successfully to the analysis of very high microwave frequencies (e.g. high millimeter-wave range), where they are referred to as "quasi-optical". Of course, it should be noted that further application of Maxwell's equations leads to an advanced field of optics called "physical optics or Fourier optics", which treats light as a wave and explains such phenomena as diffraction and interference, where geometrical fails completely.

The important conclusion to be drawn from this discussion is that Maxwell's equations present a unified theory of analysis for any system at any frequency, provided one uses appropriate simplifications when the wavelengths involved are a) much larger, b) comparable to, or c) much smaller than the circuit dimensions.

1.1.3 Properties of RF and Microwaves

An important property of signals at RF, and particularly at higher microwave frequencies is their great capacity in carrying information. This is due to the existence of large bandwidths that are available at these high frequencies. For example a 10% bandwidth at 60 MHz carrier signal is 6 MHz which is approximately one TV channel of information; on the other hand 10% of a microwave carrier signal at 60 GHz is 6 GHz which is equivalent to 1000 TV channels.

Another property of microwaves is that they travel by line of sight, very much like traveling of light rays as described in the field of geometrical optics. Furthermore, unlike the lower frequency signals, the microwave signals are not bent by the ionosphere. Thus the use of line-of-sight communication towers or links on the ground and orbiting satellites around the Globe are a necessity for local or global communications.

A very important civilian as well as military instrument is Radar. The concept of Radar is based upon Radar cross-section which is the effective reflection area of the target. Target's visibility greatly depends on the target's electrical size which is a function of the incident signal's wavelength. Microwave frequencies form the ideal signal band for Radar applications.

Of course, another important advantage of using microwaves in Radars is the availability of higher antenna gains as the frequency increases for any given physical antenna size. This is because the antenna gain is proportional to the electrical size of the antenna, which becomes larger as the frequency is increased in the microwave band. The key factor in all this is that microwave wavelengths are comparable to the physical size of the transmitting antenna as well as the target.

There is a fourth and yet a very important property of microwaves and that is the molecular, atomic, and nuclear resonance of conductive materials and substances when exposed to microwave fields. This property creates a wide variety of applications.

For example, since almost all biological units are composed of water predominantly, and as we know water is a good conductor, thus microwave gains tremendous importance in the field of detection, diagnostics, and treatment of biological problems or investigations as in medicine (e.g. diathermy, scanning, etc.). There are other areas that this basic property would create a variety of applications such as remote sensing, heating (e.g. industrial purification, cooking, etc.) and many others which are listed in a later section.

1.2 REASONS FOR USING RF/MICROWAVES

Over the past several decades, there has been a growing trend toward the use of RF/Microwaves in system applications. The reasons are many, amongst which the following are prominent:

a. Wider bandwidths due to higher frequency
b. Smaller component size leading to smaller systems
c. More available and uncrowded frequency spectrum
d. Better resolution for Radars due to smaller wavelengths
e. Lower interference due to a lower signal crowdedness
f. Higher speed of operation
g. Higher antenna gain possible in a smaller space

On the other hand, there are some disadvantages in using RF/Microwaves such as use of more expensive components; availability of lower power levels; existence of higher signal losses and use of high-speed semiconductors (such as GaAs or InP) along with their corresponding less-mature technology relative to the traditional Silicon technology which is quite mature and less expensive at this time.

In many RF/Microwave applications the need for a system operating at these frequencies with all the above advantages, is so great that it outweighs these disadvantages aside and spurs the engineer forward into a high-frequency design.

1.3 RF/MICROWAVE APPLICATIONS

The major applications of RF/Microwave signals can be categorized as follows:

A. Communication

This application includes satellite, space, long-distance telephone, marine, cellular telephone, data, mobile phone, aircraft, vehicle, personal and Wireless Local Area Network (WLAN), and so on. There are two important sub-categories of applications that need to be considered as follows:

A1. TV and Radio broadcast

In this application, RF/Microwaves are used as the carrier signal for the audio and video signals. An example is the Direct Broadcast Systems (DBS) which is designed to link satellites directly to home users.

A2. Optical Communications

In this application a microwave modulator is used in the transmitting side of a low-loss optical fiber with a microwave demodulator at the other end. The microwave signal acts as a modulating signal with the optical signal as the carrier. Optical communications is useful in cases where a much larger number of frequency channels as well as lower interference from outside electromagnetic radiation are desired. Current applications include telephone cables, computer network links, low-noise transmission lines, etc.

B. Radar

This application includes air defense, aircraft/ship guidance, smart weapons, police, weather, collision avoidance, imaging, etc.

C. Navigation

This application is used for the orientation and guidance of aircraft, ships and land vehicles. Particular applications in this area are:

C1. Microwave Landing System (MLS), which is used to guide aircraft to land properly in airports.

C2. Global Positioning Systems (GPS) which is used to find one's exact coordinates on the Globe.

D. Remote Sensing
In this application many satellites are used to monitor the Globe constantly for weather conditions, meteorology, ozone, soil moisture, agriculture, crop protection from frost, forests, snow thickness, icebergs, and other factors such as natural resources monitoring and exploration, etc.

E. Domestic and industrial applications
This application includes microwave Ovens, microwave clothes dryer, fluid heating, moisture sensors, tank gauges, automatic door openers, automatic toll collection, Highway traffic monitoring and control, chip defect detection, flow meters, power transmission in space, food preservation, pest control, etc.

F. Medical applications
This application includes cautery, selective heating, Heart stimulation, Hemorrhaging control, sterilization, imaging, etc.

G. Surveillance
This application includes security systems, intruder detection, Electronic warfare (EW) receivers to monitor signal traffic, etc.

H. Astronomy and space exploration
In this application, gigantic dish antennas are used to monitor, collect and record incoming microwave signals from outer space, providing vital information about other planets, stars, meteors, etc., in this or other galaxies.

I. Wireless applications
Short-distance communication inside as well as between buildings in a local area network (LAN) arrangement can be accomplished using RF and Microwaves.

Connecting buildings via cables (e.g. coax or fiber optic) creates serious problems in congested metropolitan areas, since the cable has to be run underground from upper floors of one building to upper floors of the other. However, this problem can be greatly alleviated using RF and microwave transmitter/receiver systems which are mounted on rooftops or in office windows (see Figure 1.1).

Inside buildings, RF and Microwaves can be used effectively to create a wireless LAN in order to connect telephones, computers, and various LANs to each other. Using wireless LANs has a major advantage in office re-arrangement where phones, computers, and partitions are easily moved with no change in the wiring in the wall outlets. This creates enormous flexibility and cost-saving features for any business entity.

A summary of RF and microwave applications is shown in table 1.1.

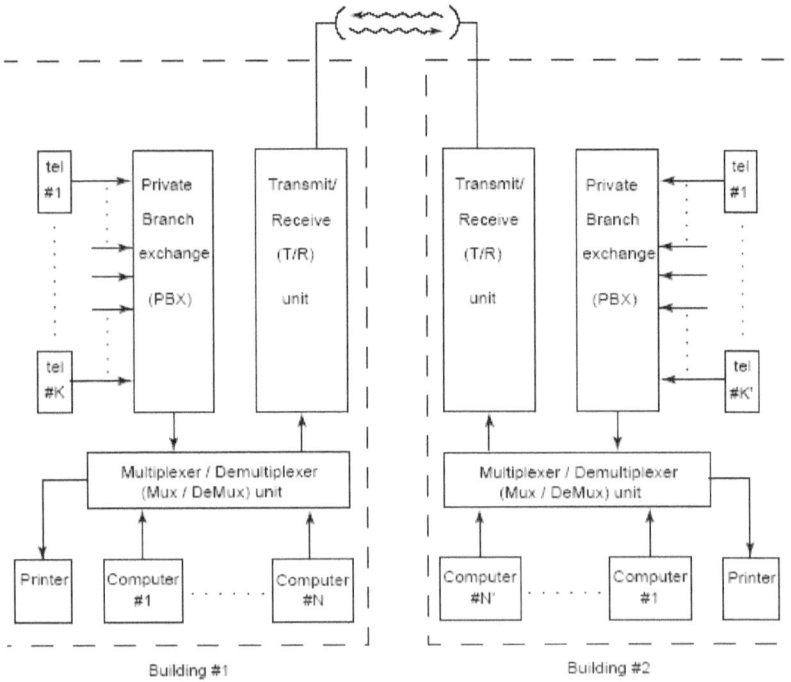

Figure 1.1 A typical local area network (LAN) for connectivity.

Table 1.1 Summary of applications of RF and Microwaves.

Category of Application	Description
Astronomy and space exploration	Deep space probes Galactic explorations

Communication	Optical communications
	    Telephone systems
	    Computer networks
	    Low-noise transmission media
	TV and radio broadcast
	    Direct broadcast satellite
	    High-definition TV

Domestic & industrial applications	Agriculture
	    Moisture detection and soil treatment
	    Pesticides
	    Crop protection from freezing
	Automobiles
	    Anti-theft radar or sensor
	    Automotive telecommunication
	    Blind spot radar
	    Collision avoidance radar
	    Near-obstacle detection
	    Radar speed sensors
	    Road-to-vehicle communication
	    Vehicle-to-vehicle communication
	Highway
	    Automatic toll collection
	    Highway traffic control and monitoring
	    Range and speed detection
	    Structure inspection
	    Vehicle detection
	Microwave Heating
	    Home microwave ovens
	    Microwave clothes dryer
	    Industrial heating
	Microwave Imaging
	    Hidden weapon detection
	    Obstacle detection & Navigation
	Office
	    Mail sorting
	    Wireless phones and computers
	Power
	    Beamed power propulsion
	    Power transmission in space
	Preservation
	    Food preservation
	    Treated manuscript drying
	Production control
	    Etching system production
	    Industrial drying
	    Moisture control

Table 1.1 Summary of applications of RF and Microwaves (continued).

Category of Application	Description
Medical applications	Cautery
	Heart stimulation
	Hemorrhaging control
	Hyperthermia
	Microwave imaging
	Sterilization
	Thermography
Radar	Air defense
	Navigation & position information
	Airport traffic control
	Global positioning system (GPS)
	Microwave landing system (MLS)
	Police patrol (velocity measurement)
	Smart weapons
	Tracking
	Weather forecast
Remote sensing	Earth monitoring
	Meteorology
	Pollution control
	Natural resources and exploration
Surveillance	Security system
	Intruder detection
	Security system
	Signal traffic monitoring
Wireless applications	Wireless local area networks (LANs)

1.4 RADIO FREQUENCY (RF) WAVES

Having briefly reviewed many of the current applications of RF/Microwaves, we can see that this rapidly advancing field has great potential to be a fruitful source of many future applications.

As discussed earlier, electromagnetic (EM) waves are generated when electrical signals pass through a conductor. EM waves start to radiate more effectively from a conductor when the signal frequency is higher than the highest audio frequency which is approximately 15 to 20 kHz. Because of this radiating property, signals of such or higher frequencies are often known as radio frequency (RF) signals.

1.4.1 RF Bands

Since it is not practical either a) to design a circuit that covers the entire frequency range, or b) to use all radio frequencies for all purposes, therefore the RF spectrum is broken down into various bands. Each band is used for a specific purpose and in general, RF circuits are designed to be used in one particular band. Table 1.2 shows the most common assignment of RF commercial bands.

Table 1.2 Commercial Radio Frequency Bands

Name of Band	Abbreviation	Frequency Range
Very low frequency	VLF	3–30 kHz
Low frequency	LF	30–300 kHz
Medium frequency	MF	300 kHz-3 MHz
High frequency	HF	3–30 MHz
Very high frequency	VHF	30–300 MHz
Ultra-high frequency	UHF	0.3–3 GHz
Super-high frequency	SHF	3–30 GHz
Extra-high frequency	EHF	30–300 GHz

1.4.2 Definition of Microwaves

When the frequency of operation starts to increase toward approximately 1GHz and above, a whole set of new phenomena occurs that is not present at lower frequencies. The radio waves at frequencies ranging from 1 GHz to 300 GHz are generally known as **microwaves.** Signals at these frequencies have wavelengths that range from 30 cm (at 1 GHz) to 1 millimeter (at 300 GHz). The special frequency range from 30 GHz to 300 GHz has a wavelength in the millimeter range thus is generally referred to as millimeter-waves.

NOTE: *It should be noted that in some texts, the range 300 MHz to 300 GHz is considered to be the microwave frequency range. This is in contrast with the microwave frequency range defined above, where the frequency range from 300 MHz to 1 GHz is referred to as the "RF" and waves at frequencies ranging from 1 GHz to 300 GHz are generally known as "microwaves."*

1.4.3 Microwave Bands

The microwave frequency range consisting of the three main commercial frequency bands (UHF, SHF, and EHF) can further be subdivided into several specific frequency ranges each with its own band designation. This band subdivision and designation facilitates the use of microwave signals for specific purposes and applications. In electronics industries and academic institutions, the most commonly used microwave bands are as set forth by the Institute of Electrical and Electronics Engineers (IEEE) and is shown in table 1.3. In this table the "Ka" to "G" are the millimeter-wave (mmw) bands.

Table 1.3 IEEE and commercial radio band designations

Band Designation	Frequency Range (GHz)
L Band	1.0-2.0
S band	2.0-4.0
C band	4.0-8.0
X band	8.0-12.0
Ku band	12.0-18.0
K band	18.0-26.5
Ka band (mmw)	26.5-40.0
Q band (mmw)	33.0-50.0
U band (mmw)	40.0-60.0
V band (mmw)	50.0-75.0
E band (mmw)	60.0-90.0
W band (mmw)	75.0-110.0
F band (mmw)	90.0-140.0
D band (mmw)	110.0-170.0
G band (mmw)	140.0-220.0

1.5 RF AND MICROWAVE (MW) CIRCUIT DESIGN

Because of the behavior of waves at different frequencies, basic considerations in circuit design has evolved greatly over the last few decades and generally can be subdivided into two main categories:

a. Radio Frequency (RF) circuit design considerations, and

b. Microwave (MW) circuit design considerations

Each category is briefly described next.

1.5.1 Low RF Circuit Design Considerations

RF circuits have to go through a four-step design process. In this design process, the effect of wave propagation on the circuit operation is negligible and the following facts can be stated:

a. The length of the circuit (ℓ) is generally much smaller than the wavelength (i.e. $\ell << \lambda$)
b. Propagation delay time (t_d) is approximately zero (i.e. $t_d \approx 0$).
c. Maxwell's Equations simplify into all of the low frequency laws such as KVL, KCL, Ohm's law, etc. Therefore at RF frequencies (f<1 GHz), the delay time of propagation (t_d) is zero when $\ell << \lambda$ and all elements in the circuit can be considered to be lumped.

The design process has the following four steps:

Step 1. The design process starts with the selection and of a suitable device and performing a DC design to obtain a proper Q-point.

Step 2. Next, the device will be characterized (either through measurement or calculations) to obtain its AC small-signal parameters based on the specific DC operating point selected earlier.

Step 3. The third step consists of designing two matching circuits that transition this device to the outside world which are the signal source at one end and the load at the other. Various design considerations and criteria such as stability, gain, noise, etc. are included at this stage and must be incorporated in the design of the final matching networks.

Step 4. In this final step, the entire circuit is put together in one seamless design to create a functional circuit. This circuit is now packaged properly by enclosing it in an appropriate box with correct connectors or terminals for communication to the outside world.
The design process for RF circuits is summarized and shown in Figure 1.2.

1.5.2 High RF and Microwave Circuits
To understand microwave circuits we should know that microwave circuits may have one or more lumped elements but should at least contain one distributed element. This last needs to be defined at this point:

DEFINITION- DISTRIBUTED ELEMENT: *Is defined to be an element whose property is spread out over an electrically significant length* *or area of a circuit instead of being concentrated at one location or within a specific component.*

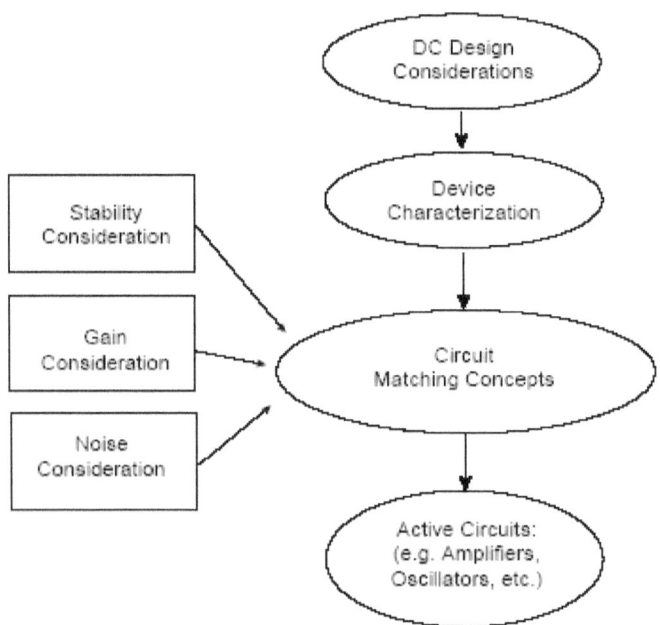

Figure 1.2 RF circuit design steps

EXAMPLE 1.1
Describe what a distributed inductor is?

Answer:
A distributed inductor would be an element whose inductance is spread out along the entire length of a conductor (such as self-inductance) as distinguished from an inductor whose inductance is concentrated within a coil.

EXAMPLE 1.2
Describe what a distributed capacitor is?

Answer:
A distributed capacitor is an element whose capacitance is spread out over a length of wire and not concentrated within a capacitor,

such as the capacitance between the turns of a coil or between adjacent conductors of a circuit.

Working with distributed circuits, we need to know the following facts about them:

a. The wave propagation concepts as set forth by the Maxwell Equations fully apply and,

b. The circuit has a significant electrical length, i.e. its physical length is comparable to the wavelength of the signals propagating in the circuit.

This fact brings the next point into view:

c. The time delay (t_d) due to signal propagation can no longer be neglected (i.e. $t_d \neq 0$).

To illustrate these points we will consider the following example.

EXAMPLE 1.3

How does a two-conductor transmission line (Such as a coaxial line, etc.) behave at low and high frequencies?

Answer:

At low frequencies this transmission line is considered to be a short piece of wire with a negligibly small distributed resistance which can be considered to be lumped for the purpose of analysis (since $t_d \approx 0$).

However at higher frequencies, the resistive, capacitive, and inductive properties can no longer be separated and each infinitesimal length (Δx) of this transmission line exhibits these properties as shown in Figure 1.3.

From this Figure, we can see that the elements are series elements (R, L) and shunt elements (G ,C) which are defined as:

R= resistance per unit length in Ω/m

L = inductance per unit length in H/m

G = conductance per unit length in S/m

C = capacitance per unit length in F/m

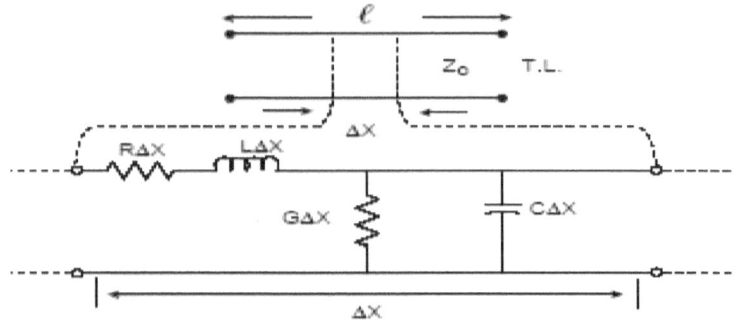

Figure 1.3 An Infinitesimal portion of a transmission line (TL).

This equivalent circuit is referred to as a *distributed circuit model* of a two-conductor transmission line and will be used in the next example to derive the equivalent circuit model of a transmission line.

EXAMPLE 1.4
Using KVL and KCL derive the relationship between voltage and current in a transmission line at:
a.   Low frequencies
b.   High frequencies (i.e., RF/Microwave frequencies)

Solution:
   a. At low frequencies a transmission line (which can be lossy in general), can be represented as shown in Figure 1.4.

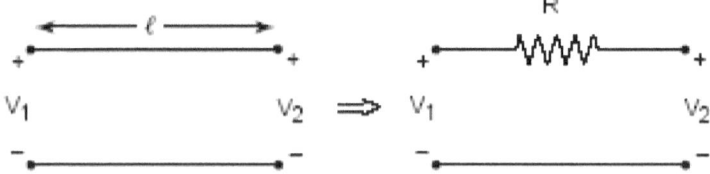

Figure 1.4 The equivalent circuit of a TL at low frequencies.

In this Figure, "R" is the distributed loss resistance of the line, which can be modeled as a lumped element. The voltage and current relationship can be written as:
$V_1=V_2+IR$

Note: *If the line is lossless, then we have:* $V_1=V_2$

b. At high frequencies, based on Figure 1.3 a transmission line can be modeled as a distributed element as shown by the equivalent circuit in Figure 1.5.

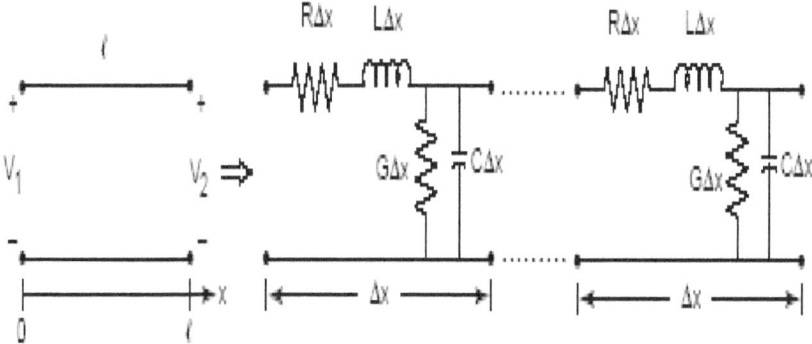

Figure 1.5 The equivalent circuit of a TL at high frequencies.

The analysis of this equivalent circuit will be postponed until the next chapter where we will examine one Δx section of a transmission line and will develop the governing equations of a transmission line in great depth.

1.5.3 High RF and Microwave Circuit Design Process

The microwave circuit design process is very similar to the RF circuit design except for the wave propagation concepts which should be taken into account.

The design process has the following four steps:
Step 1: The design process starts with the design of the DC circuit to establish a stable operating point.

Step 2: The next step is to characterize the device at the operating point (Q-point), using electrical waves to measure the percentage of reflection and transmission that the device presents at each port.

Step 3: The third step consists of designing the matching networks that transition the device to the outside world such that the required specifications such as stability, overall gain, etc. are satisfied.

Step 4. In this final step, the entire circuit is put together in one seamless design to create a functional circuit. This circuit is now packaged properly by enclosing it in an appropriate box with correct connectors or terminals for communication to the outside world.

Except for the fact that one's familiarity with wave propagation concepts becomes crucial, "Microwave Circuit Design" process is similar to the RF circuit design steps as delineated in Figure 1.6.

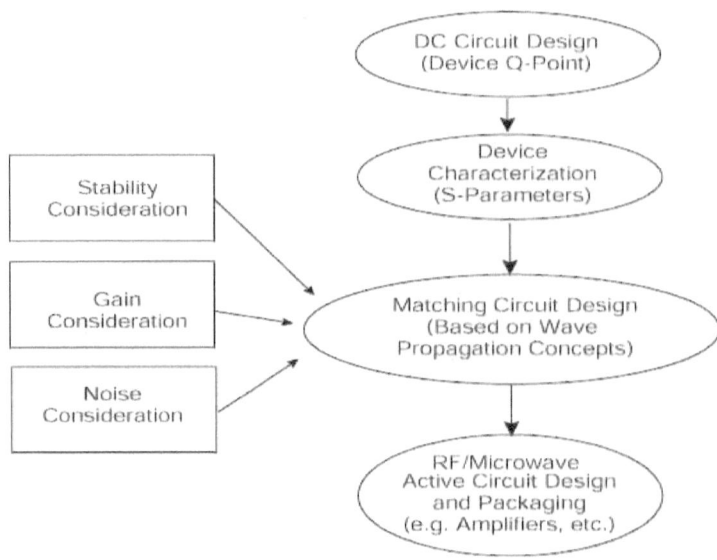

Figure 1.6 Microwave Circuit Design Steps

1.6 FUNDAMENTALS VERSUS STRUCTURE

Before we get into specific analysis and design of RF and microwave circuits, it is worthwhile first to examine a general communication system in which each circuit or component has a specific function in a bigger scheme of affairs. In general, any communication system is based upon a very simple and yet extremely fundamental truth, commonly referred to as the "universal communication principle".

The "Universal Communication Principle" is a fundamental concept which is at the heart of a wide sphere of existence called "life and livingness", or for that matter any of its subsets particularly the field of RF/Microwaves. This principle is intertwined throughout the entire field of RF/microwaves and thus plays an important role in our understanding of this subject. Therefore it behooves us well to define it at this juncture.

THE UNIVERSAL COMMUNICATION PRINCIPLE: *This principle states that communication is the process whereby information is transferred from one point in space and time (X_1, Y_1, Z_1, t_1), called the source point, to another point in space and time (X_2, Y_2, Z_2, t_2), called the receipt point. Usually, the receipt point at location (X_2, Y_2, Z_2) is separated by a distance (d) from the source point location (X_1, Y_1, Z_1).*

The physical embodiment of the universal communication principle is a "communication system", which takes the information from the source point and delivers an exact replica of it to the receipt point (see Figure 1.7).

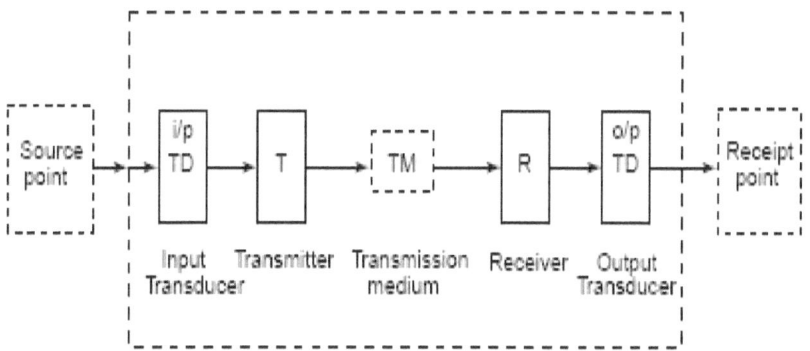

Figure 1.7 Depiction of a communication system.

Thus in general, it can be seen that any communication system can be broken down into three essential elements:

1. Source point: *A point of emanation or generation of information.*
2. Receipt point: *A point of receipt of information.*

3. Distance (or Imposed space): *The space existing between the "Source point" and "Receipt point" where the information travels through.*

Furthermore, it can be observed that in order to achieve effective communication between two systems, we need to have three more factors present: a) There must be an intention on the part of the source point and the receipt point to emit and to receive the information, respectively, b) source and receipt points must have attention on each other (i.e. both being ready for transmission and reception), and c) duplication (i.e. an exact replica) must occur at the receipt point of what emanated from the source point.

Use of the universal communication principle in practice creates a one-way communication system (such as radio and TV broadcast, etc.), and forms one leg of a two-way communication system (such as CB radio, telephone, etc.), where this process is reversed to create the second leg of the communication action.

An important application of the universal communication principle is in a radar communication system where the source point (X_1, Y_1, Z_1) is at the same physical location as the receipt point (X_2, Y_2, Z_2), i.e., $X_1=X_2$, $Y_1=Y_2$, $Z_1=Z_2$; however the times of sending and reception are different ($t_1 \neq t_2$). Otherwise no communication would take place. This brings us to the obvious conclusion that one can not have a condition where the source and the receipt points are the same, simultaneously!

Based on this simple concept of communication, the most complex communication systems can be understood, analyzed, and designed. Figure 1.8 is a simple and yet a very generalized block diagram of such a practical communication system in use today.

It should be noted that the design and structure of this communication system can change and evolve into a more efficient system with time whereas the universal communication principle will never change.

Of course, this should be no surprise to the workers in the field because as it turns out the foundation (which consists of fundamental postulates, axioms, and natural laws) along with fundamental concepts (i.e., theorems, analytical techniques, the theory of operation, etc.) of any science is far superior in importance to any designed circuitry, machinery, network, etc. This observation makes us realize that *the fundamentals are unchanging whereas the structure exists on a constant-change basis and is always evolving.* This brings us to the following conclusion:

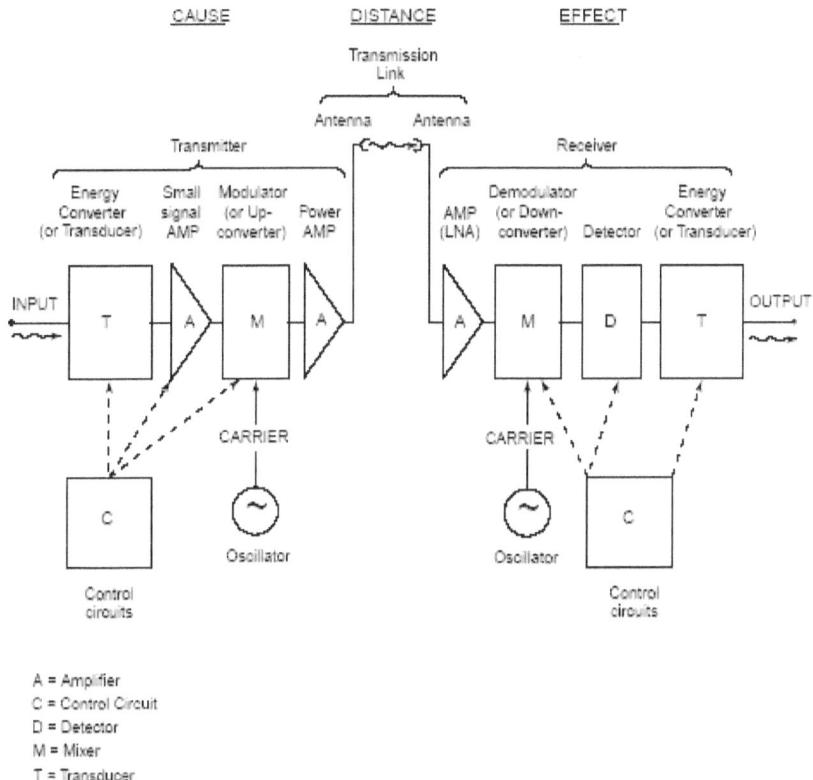

Figure 1.8 Block diagram of a general communication system.

Fundamentals of any science are superior and dictate the designed forms, structures or in general the entire application mass of that science, and not vice versa.

This is true in all aspects of design, i.e., while the underlying principle remains constant, the structure which is the electronic circuit constantly undergoes improvements with new designs and evolves in time toward a more efficient circuitry.

This can best be described as "engineering principle as a constant" vs. "the application mass as a constantly evolving structure" where it approaches closer and closer to the underlying principle with each improvement.

Even though rarely new discoveries may bring about new underlying fundamentals to the forefront, nevertheless the fundamentals, as a general rule of thumb, remain invariant.

For example, circa 1864-1873 James clerk Maxwell interrelated all of the known data about electricity and magnetism, formulated and presented the classical laws of Electromagnetics. Since that time, which is over a century, tremendous technological changes and advances have happened all over the Globe and yet Maxwell's equations have not changed an iota. **This set of celebrated equations have remained timeless!**

Of course it should be noted that quantum mechanics, dealing with sub-atomic particles may be considered by some, to have generalized these equations and shown that energy is not continuous but quantized. Nevertheless, Maxwell's equations at the classical level of observation have not been surpassed and are still true today and currently form the foundation of the "Electromagnetics" as a science -- the backbone of electronics and electrical engineering.

Now to build a communication system in the physical universe that works and is practical, one must satisfy two conditions:

1. First, it must be based on the fundamental concept of a) " the universal communication principle" and then b) "Maxwell's Equations"-- both in combination form a static which is unchanging!

2. Secondly, it must follow and conform to the current state of technology in terms of manufacturing, materials, device fabrication, circuit size, and structure -- a kinetic and constantly evolving!

These two pre-requisites, in essence, clearly demonstrate and confirm the interplay of "static vs. kinetic" which is interwoven throughout our entire world of science and technology.

The above two steps of system design set up the "Blue Print" for any "general engineering system design". One must heed these points carefully before one has gone very far in the quest for workable knowledge.

1.7 ACTIVE CIRCUIT BLOCK DIAGRAMS

Considering Figure 1.8, we note several stages:

1. **Energy Conversion Stage:** This is a simple transducer casing the incoming energy (e.g. sound, etc.) to be converted to electrical energy. An example for this stage could be a microphone.

2. **Amplification Stage:** This is a high gain small-signal amplifier causing a higher signal to compensate for losses in the energy conversion stage.

3. **Frequency Conversion Stage (also called Modulation or Up-Conversion):** This causes a carrier wave to be modulated by the amplified signal of stage 2. This is the stage that prepares the signal for transmission for long distance by increasing its frequency, since higher frequency signals travel longer and require smaller antennas. Needless to say that a local oscillator is needed to produce the carrier wave before the modulation process can take place.

4. **Power Amplification Stage:** This is the stage where the signal power level is boosted greatly so that a higher range of reception is allowed.

5. **Transmission Link**: This is the transmission media in which the modulated signal is transported from "cause or source point" to the "effect or the receipt point".

6. **Low-Noise Amplification Stage:** This is the first stage (or Front-end) of the receiver wherein the modulated signal is amplified and prepared by a low-noise amplifier (LNA) in such a

way that the effect of noise which could possibly be added to the signal by later stages, is minimized.

7. **Frequency Conversion Stage (Demodulation Or Down-Conversion):** This stage demodulates the signal and brings the carrier frequency down to workable levels. Just like the modulation stage, a local oscillator of a certain frequency is needed to make the demodulation process effective. **Note:** If the local oscillator is tunable, then the same receiver can be used to receive signals from other sources at other frequencies (a Heterodyne receiver!).

8. **Detector Stage:** This stage removes the carrier wave and reconstructs the original signal.

9. **Energy Conversion Stage:** This stage converts the electrical signal back to its original form (e.g. sound). An example for this stage could be a speaker.

10. **Control Stage:** This is where all the decisions with regard to circuit connection/disconnection, routing, switching, etc. take place and are present at both the source and the receipt points of the communication system.

To gain a full conceptual understanding of different types of circuit designs one needs to have an overall idea of "how different components fit together". To bring this point into a realm of practicality each specific type of microwave circuit has been cast into an exact block diagram that clearly depicts the relationship of the device with other circuit components and sections.

The circuits considered for the purpose of the block diagram are as follows:

A. **Amplifier:** is defined to be an electronic circuit capable of increasing the magnitude or power level of an electrical signal without distorting the wave-shape of the quantity. The block diagram for this circuit is shown in Figure 1.9.

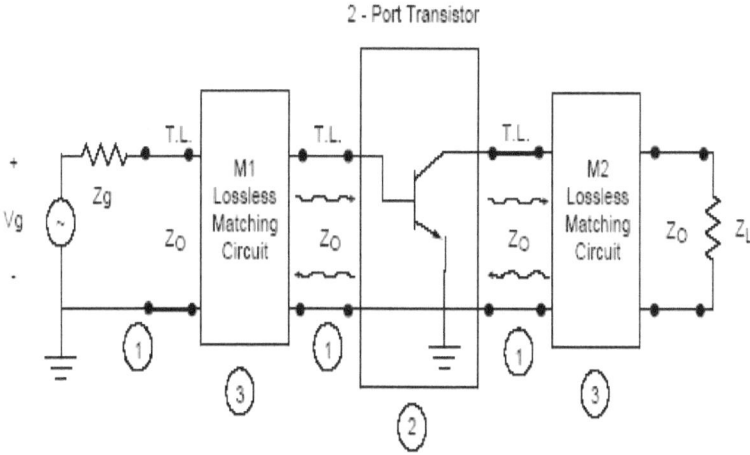

FIGURE 1.9 An amplifier circuit block diagram.

B. **Oscillator**: is defined to be an electronic circuit that converts energy from a DC source to a periodically varying electrical signal. The block diagram for this circuit is shown in Figure 1.10.

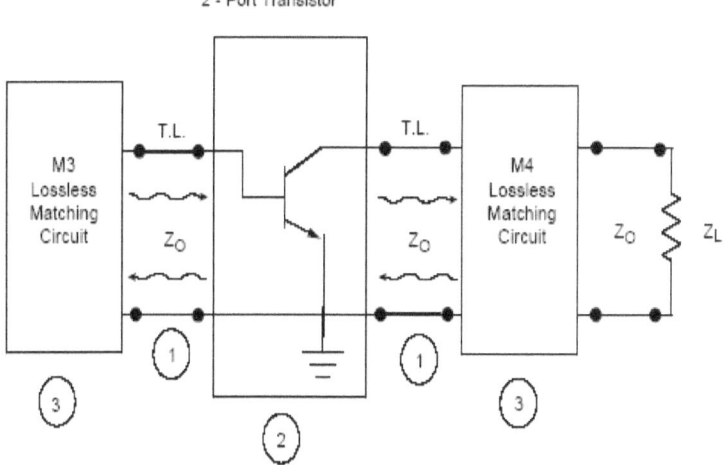

FIGURE 1.10 An oscillator circuit block diagram.

C. **Mixer**: is defined to be an electronic circuit that generates an output frequency equal to the sum and difference of two input frequencies or in short a frequency converter. The block diagram for this circuit is shown in Figure 1.11.

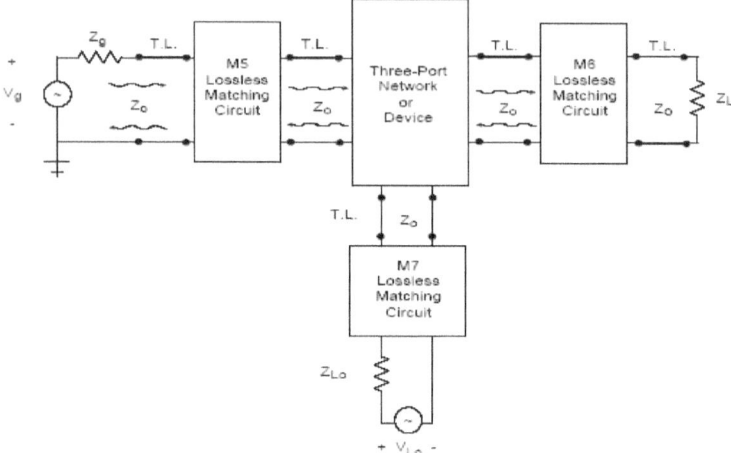

FIGURE 1.11 A mixer circuit block diagram.

D. Detector: is defined to be an electronic circuit concerned with demodulation, i.e., extracting a signal which has modulated a carrier wave. The block diagram for this circuit is shown in Figure 1.12.

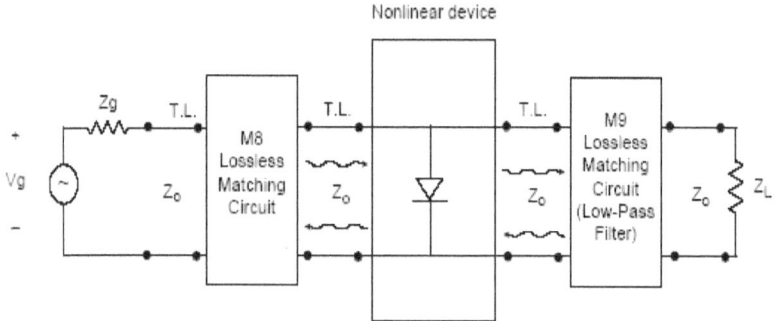

FIGURE 1.12 A detector circuit block diagram.

From these block diagrams we can see that the device forms the "heart" or "engine" of the circuit around which all other circuit components should be properly designed in order to control the input/output flow of signals and eventually obtain an optimum performance.

Furthermore, these four block diagrams show the irresistible fact that the knowledge gained in earlier chapters is essential in the design of these complicated circuits.

1.8 SUMMARY

To be proficient at higher frequency circuits (analysis or design), one needs to master, on a gradient scale, all of the underlying principles and develop a depth of knowledge before one can be called a skilled microwave practitioner.

Figure 1.13 depicts the gradient scale of concepts that need to be fully understood in order to achieve a mastery of circuit design skills at higher frequencies.

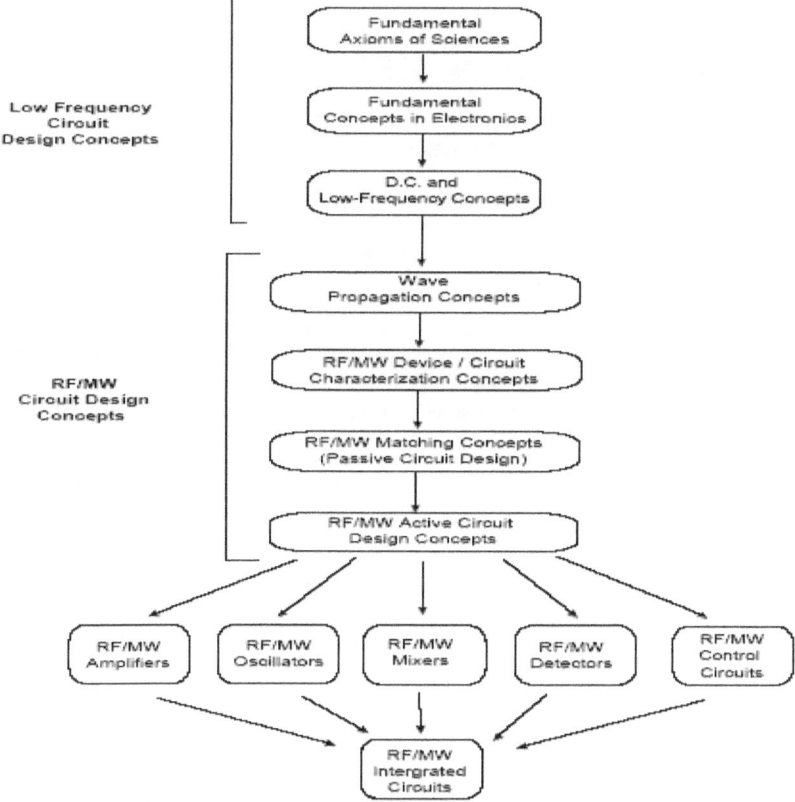

Figure 1.13 The gradient scale of concepts in RF/MW circuit design.

As shown in Figure 1.13, one starts with the fundamental axioms of sciences, fundamental concepts in electronics, and progresses toward high frequency electronic circuit design by learning the DC and low frequency concepts at first, then wave propagation concepts, device-circuit characterization, matching concepts and eventually arrives at the final destination of RF/MW active circuit design concepts, which was originally set forth as the goal of this book.

Knowing this progressive series of concepts will enable one to design amplifiers, oscillators, mixers, detectors, control circuits, and integrated circuits with relative ease and proficiency at RF/MW frequencies.

CHAPTER 1- SYMBOL LIST

A symbol will not be repeated again, once it has been identified and defined in an earlier chapter, with its definition remaining unchanged.

ℓ – Length of the circuit

t_d – Time delay

λ - Wavelength

CHAPTER -1 PROBLEMS

1.1)   What is the difference between a lumped element and a distributed element?

1.2)   How many steps are required to design a) an RF circuit? b) A microwave circuit? Describe the steps.

1.3)   What are the similarities and difference(s) between an RF and a microwave circuit design procedures?

1.4)   Describe: a) What is meant by "fundamentals vs. application mass"? b) What is meant by the timelessness of a fundamental truth? Give an example, c) What part of a system constantly evolves? and d) What are the pre-requisites for any general system design?

1.5)   What is at the heart of an amplifier, an oscillator, a mixer, or a detector block diagrams?

1.6)   What are the main concepts one needs to master in order to design an RF or a microwave circuit?

1.7)   Why is it necessary to understand the low frequency electronics fully before trying to master RF/microwave electronics?

REFERENCES

[1.1] Carlson, A. B. *Communication Systems: An Introduction to Signals and Noise in Electrical Communication.* New York: McGraw-Hill, 1968.

[1.2] Cheung, W. S. and F. H. Levien, *Microwave Made Simple,* Dedham: Artech House, 1985.

[1.3] Gardiol, F. E. *Introduction to Microwaves,* Dedham: Artech House, 1984.

[1.4] Ishii, T. K. *Microwave Engineering.* 2nd ed., Orlando: Harcourt Brace Jovanovich, publishers, 1989.

[1.5] Laverghetta, T. *Practical Microwaves,* Indianapolis: Howard Sams, 1984.

[1.6] Lance, A. L. *Introduction to Microwave Theory and Measurements,* New York: McGraw-Hill, 1964.

[1.7] Radmanesh, M. M. *Applications and Advantages of Fiber Optics as Compared with other Communication Systems,* Hughes Aircraft Co., Microwave Products Div. , pp. 1-11, April 1988.

[1.8] Radmanesh, M. M. *Radiated and Conducted Susceptibility Induced Current in Bundles: Theory and Experiment,* Boeing Co., HERF Div., pp. 1–115, Sept. 1990.

[1.9] Radmanesh, M. M. *The Gateway to Understanding: Electrons to Waves and Beyond,* AuthorHouse, 2005.

[1.10] Radmanesh, M. M. *Cracking the Code of Our Physical Universe,* AuthorHouse, 2006.

[1.11] Saad, T. *Microwave Engineer's Handbook,* Vols I, II. Dedham: Artech House, 1988.

[1.12] Scott, A. W. Understanding Microwaves. *New York: John Wiley & Sons, 1993.*

CHAPTER 2

RF Electronics

2.1 INTRODUCTION

It is important to set the stage properly for the introduction of microwave circuits. To that end we will introduce RF circuit analysis and design to serve as a platform of fundamental information in order to catapult us into the world of microwave circuit design. Therefore this chapter will primarily deal with the world of RF circuit design with the intention of preparing the reader for a much broader field of study, namely, microwave circuit analysis and design presented in the future chapters.

2.2 RF/MICROWAVES VERSUS DC OR LOW AC SIGNALS

There are several major differences between signals at higher radio frequency (RF) or microwaves (MW) and their counterpart at low AC frequency or DC.   These differences which influence the electronic circuits and their operation greatly, become increasingly important as the frequency is raised. The following four effects provide a brief summary of the effects of RF/MW signals in a circuit which are not present at DC or low AC signals:

EFFECT #1. PRESENCE OF STRAY CAPACITANCE - This is the capacitance that exists:

    a. Between conductors of the circuit

    b. Between conductors or components and ground

c. Between components

This effect is shown in Figure 2.1.

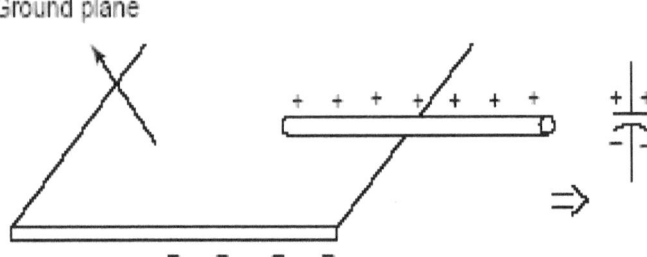

Figure 2.1 The stray capacitance effect in a circuit.

EFFECT #2. PRESENCE OF STRAY INDUCTANCE- This is the inductance that exists due to:

a. The inductance of the conductors that connect components, and

b. The parasitic inductance of the components themselves.

These stray parameters are not usually important at DC and low AC frequencies but as frequency increases, they become a much larger portion of the total. This concept is shown in Figure 2.2.

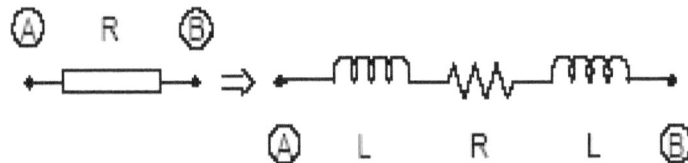

Figure 2.2 The stray inductance effect in a circuit.

EFFECT #3. SKIN EFFECT- This refers to the fact that AC signals penetrate a metal partially and flow in a narrow band near the outside surface of each conductor. This is in contrast to the DC signals where they flow through the whole cross-section of the conductor as shown in Figures 2.3a and 2.3b.

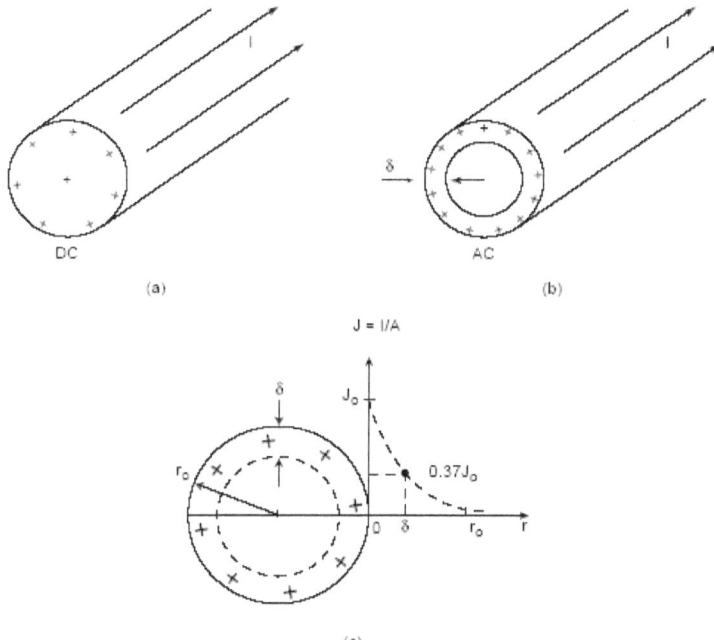

Figure 2.3 Skin effect and current flow for (a) DC signal, (b) AC signal, and (c) Skin effect.

For AC signals, the current density falls off exponentially from the surface of the conductor toward the center. At a critical depth (δ), called the skin depth or depth of penetration, the signal amplitude is 1/e or 36.8% of its surface amplitude (see Figure 2.3c) which is given by:

$$\delta = \sqrt{\frac{1}{\pi f \mu \sigma}} \qquad (2.1)$$

where μ is the permeability (H/m), and σ is the conductivity of the conductor.

EXAMPLE 2.1
Considering copper as the conductive medium, what is the skin depth at 60 Hz and 1 MHz?
Solution:
For copper we have:
$\mu = 4\pi \times 10^{-7}$ H/m

$\sigma = 5.8 \times 10^7$ S/m

At f=60 Hz $\Rightarrow$ $\delta = (1/\pi \times 60 \times 4\pi \times 10^{-7} \times 5.8 \times 10^7)^{1/2} = 0.85$ cm

While on the other hand for f=1 MHz, we calculate δ to be :

$\delta = 0.007$ cm

which is a substantial reduction in penetration depth.

___

As seen from example 2.1, we can observe that as frequency increases, skin effect produces a smaller zone of conduction and a correspondingly higher value of AC resistance compared with DC resistance.

EFFECT #4. RADIATION- This is caused by the leakage or escape of signals into the air. This, in essence, means that the signals bypass the conducting medium and not all of the source energy is reaching the load. Radiation can occur outside or within a circuit as shown in Figure 2.4

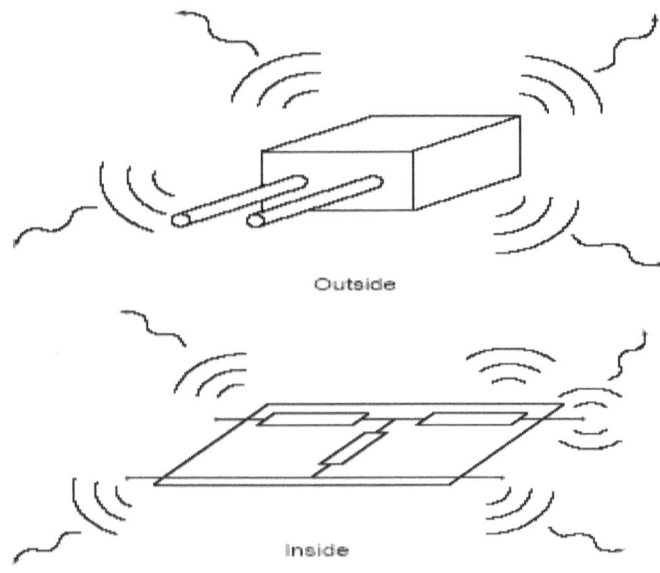

Figure 2.4 Radiation of a circuit (a) outside or (b) inside.

The radiation factor causes coupling effects to occur as follows:

a. Coupling between elements of the circuit,

b. Coupling between the circuit and its environments, and
c. Coupling from the environment to the circuit.

"Electromagnetic interference" (EMI), also called "Radio Frequency Interference" (RFI) or "RF-noise", which is due to signals at RF/MW frequencies, is missing in DC circuits and is considered to be negligible in most low-frequency AC circuits.

2.3 EM SPECTRUM

When an RF/MW signal radiates, it becomes an EM wave that is propagating through a medium such as air. The range of frequencies of electromagnetic waves known as the EM spectrum is shown in Figure 2.5.

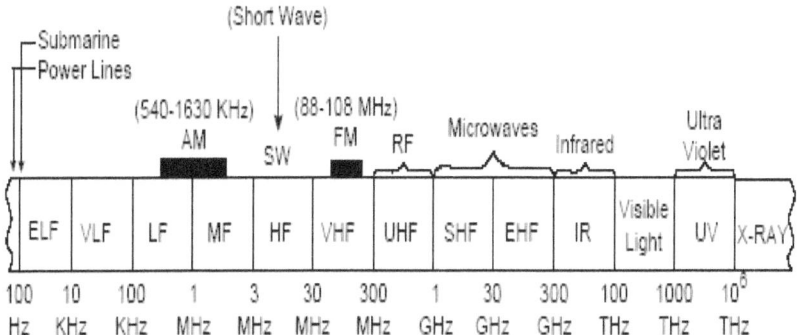

Figure 2.5 The EM spectrum.

Looking at this spectrum one may wonder, "How do microwaves differ from other EM waves?" The answer lies in the fact that microwaves is a separate topic all by itself because at these frequencies the wavelength (λ) approximates the physical size of the ordinary electronic components as discussed earlier in Chapter 5. Therefore components behave differently at microwave frequencies than they do at lower frequencies.

EXAMPLE 2.2

How does an ordinary resistor element behave at microwave frequencies?

Solution:
An ordinary carbon resistor at microwave frequencies (e.g. at $f = 10$ GHz) has a stray capacitor and a stray inductor as well as a higher resistance due to the skin effect (since the cross section is reduced) and radiation (since part of the power is lost in the air).

These factors are added into the equivalent circuit (as shown in Figure 2.6).

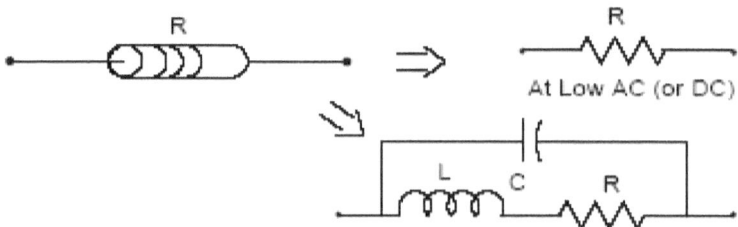

Figure 2.6 The equivalent circuit of an ordinary resistor at low Ac and at microwave frequencies.

The addition of extra parasitic elements in the equivalent circuit is due to the combined length of the leads and the physical size of the component itself which is comparable to the wavelength.

2.4 Wavelength and Frequency
When an electromagnetic wave with a certain oscillation frequency (f) propagates through the air or any other medium, it does so at a certain fixed speed or velocity (also known as the phase velocity (V_p) and a corresponding fixed wavelength (λ) as shown in Figure 2.7.
These three factors: f, V_p and λ are not independent of each other and in fact are interrelated such that the product of frequency (f) and wavelength (λ) is equal to the velocity (V_p), i.e.,
$$\lambda f = V_p \tag{2.2a}$$

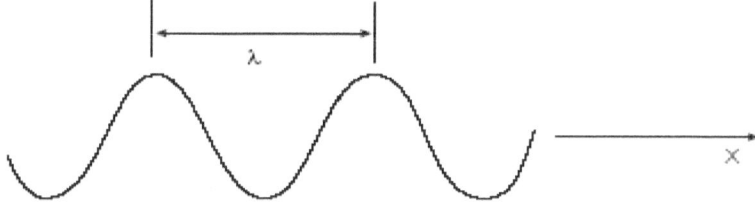

Figure 2.7 Wavelength of a wave.

It has been shown that the velocity of propagation for any and all EM waves through the air is approximately equal to the speed of light(c_o):

$c_o = 3 \times 10^8$ m/s

However, if the medium is not air the speed is lower than " c_o " and can be shown to be :

$$V_p = c = c_o / \sqrt{\varepsilon_r} \qquad (2.2b)$$

Where ε_r is the relative dielectric constant of the medium of propagation.

2.5 COMPONENT BASICS

In this section, the properties of resistors, capacitors and inductors at high radio frequencies will be studied. . But first, we will take a brief look at the most simple component of all: a piece of wire. We will consider this element first and examine its problems at radio frequencies.

2.5.1 Wire

A wire is the simplest element to study having a zero resistance which makes it appear as a short circuit at DC and low AC frequencies. Yet at RF/MW frequencies it becomes a very complex element and deserves special attention that will be studied in depth shortly. Wire in a circuit can take on many forms, such as:

- Wire-wound resistors,
- Wire-wound inductors,
- Leaded capacitors (see Figure 2.8), and
- Element-to-element interconnect applications

(a)   (b)
Figure 2.8 A loaded capacitor, a) axial, b) radial.

The behavior of a piece of wire in the RF spectrum depends to a large extent on the wire's diameter and length. A system for different wire sizes is the American Wire Gauge (AWG) System. In this system, the diameters of a wire will roughly double every six gauges.

EXAMPLE 2.3
Given that the diameter of AWG 50 is 1.0 mil (0.001 of an inch), what is the diameter of AWG 14?

Solution:
Starting from AWG 50 we descend downward by 6 gauges until we reach AWG 14 as follows:
AWG 50 $\Rightarrow$ d = 1 mil,
AWG 44 $\Rightarrow$ d = 2 mils,
AWG 38 $\Rightarrow$ d = 4 mils
AWG 32 $\Rightarrow$ d = 8 mils,
AWG 26 $\Rightarrow$ d= 16 mils,
AWG 20 $\Rightarrow$ d= 32 mils,
AWG 14 $\Rightarrow$ d = 64 mils.

Problems Associated with a Piece Of Wire
Problems associated with a wire can be traced to two major areas:
a. Skin effect, and
b. Straight-wire inductance.
These two problems are discussed next.

a. Skin Effect in a Wire
As frequency increases, the electrical signals propagate less and less

in the inside of the conductor. The current density increases near the outside perimeter of the wire and causes a higher impedance to be seen by the signal as shown in Figure 2. 9. This is because the resistance of the wire is given by:

$$R = \frac{\rho \ell}{A},$$ (2.3a)

and if the effective cross-sectional area "A" decreases, it would lead to an increase in resistance (R).

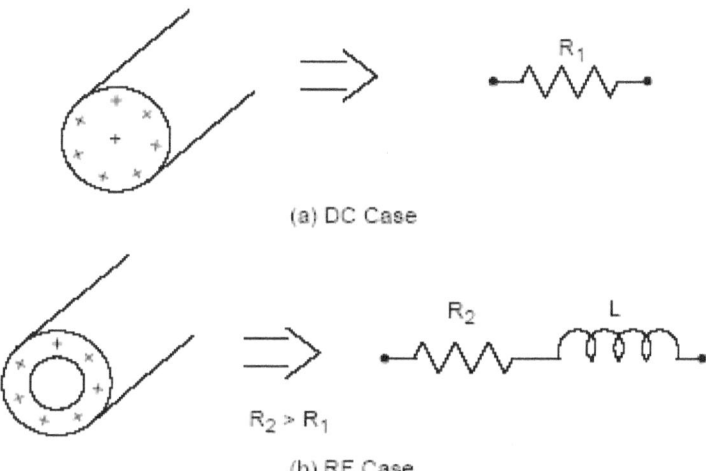

(a) DC Case

$R_2 > R_1$

(b) RF Case

Figure 2.9 The skin effect in a wire.

b. Straight-Wire Inductance

In the medium surrounding any current-carrying conductor, there exists a magnetic field. If the current (I) is AC, this magnetic field is alternately expanding and contracting (and even reversing direction if there is no DC bias present). This produces an induced voltage (as specified by the Faraday's law) in the wire which opposes any change in the current flow. This opposition to change is called "self-inductance" as shown in Figure 2.10.

The concept of inductance is important because at RF/MW, any and all conductors including hookup wires, capacitor leads, bonding wires and all interconnections tend to become inductors and exhibit the property of inductance as shown in Figure 2. 11.

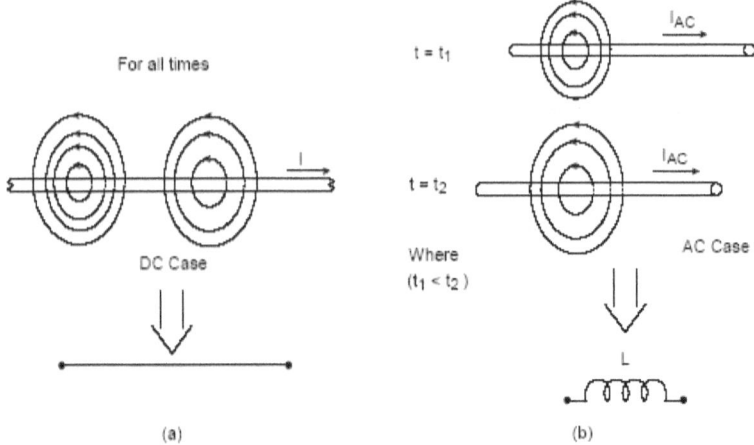

Figure 2.10 Interactive properties of a wire (a) DC case: no self inductance, b) AC case: self inductance.

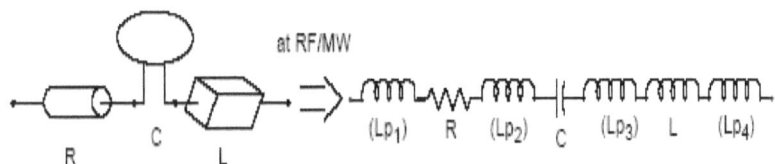

Figure 2.11 A simple RLC Circuit at RF/MW frequency.

2.5.2 Resistors

DEFINITION-RESISTOR: *Is an element specializing in the resistance property, which is the property of a material that determines the rate at which electrical energy is converted into thermal energy when an electric current passes through it.*

Resistors are used in almost all circuits for different purposes, such as:
a. In transistor bias networks, to establish an operating point,
b. In attenuators (also called pads), to control the flow of power, and
c. In signal combiners, to produce a higher output power.

Once we depart from the world of DC, resistors start to behave differently, i.e.,

At DC :   **V = RI (Ohm's law),**

At low AC: **V ≈ RI ,**
At high RF/MW: **V ≠ RI.**

At RF/MW frequencies, a resistor (R) appears like a combination of several elements as shown in Figure 2.12.

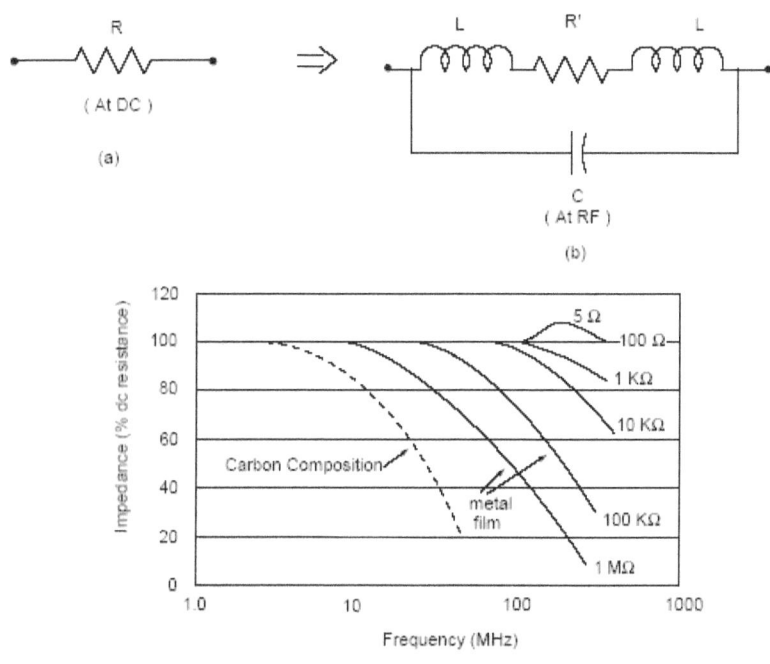

Figure 2.12 A simple resistor at DC, at RF and its graph vs. frequency.

Figure 2.12a shows a simple resistor at DC, and as frequency increases the lead wire inductances (L) bring about a higher resistor value (R'>R) due to skin effect as well as parasitic capacitances, both becoming prominent as shown in Figure 2.12b. The net effect of all these parasitic elements, on the average, is a decrease in value of the carbon-composition as well as metal resistors as shown in Figure 2.12c

NOTE: *The 5 Ω resistor graph in Figure 2.12c shows a slight resonance due to the parallel combination of lead inductance and capacitance which causes a small increase in the resistor value with a subsequent decline as the frequency is increased further.*

There are several types of resistors, which can be briefly summarized as follows:

a. Carbon composition type resistors, which have a high capacitance due to carbon granule's parasitic capacitance,

b. Wirewound resistors, which have high lead inductance,

c. Metal film resistors, which are usually made up of highly resistive films such as NiCr, etc.

d. Thin-film chip resistors, which are produced on an Alumina or Beryllia substrate and thus reduce the parasitic reactances greatly. These four types of resistors are shown in Figure 2.13 a, b, c, and d.

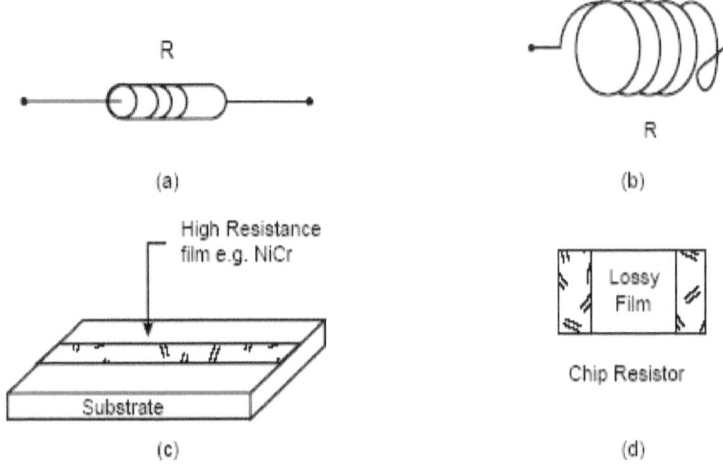

Figure 2.13 Various types of resistors: a) carbon resistor, b) wire-wound resistor, c) metal film resistor, and d) chip resistor.

2.5.3 Capacitors

A capacitor was defined earlier in Chapter 3 and basically is any device, which consists of two conducting surfaces separated by an insulating material or a dielectric. The dielectric is usually ceramic, air, paper, mica, etc.

The capacitance is that property which permits the storage of charge when a potential difference exists between the conductors, and is measured in Farads (see Figure 2.14).

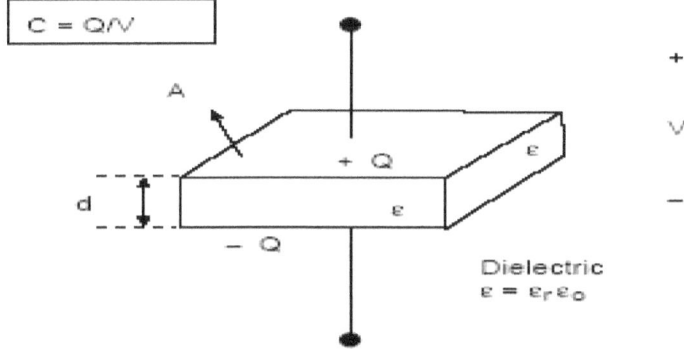

Figure 2.14 A parallel-plate capacitor.

The performance of a capacitor is primarily dependent on the characteristic of its dielectric. It determines voltage and temperature extremes of the capacitor at which it can be used. Thus any losses or imperfections in the dielectric have an enormous effect on the circuit operation. A few examples of different types of dielectric materials is shown in Figure 2.15.

Figure 2.15 effect of dielectric on capacitance value.

A practical capacitor has several parasitic elements which become important at higher frequencies. The equivalent circuit of a real capacitor is shown in Figure 2. 16.

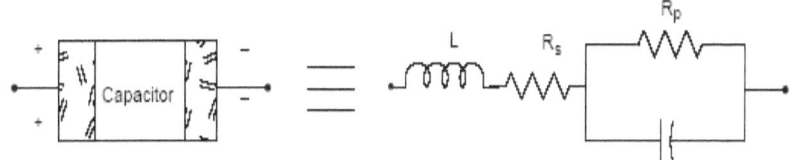

Figure 2.16 parasitic elements in a capacitor.

In Figure 2.16 the elements are defined as follows:
C is the actual capacitance, L is the lead inductance, R_S is the series

resistance, and R_P is the insulation resistance (both creating heat and loss).

The existence of parasitic elements, as shown in Figure 2.16, brings the concept of real-world capacitors to the forefront which needs further explanation:

PERFECT CAPACITORS: In a perfect capacitor, the current will lead the applied voltage in phase by 90 degrees. In phasor notation this can be written as:

$$I = j\omega\ CV = \omega\ C\ Ve^{j90°} \qquad\qquad (2.4a)$$

Practical capacitors: In a real-world capacitor, the phase angle (ϕ) will be less than 90 degrees (i.e. $\phi < 90°$) as shown in Figure 2.17. The reason $\phi < 90°$ is due to the existence of R_S and R_P (parasitic resistances are shown in Figure 2.16) which combine into one equivalent resistor (R_{EQ}) as shown in Figure 2.18.

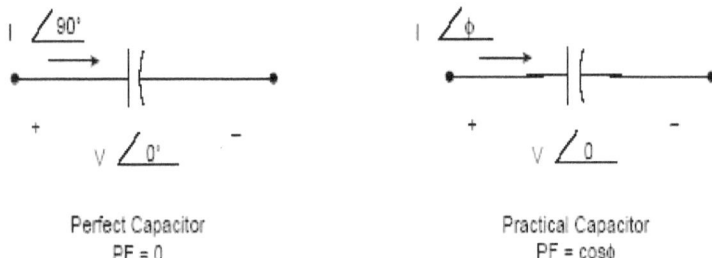

Perfect Capacitor
PF = 0

Practical Capacitor
PF = cosϕ

Figure 2.17 A capacitor current-voltage relationship.

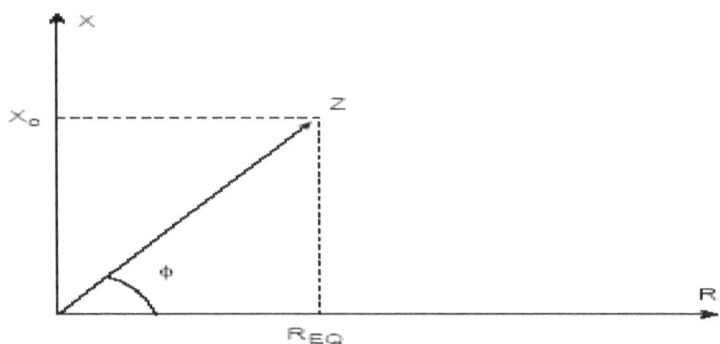

Figure 2.18 Phase angle.

In a practical capacitor cos(ϕ), called the power factor (PF), can be written as (see Figure 2.18):

$$PF=\cos(\phi)=\frac{R_{EQ}}{\sqrt{X_C{}^2+R_{EQ}{}^2}} \qquad (2.4b)$$

Usually $R_{EQ} \ll X_C$ where $X_C = 1/\omega C$. Therefore we can write:
$$PF=\cos(\phi)\approx R_{EQ}/X_C \qquad (2.5)$$

An important factor in practical capacitors or in general any imperfect element is the Quality Factor (Q):

DEFINITION-QUALITY FACTOR (Q): *Is a measure of the ability of an element (or circuit) with periodic behavior to store energy equal to 2π times the average energy stored divided by the energy dissipated per cycle.*

Q is a "Figure of Merit" for a reactive element and can be shown to be the ratio of element's reactance to its effective series resistance. Thus For a capacitor, Q is given by:

$$Q = X_C/R_{EQ} = \frac{1}{\omega C R_{EQ}} \approx 1/PF \qquad (2.6)$$

From Equation (2.6), we can observe that for a practical capacitor, as the effective series resistance (R_{EQ}) decreases, Q will increase until $R_{EQ}=0$ which corresponds to a perfect capacitor having $Q=\infty$, i.e.,
$$R_{EQ}=0 \Rightarrow PF=0, \ Q=\infty \quad \text{(a perfect capacitor)} \qquad (2.7)$$

The effect of these imperfections in a capacitor is shown in Figure 2.19.

From Figure 2.19, two distinct regions in the frequency response plot of a capacitor can be identified. These two regions straddle the resonance frequency (f_r)as follows:

a. $f < f_r$
In this region as frequency increases, the lead inductance's reactance goes up gradually toward resonance (f_r).

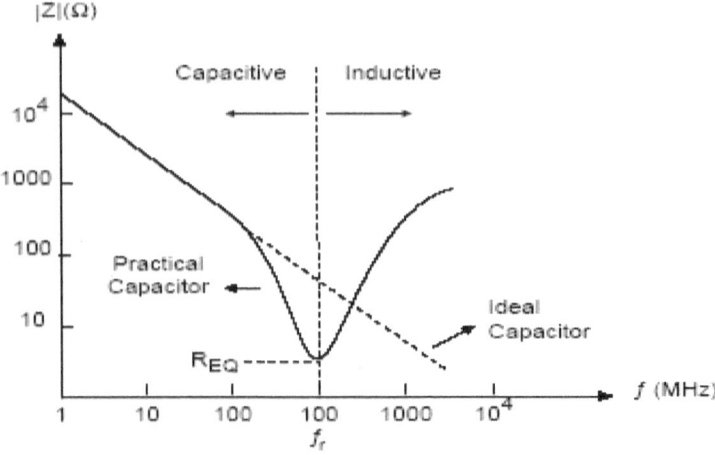

Figure 2.19 The behavior of a capacitor vs. frequency.

b. $f > f_r$

In this region the capacitor acts like an inductor and is no longer performing its intended function.

From Figure 2.19 we can conclude that we need to examine the capacitor at RF/MW frequencies before final design and production. This concept is shown in Figure 2.20 where the distinction between the low AC and RF/MW is clearly shown.

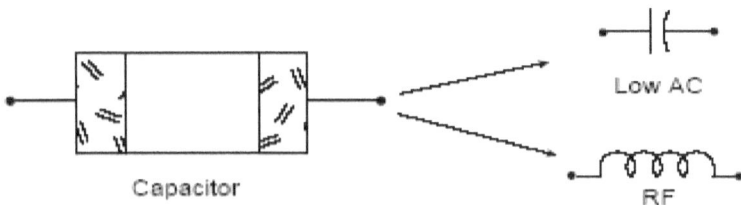

Figure 2.20 A capacitor performance at low AC and at RF/MW frequency.

2.5.4 Inductors

Definition-Inductor: is a wire which is wound (or coiled) in such a manner as to increase the magnetic flux linkage between the turns of the coil.   The increased flux linkage increases the wire's self-inductance, as shown in Figure 2.21.

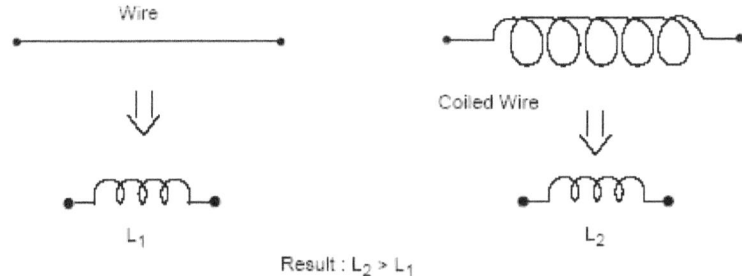

Figure 2.21 A simple piece of wire vs. a coiled wire.

Inductors have a variety of applications in RF circuits such as in resonance circuits, filters, phase shifters, delay networks, and RF chokes as shown in Figure 2.22.

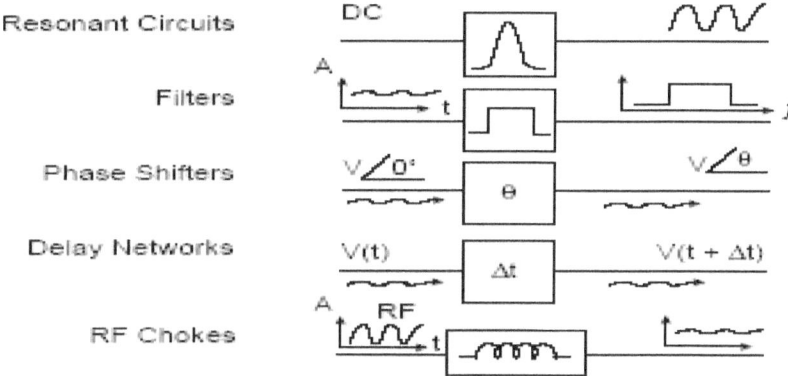

Figure 2.22 Applications of an inductor.

Since there is no such thing as a perfect component, it is found that amongst all components, inductors are most prone to very drastic changes over frequency. This is due to the fact that the distributed capacitance (C_d) and series resistance (R) in an inductor at RF/MW plays a major role in the performance of an inductor as shown in Figure 2.23.

From Figure 2.23, we can see that C_d exists due to a voltage drop in the coil caused by internal resistance. The voltage drop causes a voltage difference between two turns of the coil separated from each other (with air as the dielectric). The aggregate of all small C_d's and r's provides the equivalent circuit shown in Figure 2.24.

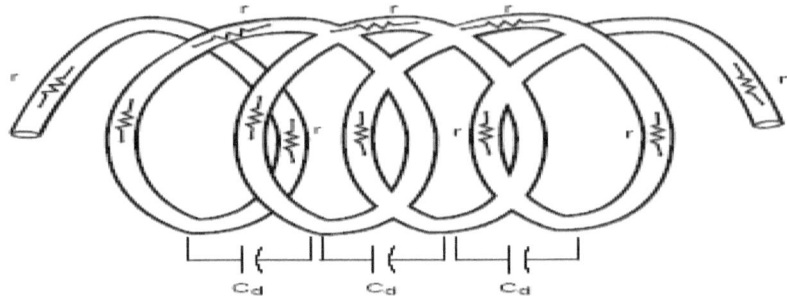

Figure 2.23 The distributed parasitic elements of an inductor.

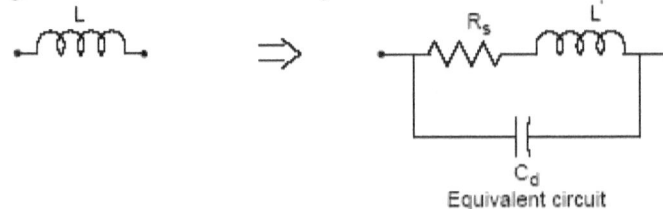

Figure 2.24 The equivalent circuit of an inductor at RF/MW frequencies.

The effect of C_d on an inductor's frequency response is shown in Figure 2.25. From this Figure (just like a capacitor) there are two regions that straddle the resonant circuit.

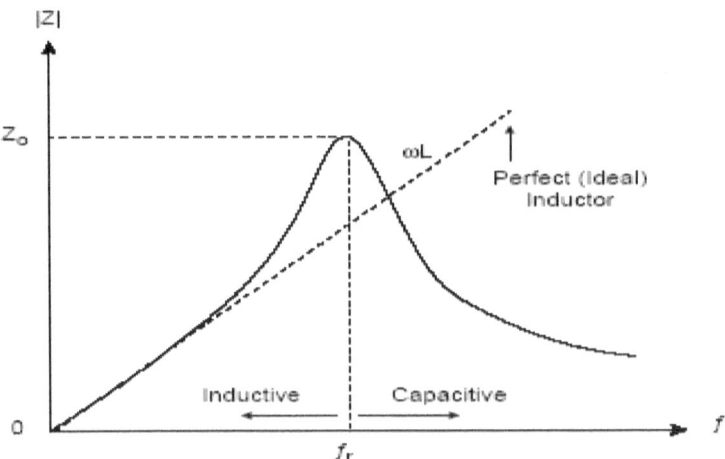

Figure 2.25 Effect of parasitic C_d on an inductor's reactance. These two regions can be identified as:

a. $f < f_r$

In this region, the inductor's reactance ($X_L=\omega L$) increases as frequency is increased.

b. $f > f_r$

In this region the inductor behaves like a capacitor and as the frequency is increased the reactance decreases.

At $f = f_r$ resonance takes place in an inductor and theoretically the inductor's reactance is infinity; however, in practice, the total impedance of the element is finite due to a non-zero series resistance.

The quality factor (Q) of an inductor is defined to be:

$$Q= X_L/R_S = \omega L/R_S \tag{2.8a}$$

For a perfect inductor, the series resistance is zero, thus we have:

$$R_S= 0 \quad \Rightarrow \quad Q=\infty \quad \text{(A perfect inductor)} \tag{2.8b}$$

At low frequencies, Q is very large since R_s is very small, however, as frequency increases the skin effect and winding distributed capacitor (C_d) begin to degrade the Q of an inductor as shown in Figure 2. 26.

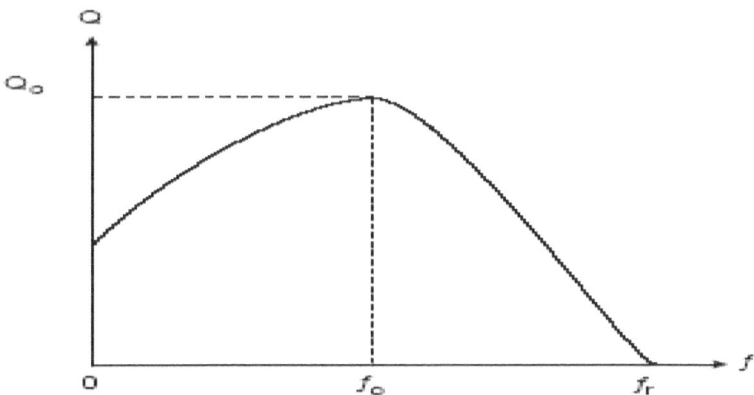

Figure 2.26 Degradation of the Q of an inductor.

From Figure 2.26, it can be seen that as frequency increases, Q will increase up to Q_o which is at $f = f_o$. However, for frequencies $f_o<f<f_r$, R_s and C_d combine to decrease the Q of the inductor toward

zero. At resonance (f=f$_r$), where the total reactance of the element is zero, the inductor is no longer useful.

To extend the frequency range of an inductor (by increasing its Q), we can do one of the following solutions:

a. Use a larger diameter for the wire which effectively reduces the resistance value, or

b. Spread the winding apart which reduces the distributed capacitance (C$_d$) between the windings, or

c. Increase the inductance (L) by increasing the permeability of the flux linkage path by using a magnetic-core material. These are shown in Figure 2.27.   Solution (c) is the most effective and practical one of all the three solutions.

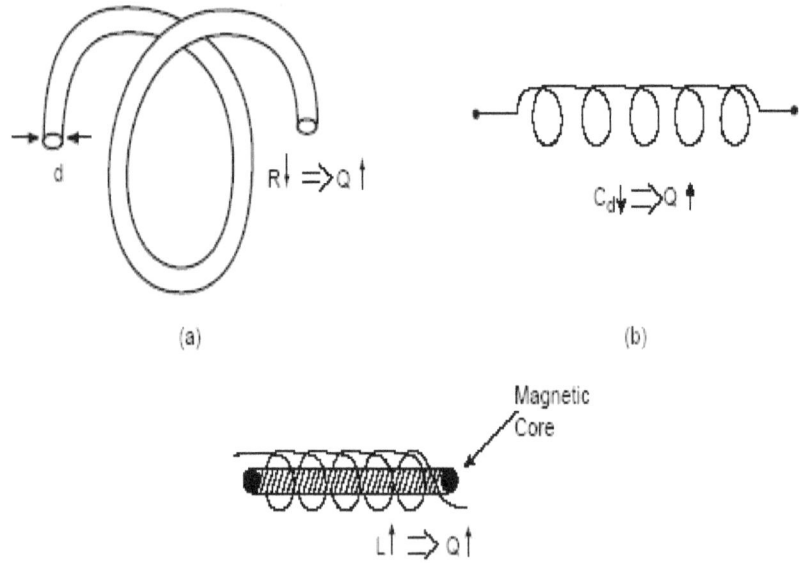

Figure 2.27 Three possible ways of increasing the Q in an inductor.

2.6 RESONANT CIRCUITS

A resonant circuit (also called a filter) is certainly not new and in fact has been and is used in practically every transmitter, receiver, or any piece of test equipment in existence.

Resonant circuit's function is to pass selectively, a certain frequency (or a frequency range) from the source to the load, while attenuating all other frequencies outside of this passband as shown in Figure 2.28.

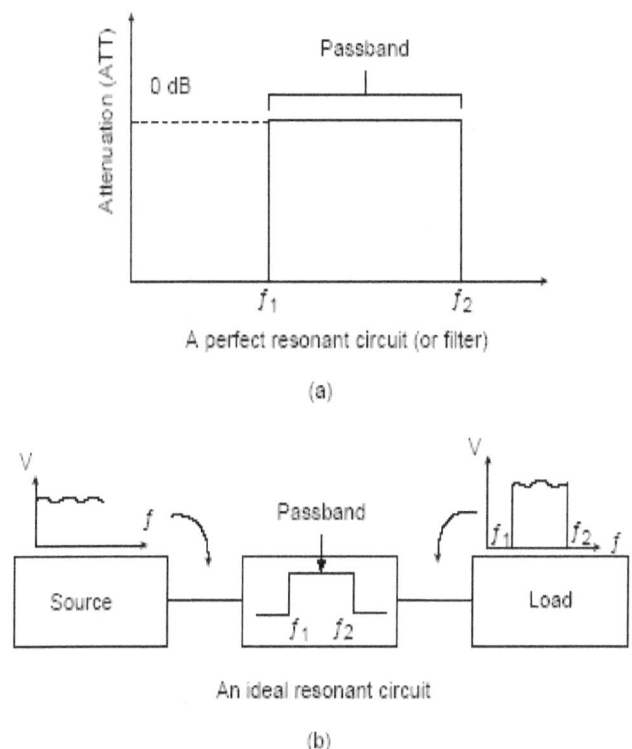

A perfect resonant circuit (or filter)

(a)

An ideal resonant circuit

(b)

Figure 2.28 An ideal resonant circuit: frequency response and an application.

Since there is no perfect component, a perfect resonant circuit does not exist and can not be built. However, knowing the mechanics of resonant circuits, an imperfect resonant circuit (or filter) can be tailored to suit our needs. A typical practical filter's frequency response is shown in Figure 2.29.

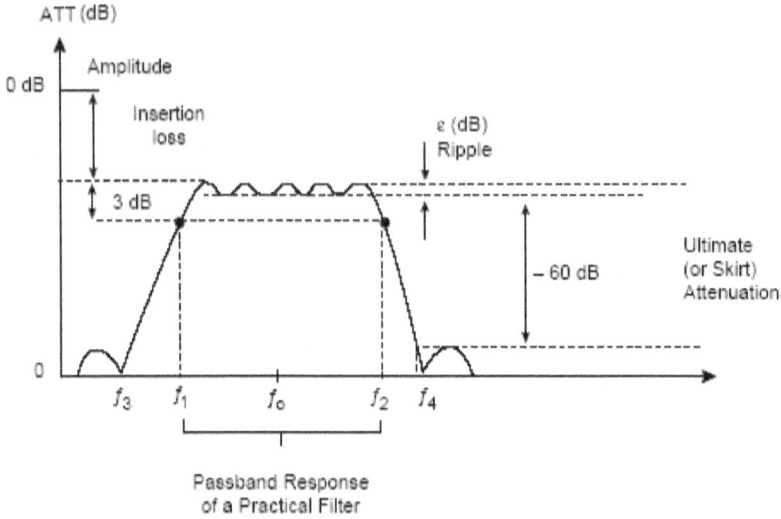

Figure 2.29 A practical filter frequency response.

From this Figure, several important features must be first defined:

a. DEFINITION-BANDWIDTH (BW): *Is the difference between upper* (ω_2) *and lower* (ω_1) *angular frequencies at which the amplitude response is 3 dB below the passband response value (also called the half-power BW). Therefore we can write: BW=(ω_2 - ω_1)*

b. DEFINITION-CIRCUIT Q: *Is the ratio of center angular frequency* (ω_o) *to the bandwidth (BW), i.e.,*

$$Q = \omega_o / BW = f_o/(f_2-f_1) = f_o/\Delta f \qquad (2.9a)$$

Where

$BW = (\omega_2 - \omega_1) = 2\pi\Delta f$,

$\Delta f = f_2-f_1$, $\omega_o = 2\pi f_o$,

$\omega_2 = 2\pi f_2$,

and $\omega_1 = 2\pi f_1$.

It is important to note that "circuit Q" should not be confused with "component Q" which is a measure of component loss, while "Circuit Q" is a measure of the selectivity of a resonance circuit which means that as the bandwidth (BW) decreases, the selectivity of the resonant circuit increases. Furthermore it should be noted that the component Q does have an effect on the circuit Q but the reverse is not true.

c. DEFINITION - SHAPE FACTOR (SF)- OF A RESONANT CIRCUIT: *is defined as being the ratio of the 60-dB bandwidth to the 3-dB bandwidth, i.e.,*

$$SF = \frac{f_4 - f_3}{f_2 - f_1} \tag{2.9b}$$

Shape factor (SF) is simply a measure of the steepness of the skirts. The smaller the SF number, the steeper the response skirts. A perfect filter has SF = 1, however, in practice SF is always greater than one (SF≥ 1). When SF is less than 1(SF<1), we have a physical impossibility as shown in Figure 2.30.

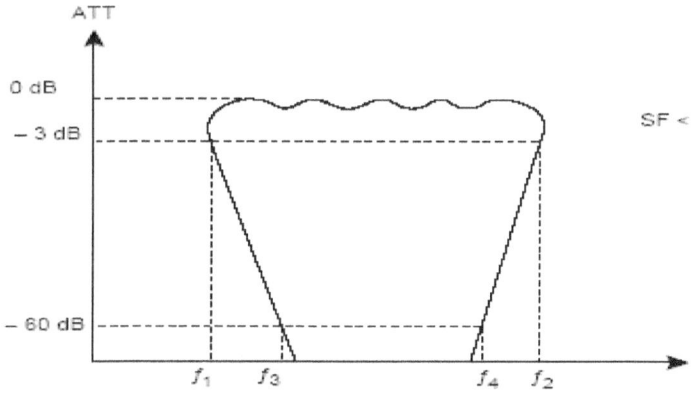

Figure 2.30 Physical impossibility when SF<1.

d. DEFINITION-ULTIMATE ATTENUATION: *Is the final minimum attenuation that the resonance circuit presents outside of the specified passband.*

A perfect resonant circuit has an ultimate attenuation of infinity. If there are response peaks outside of the passband, then this will detract from the ultimate attenuation specification of the circuit as shown in Figure 2. 31

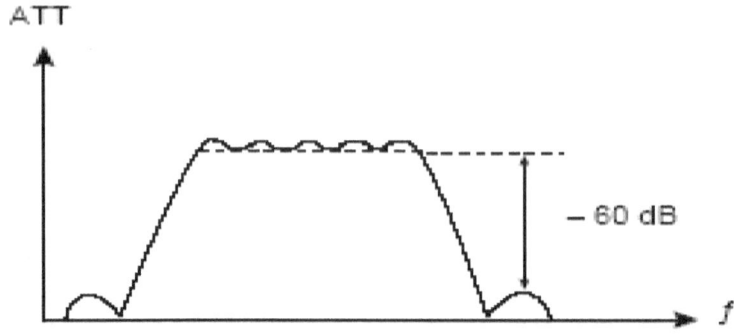

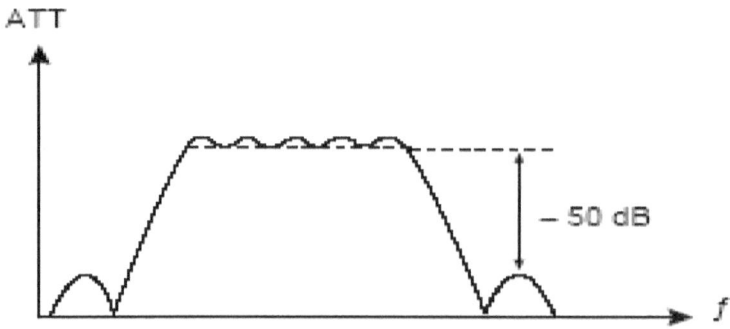

Figure 2.31 Response peaks outside the passband reduce ultimate attenuation.

e. DEFINITION-INSERTION LOSS: *is the attenuation resulting from inserting a circuit between source and load.*

This concept is depicted in Figure 2.32. Therefore the insertion loss is the attenuation that results from the insertion of a resonant circuit, usually expressed in dB.

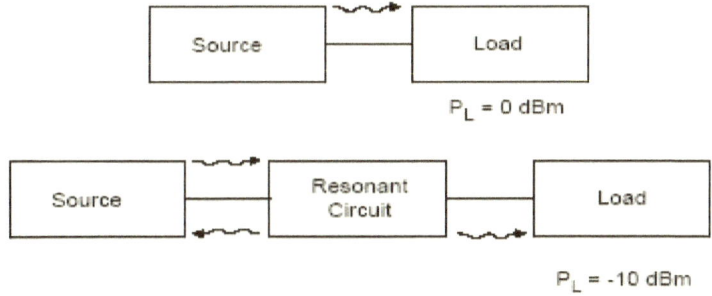

P$_L$ = 0 dBm

P$_L$ = -10 dBm

Insertion Loss = 10 dB

Figure 2.32 The concept of insertion loss.

f. DEFINITION- RIPPLE: *Is a measure of the flatness of the frequency response of the resonance circuit and is defined to be the attenuation difference (in dB) of the maximum value from the minimum in the passband, i.e.,*

ε=|max. attenuation - min. attenuation| *(in dB)*

This concept is shown in Figure 2.33a.

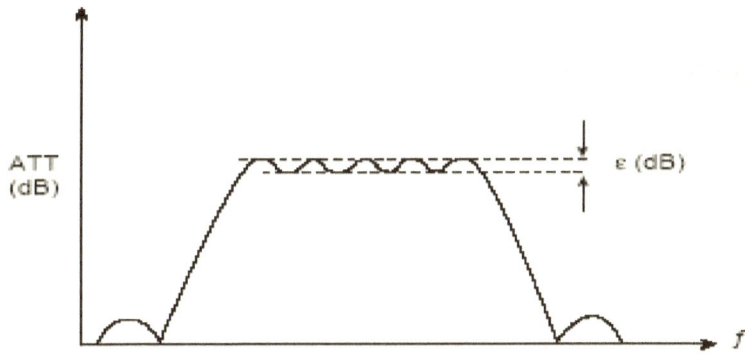

Figure 2.33a the concept of ripple in the passband.

2.7 RESONANCE

The term "resonance" and "resonant frequency" have exact meanings, which we need to define at this point as:

DEFINITION-RESONANCE: *Is a condition that occurs when the reactive part of Z or Y of a circuit vanishes at one or more frequencies. In this case Z or Y becomes purely a real number.*

RESONANT FREQUENCY- *The frequency at which resonance occurs is called the resonant frequency.*

Resonant circuits are used to separate out the wanted signals from spurious or unwanted signals, thus they can be considered to be excellent filters. At resonance, the impedance of most circuits goes through a sharp minimum or maximum, thus the selectivity of such circuits is often defined in terms of the half-power bandwidth centered around the resonant frequency, which effectively defines the width of the peak (when |Z| is max.) or notch (when |Z| is min.).

2.7.1 RLC Series Resonance

The series RLC circuit is shown in Figure 2.33b.

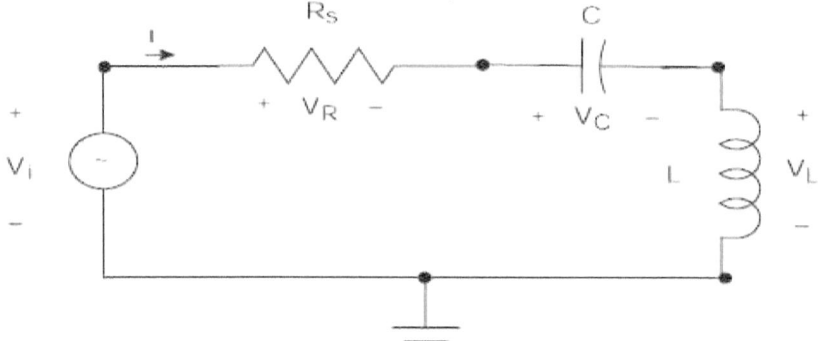

Figure 2.33b Series RLC Circuit

$$Z(j\omega)=V_i/I=R_S+j(\omega L-1/\omega C)=R_S+jX \qquad (2.10a)$$
Where $X=(\omega L-1/\omega C)$.

By definition, at resonance we should have $X=0$, which makes |Z| achieve a minimum value of R_S and thus makes it possible to obtain the resonant frequency (ω_o) as follows:
$$(\omega_o L-1/\omega_o C)=0 \Rightarrow \omega_o^2 LC=1$$
Or,
$$\omega_o = 2\pi f_o = 1/\sqrt{LC} \text{ rad/s}$$
As defined earlier,
$$Q=X_S/R_S=\omega_o L/R_S=1/R_S\omega_o C$$
Thus we can write Equation (2.10a) as:
$$Z(j\omega)=R_S[1+j(\omega L/R_S-1/R_S\omega C)]= R_S[1+jQ(\omega/\omega_o - \omega_o/\omega)] \quad (2.10b)$$

Using the half power frequencies $(\omega_1$ and $\omega_2)$, we can write:

$BW=\Delta\omega=f_2-f_1$.

$\omega_1=\omega_0-\Delta\omega \Rightarrow \omega_1/\omega_0 =1-\Delta\omega/\omega_0=1-k$

$\omega_2=\omega_0+\Delta\omega \Rightarrow \omega_2/\omega_0 =1+\Delta\omega/\omega_0=1+k$

where $k=\Delta\omega/\omega_0$

$|Z(j\omega_0)|= R_S$ (2.10c)

$|Z(j\omega_1)|=|Z(j\omega_2)|= \sqrt{2}|Z(j\omega_0)| =\sqrt{2} R_S$ (2.10d)

$|Z(j\omega_2)|= R_S|1+jQ[(1+k)-1/(1+k)]|$ (2.10e)

If $\Delta\omega<<\omega_0$, $\Rightarrow k<<1 \Rightarrow 1/(1+k) \approx 1$

Therefore, using (2.10d) and (2.10e) we can write:

$R_S (1+Q^2k^2)^{1/2} =\sqrt{2} R_S \Rightarrow Q^2k^2=1 \Rightarrow Q=1/k = \omega_0/\Delta\omega$

2.7.2 RLC Parallel Resonance

The parallel RLC circuit is shown in Figure 2.33c.

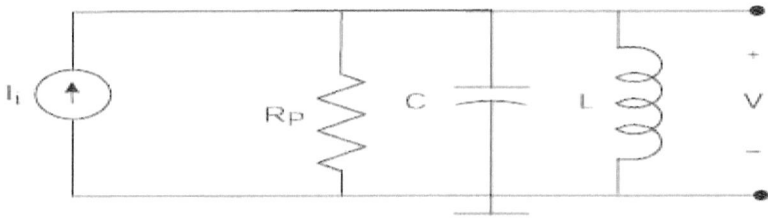

Figure 2.33c Parallel RLC Circuit

Dealing with a parallel circuit, we need to use the concept of admittance as follows:

$Y(j\omega)= I /Vi=R_P+j(\omega C-1/\omega L)=G_P+jB$ (2.11a)

Where $G_P=1/R_P$ and $B=(\omega C-1/\omega L)$.

By definition, at resonance we should have B=0, which makes $|Y|$ achieve a minimum value of G_P (or $|Z|$ achieve a maximum value of R_P). Thus we can obtain the resonant frequency (ω_0) as follows:

$(\omega_0 C-1/\omega_0 L)=0 \Rightarrow \omega_0^2 LC=1$

Or,

$\omega_0 = 2\pi f_0 = 1/\sqrt{LC}$ rad/s

As defined earlier, for a parallel circuit we have:

$Q=R_P/X_P=R_P\omega_0 C=R_P/\omega_0 L$

Thus we can write Equation (2.11a) as:

$Y(j\omega) =G_P[1+j(\omega C/G_P-1/G_P\omega L)]$

Or,

$$Y(j\omega) = G_P \left[1 + jQ(\omega/\omega_0 - \omega_0/\omega)\right] \tag{2.11b}$$

Using the half power frequencies (ω_1 and ω_2) for the admittance:

$\omega_1 = \omega_0 - \Delta\omega,$

$\omega_2 = \omega_0 + \Delta\omega,$

$$|Y(j\omega_1)| = |Y(j\omega_2)| = |Y(j\omega_0)|/\sqrt{2} \tag{2.11c}$$

Using a similar procedure to the series case, Equation (2.11c) yields:

$Q = \omega_0 / BW = \omega_0/\Delta\omega$

Where $BW = \Delta\omega = f_2 - f_1.$

Equations (2.11a) and (2.11b) for the parallel RLC circuit are the dual of the equations obtained for the series case (i.e., Equations 2.10a and 2.10b). They are dual of each other since one can replace Z with Y, L with C and R with G in the series equations and obtain the actual equations for the parallel case.

2.7.3 Analysis of a Simple Circuit in Phasor Domain

Consider the circuit shown in Figure 2.34 which consists of a series resistance (R_S) and shunt element with impedance Z_P.

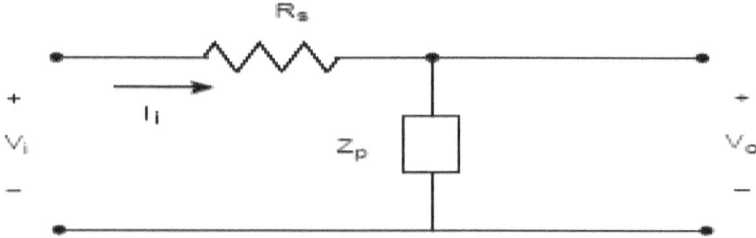

Figure 2.34 A simple series-shunt circuit.

We wish to calculate the total voltage gain (V_o/V_i) of the circuit. Using KVL in the phasor domain, we have:

$$V_o = Z_P\, I_i, \tag{2.12}$$
$$V_i = (R_S + Z_P)\, I_i \tag{2.13}$$

Dividing (2.12) by (2.13) we obtain the total voltage gain as:

$$V_o/V_i = H(\omega) = |H|\, e^{j\phi} = |H| \angle\phi = \frac{Z_P}{R_S + Z_P} \tag{2.14}$$

The gain magnitude in dB would be given by:

$$20 \log_{10}|H| = 20 \log_{10} |Vo/V_i| = 20 \log_{10}\left(\frac{Z_P}{R_S + Z_P}\right) \quad (dB) \quad (2.15)$$

NOTE 1: *From Equation (2.15) we can observe that output voltage magnitude will always be less than or at best equal to the input voltage magnitude which will be true for all passive circuits, i.e., for all passive circuits:*

$$|V_o| \le |V_i| \quad\quad\quad (2.16)$$

NOTE 2: *If the shunt element (in Figure 2.34) contains a capacitor or an inductor then the impedance of the shunt element (Z_P) will be frequency dependent and so would the output voltage(V_o) or the total voltage gain (V_o/V_i) as the following examples illustrate this point.*

EXAMPLE 2.4
If the shunt element in Figure 2.34 is a perfect capacitor, calculate and plot the voltage gain magnitude and phase.
Solution:
From the circuit shown in Figure 2.35 and Equation (2.14), we can write the following:

$Z_P = 1/j\omega C$

$$H(\omega) = |H|\angle\phi = V_o/V_i = \frac{(1/j\omega C)}{(R + 1/j\omega C)} = \frac{1}{1 + j\omega RC}$$

$$= \frac{1}{[1 + (\omega RC)^2]^{1/2}} \angle - \tan^{-1}(\omega RC)$$

Or,

$|H(\omega)| = 20 \log_{10}|H|$
$\quad\quad = -10 \log_{10}[1 + (\omega RC)^2] \quad (in\ dB)$
$\phi = -\tan^{-1}(\omega RC)$

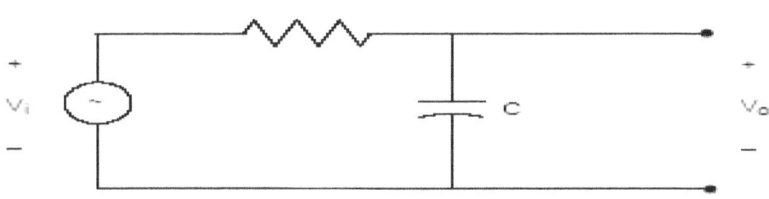

Figure 2.35 Circuit for example 2.4.

The magnitude and the phase are plotted in Figure 2.36a,b. From the magnitude plot we can see that this circuit performs like a low-pass filter.

NOTE: *Attenuation is 6 dB for every octave increase of frequency (i.e. doubling the frequency). This is due to a single reactive element. In general, for each significant reactive element added in the circuit, the slope will increase by an additional 6 dB.*

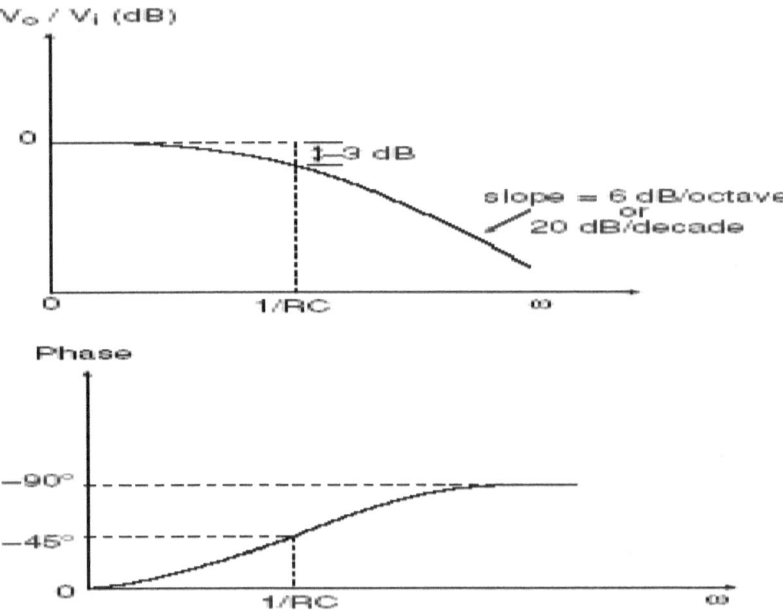

Figure 2.36 Frequency response: a) magnitude plot, b) phase plot.

EXAMPLE 2.5

If the shunt element in Figure 2.34 is a perfect inductor, calculate and plot the voltage gain magnitude and phase.

Solution:

From the circuit shown in Figure 2.37 and Equation (2.14), we can write the following:

$Z_P = j\omega L$

$$H(\omega)=V_o/V_i=\frac{j\omega L}{R_S + j\omega L}=\frac{1}{1- jR_S / \omega L}$$

$$=\frac{1}{\sqrt{1+ (R_S / \omega L)^2}} \angle \tan^{-1}(R_S / \omega L)$$

Or,

$|H| = 20 \log_{10}|H| = -20 \log_{10} [1+(R_S/\omega L)^2]^{1/2}$ (dB)

And,

$\phi = \tan^{-1}(R_S/\omega L)$

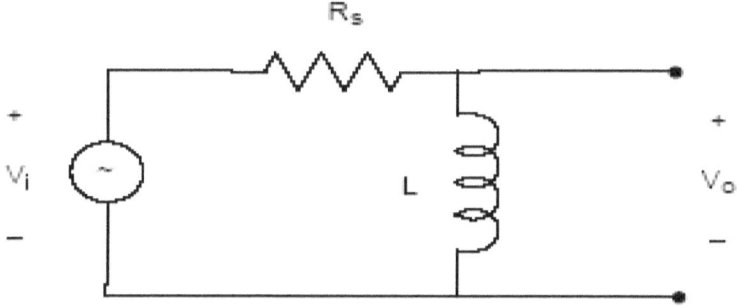

Figure 2.37 Circuit for example 2.5 .

The magnitude and the phase are plotted in Figure 2.38a,b. From the magnitude diagram we can see that this circuit performs like a high-pass filter.

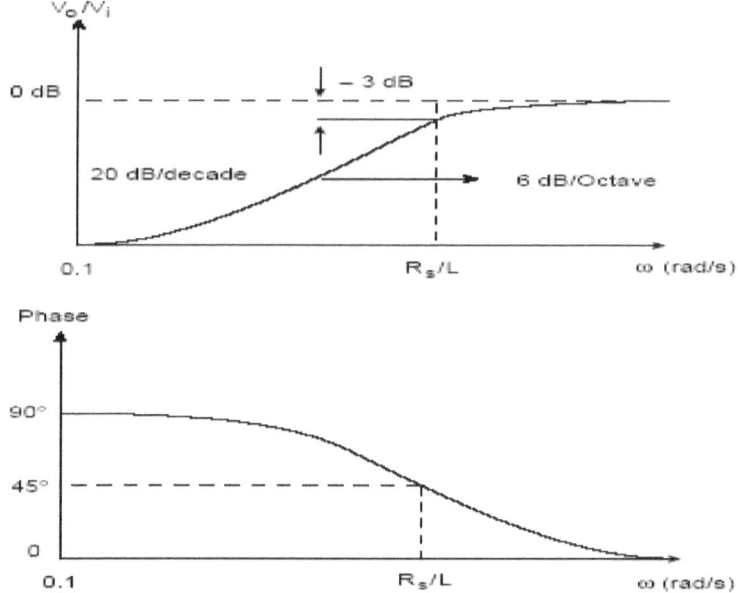

Figure 2.38 Frequency response: a) magnitude plot, b) phase plot.

EXAMPLE 2.6

If the shunt element in Figure 2.39 is a combination of a perfect capacitor in parallel with a perfect inductor, calculate and plot the voltage gain magnitude and phase.

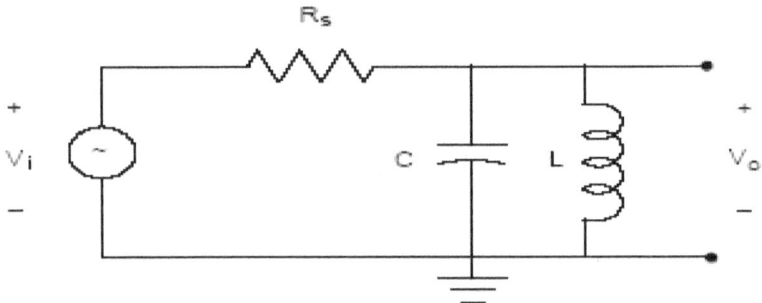

Figure 2.39 Circuit for example 2.6.

Solution:

From the circuit shown in Figure 2.39, we can write Z_P from Equation (2.14) as the parallel combination of the capacitor and inductor as follows:

$$Z_P= j\omega L \parallel -j/\omega C = \frac{j\omega L}{(1 - \omega^2 LC)}$$

$$H(\omega)=V_o/V_i= \frac{Z_P}{R_S + Z_P} = \frac{j\omega L}{R_S - \omega^2 R_S LC + j\omega L}$$

$$|H| = 20\ \log_{10}|H| = 20\log_{10}\left(\frac{\omega L}{\sqrt{R_S^2\left(1 - \omega^2 LC\right)^2 + (\omega L)^2}}\right) \quad \text{(dB)}$$

The magnitude is plotted in Figure 2.40.

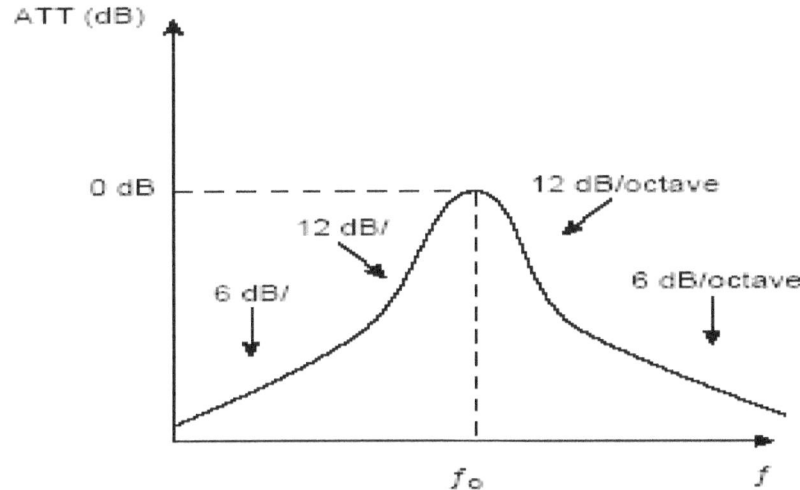

Figure 2.40 magnitude plot of the frequency response.

NOTE 1: *Near the resonance frequency of the tuned circuit, the slope of the resonance curve increases to 12 dB/octave, because there are now two significant reactances present and each one is changing at the rate of 6 dB/octave (Therefore 12 dB/octave slope.)*

NOTE 2: *Away from resonance, only one reactance becomes significant, therefore there would be only a 6 dB/octave of slope in effect.*

2.7.4 Loaded Q

The Q of a resonant circuit was defined earlier as:

$Q = \omega_0/\Delta\omega = \omega_0/BW$

The "circuit Q" is often called the "Loaded Q", because it describes the passband characteristics of the resonant circuit under actual "in-circuit" or "Loaded condition."

In general, the "loaded Q" depends on three main factors as follows:

a. The source resistance (R_S)

b. The load resistance (R_L)

c. The component Q (of each of the reactive elements)

Figure 2.41 shows a block diagram of a resonant circuit and its frequency performance.

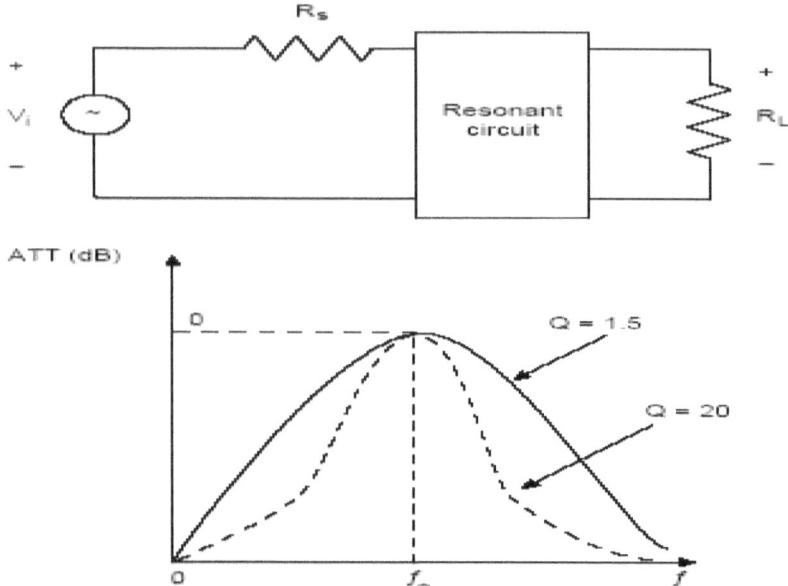

Figure 2.41 A resonant circuit: a) block diagram, b) frequency response.

From Figure 2.41, we can observe that:
a. The resonant circuit sees an equivalent resistance of R_S in parallel with R_L as its true load. This is shown in Figure 2.42.

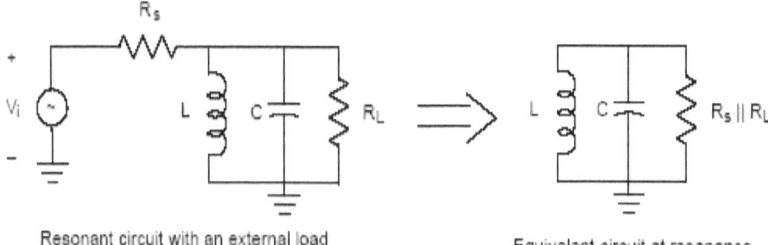

Resonant circuit with an external load   Equivalent circuit at resonance

Figure 2.42 A resonant circuit at resonance.

The loaded Q can be calculated by noting that:
$$R_P = R_S \| R_L, \tag{2.17}$$
X_P = the inductive or capacitive reactance of either of the reactive components (since they are equal at resonance!)

Therefore:

$$Q = R_P/X_P \qquad (2.18)$$

b. If R_S or R_L increases, then the equivalent resistance increases, which will reduce the energy losses and thus will narrow the curve. This will increase the selectivity and as a result the "loaded Q" .

c. For a fixed R_P, if X_P is decreased by choosing a smaller "L" or a larger "C", Q will increase. This point is illustrated in the next example.

EXAMPLE 2.7
Design a resonant circuit with a loaded Q=1.1 at f=142.4 MHz that operates between a source resistance of 100 Ω and load resistance of 100 Ω. Discuss how to increase Q. Use perfect components.
Solution:
R_P=100 ‖ 100 = 50 Ω
X_P =R_P/Q= 50/1.1= 45.45 Ω = ωL = 1/ωC
Choose: L=50 nH, C=25 pF ,
Given fixed R_S and R_P, we can increase Q by 20 times through scaling up the capacitor value by 20 while scaling down the inductor value by 20, i.e.,
Q=22 ⇒ C_{new}=500 pF, L_{new}= 2.5 nH
These two cases are shown in Figure 2.42.
Figure 2.43 shows two equivalent circuits for two different Qs obtained by scaling the inductor's and the capacitor's values appropriately.

Q = 1.1, f_0 = 142.4 MHz   Q = 22, f_0 = 142.4 MHz

Figure 2.43 Changing the Q value.

Therefore a circuit designer has two design approaches in designing resonant circuits:

a.   Select an optimal value of R_S and R_L to get the specified Q, or
b.   Given R_S and R_L, select component values of "L" and "C" to optimize Q.

NOTE 1: *If poor quality reactive components (i.e. low Q) are used in highly selective resonant circuits, the net result is that we effectively place a low-value shunt resistor directly across the circuit which will drastically reduce its loaded Q and increase the bandwidth.*

NOTE 2: *At resonance, an ideal LC parallel circuit has a very high (ideally infinite) total impedance as shown in Figure 2.44*

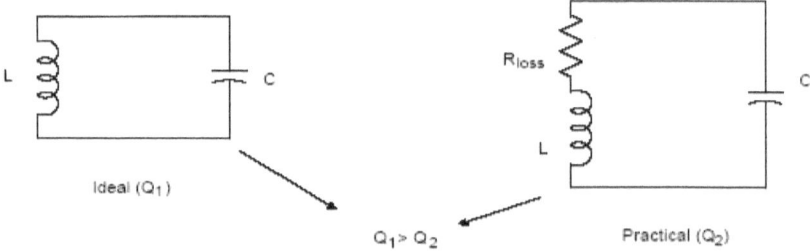

Figure 2.44 Ideal vs. practical resonant circuit.

NOTE 3: *Usually we only need to involve the Q of the inductors in the loaded-Q calculations, since the Q of most capacitors are quite high over their useful frequency range which means that they have a very small resistive passive element.*

2.7.5 Impedance Transformation
The most common type of impedance transformation is from the series elements to shunt elements as shown in Figure 2.45

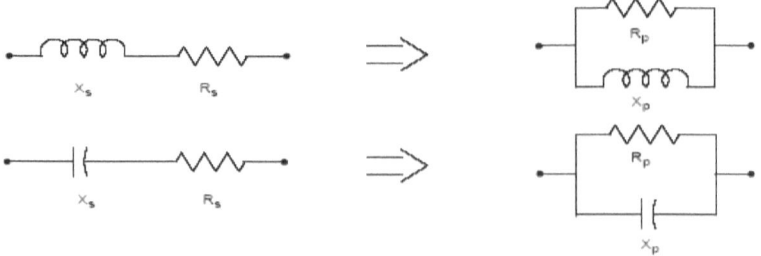

Figure 2.45 Impedance transformation.

We first define the component "Q" (represented by Qc) as:
$$Q_C = Q_S = Q_P, \quad \text{(component Q)} \tag{2.19}$$

Equation (2.19) states that the Q remains the same in the process of series-to-shunt transformation which is a correct assumption since we are still dealing with the same element even though we are changing its equivalent circuit.

Through simple mathematical manipulation, we can write:
$$R_P = (Q_C^2 + 1) R_S \tag{2.20}$$

$$X_P = R_P/Q_C \tag{2.21}$$
Using (2.21) for a shunt configuration, Q_C is defined as:
$$Q_C = R_P/X_P \quad \text{(shunt)} \tag{2.22}$$
which is in contrast with the series configuration earlier defined as:
$$Q_C = X_S/R_S \quad \text{(series)} \tag{2.23}$$

EXERCISE 2.1
It will be a worthwhile exercise to derive the above impedance transformation equations, i.e. Equations (2.20) and (2.21).
HINT: Set Z_{in} of each circuit (Real and Imaginary), equal to each other.

EXAMPLE 2.8
An imperfect inductor has an inductance of 50 nH with a series loss resistance of 10 Ω. Find the following:
a. Q_C at 100 MHz
b. The equivalent parallel configuration at f=100 MHz

Solution:
a. $Q_C = X_S/R_S = 2\pi f\ L/R_S = 2\pi \times 100 \times 10^6 \times 50 \times 10^{-9}/10 = 3.14$
b. $R_P = (Q_C^2+1)R_S = (3.14^2+1) \times 10 = 108.7\ \Omega$
    $X_P = R_P/Q_C = 108.7\ /\ 3.14 = 34.62\ \Omega$
    $X_P = \omega L_P \Rightarrow L_P = X_P/\omega = 55.1\ nH$

The equivalent circuit is shown in Figure 2.46.

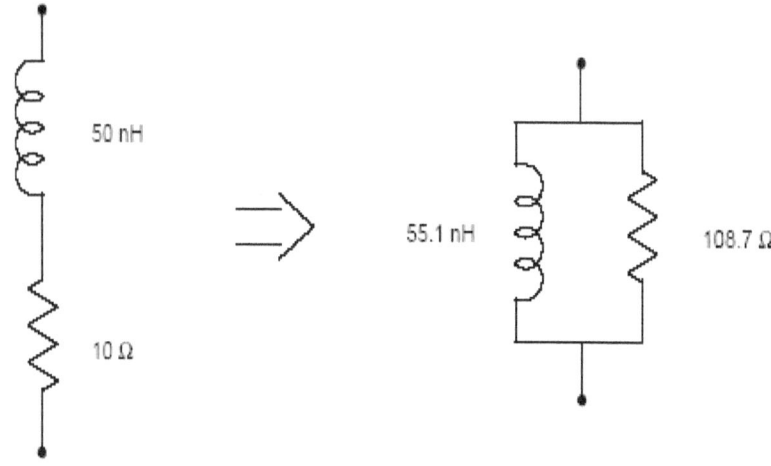

Figure 2.46 Circuit transformation.

2.7.6 Insertion Loss(IL)

If the inductor and the capacitor were perfect components with no internal loss, then the insertion loss for LC resonant circuits would have been zero dB.

In actuality, this is not the case and insertion loss is a very critical parameter in specifying a resonant circuit as shown in Figure 2.47.

DEFINITION- INSERTION LOSS : *Is a positive number expressing the difference (in dB) between the power received at the load before and after the insertion of a circuit or component in the transmission line connecting a source to a load.*

Mathematically, we can write:
Insertion Loss =IL(dB)= $-10\log_{10}(V_o'/V_o)^2$ =$-20\log_{10}(V_o'/V_o)$

Where V_o is the voltage at the load before the insertion, whereas V_o' is the voltage after the circuit is placed in the line. For example, a perfect LC circuit with no resistive loss has zero dB insertion loss as shown below.

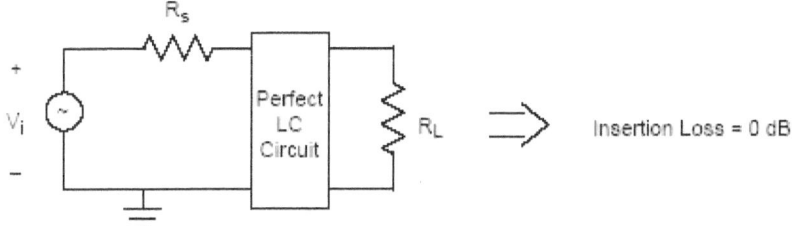

Figure 2.47 Insertion loss of a perfect LC circuit.

The following example will illustrate the concept of insertion loss further.

EXAMPLE 2.9
Calculate the insertion loss of the LC resonant circuit shown in Figure 2.48 at f=1430 MHz.
$R_S=R_L= 1 \text{ k}\Omega$
Inductor: $L= 0.05 \text{ }\mu\text{H}$, $Q_{C1}=10$
Capacitor: $C=25 \text{ pF}$, $Q_{C2}= \infty$

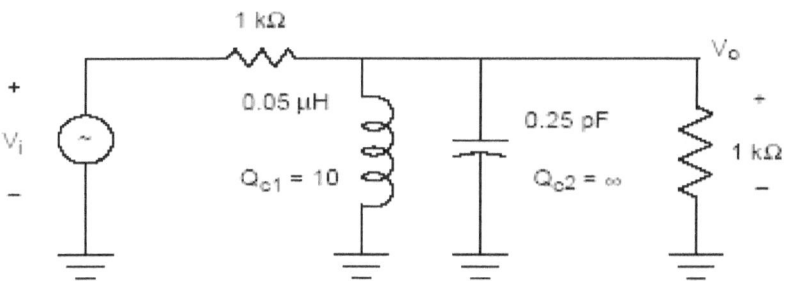

Figure 2.48 Circuit for example 6.9.

Solution:
a. Removing the LC circuit gives:
    $V_o=1000/(1000+1000) \times V_i= 0.5 \text{ } V_i$
b. Next we convert the inductor's series configuration into parallel:
    $Q_{C1}=X_{SL}/R_{SL} \Rightarrow R_{SL}=2\pi \times 1430 \times 10^6 \times .05 \times 10^{-6}/10 = 45 \text{ }\Omega$
    $R_{PL}=(Q_{C1}^2 + 1) R_{SL} \Rightarrow R_{PL}= 4.5 \text{ k}\Omega$

Therefore at resonance, we have a circuit as shown in Figure 2.49:

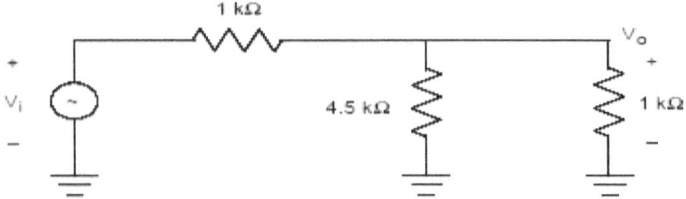

FIGURE 6.49   The equivalent circuit of Example 6.9.

Figure 2.49 The equivalent circuit for example 6.9.

$(R_L)_{EQ}$= 1 k || 4.5 k = 820 Ω

Thus the load voltage with resonant circuit in place is:

V_o' = 820/(1000+820) V_i = 0.45 V_i

Insertion Loss (IL)=V_o'/V_o = 0.45/0.5=0.9

IL(dB)= -20log$_{10}$(0.9) = 0.92 dB

An insertion loss of 0.92 dB may not appear much but can add up very quickly if we cascade several resonant circuits together.

EXAMPLE 2.10

Design a simple parallel resonant circuit to work between a source resistance (R_S) of 1 kΩ and load resistance (R_L) of 1 kΩ to provide a 3dB bandwidth of 10 MHz at a center frequency (f_o) of 100 MHz. Also calculate the insertion loss. Assume all capacitors are perfect and the inductor has a Q of 85.

Solution:

Considering the circuit shown in Figure 2.50, we have:

$Q = f_o/\Delta f = 100/10 = 10$

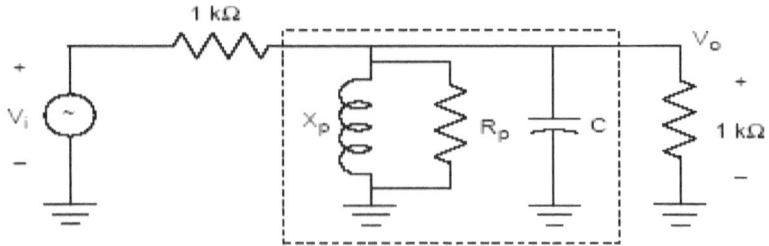

Figure 2.50 Circuit for example 2.10.

For inductor:

$Q_C=R_P/X_P = 85 \Rightarrow R_P=85\ X_P$ (1)

The loaded Q for the circuit is:

$Q=10= R_{tot}/X_P = (R_P \| 1\text{ k} \| 1\text{ k})/ X_P$ (2)
Solving (1) and (2) for R_P and X_P we obtain:
$L=X_P/\omega_o=70$ nH
$X_P=44.1\ \Omega =\omega_o L=1/\omega_o C \Rightarrow C=1/\omega_o X_P = 36$ pF
$R_P = 3.75$ kΩ
To find the insertion loss we note two cases as follows:
a) Without the resonant circuit
$V_o= 1\text{ k}/ (1\text{ k}+1\text{ k})=0.5\ V_i$
b) With the resonant circuit in place
$(R_L)_{EQ}=R_P \| R_L = 3.75\text{ k} \| 1\text{ k}=789.5\ \Omega$
$V_o' = V_i \times 789.5/(1000+789.5) = 0.44\ V_i$
I.L.(dB)= 20 $\log_{10}$ (0.44 V_i/0.5 V_i)= 1.1 dB

2.8 IMPEDANCE TRANSFORMERS

By observation, it becomes apparent that low values of R_S and R_L tend to load down a given resonant circuit, leading to a decrease of its Q and broadening of its bandwidth. Thus it is very difficult to design a high-Q simple LC resonant circuit that would function well between two low values of R_L and R_S as shown in Figure 2.51.

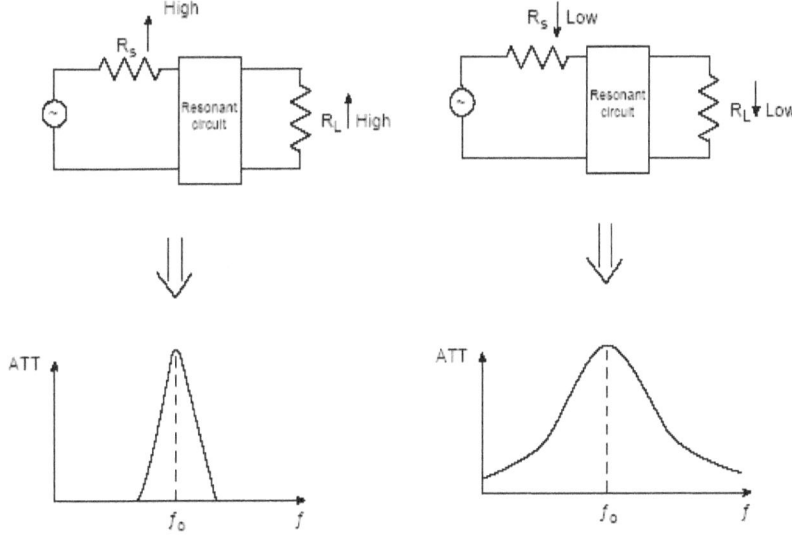

Figure 2.51 Comparison of frequency responses.

To solve this design problem, two types of impedance transformer may be used as shown in Figures 2.52 and 2.53. The impedance transformer placed between the load and the source in the circuit converts R_S or R_L to a much larger resistance. Thus the Q of the circuit is pushed to a higher value, since the resonating circuit as a whole is presented with a higher resistance value.

As shown in Figures 2.52 and 2.53 there are two types of transformers:

a. Tapped-C transformer and,
b. Tapped-L transformer.

These are described next.

2.8.1 Tapped-C Transformer

$$R_S' = R_S(1+C_1/C_2)^2, \tag{2.24a}$$

$$C_T = \frac{C_1C_2}{C_1+C_2} \tag{2.24b}$$

Where "C_T" is the equivalent capacitance that resonates with "L".

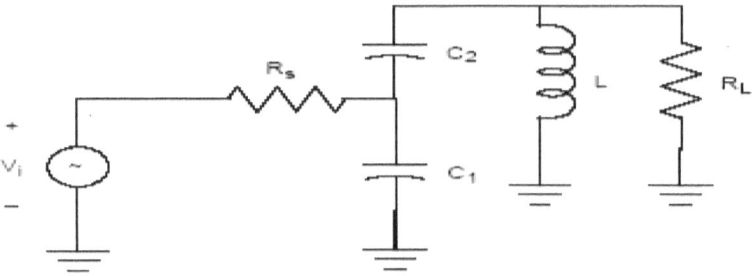

Figure 2.52 Tapped -C transformer.

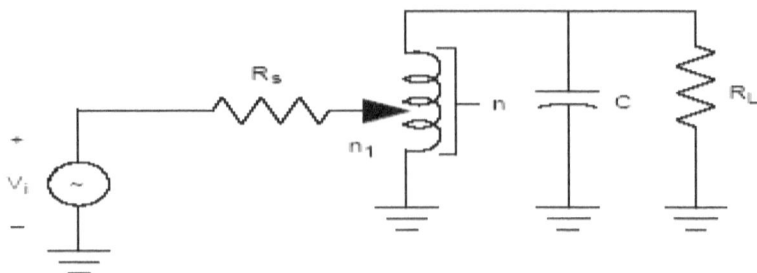

Figure 2.53 Tapped -L transformer.

2.8.2 Tapped-L Transformer

$$R_S' = R_S (n/n_1)^2 \qquad (2.25)$$
$$L_T = L \qquad (2.26)$$

The transformer circuits (a) and (b) present a much larger R_S and R_L that is actually present. For example for a circuit with $R_S = 50\ \Omega$, the transformer would turn the 50 Ω into a 500 Ω and the circuit will be able to see a higher R_S and thus its "Q" would be much higher.

EXAMPLE 2.11

Design a resonant circuit (as shown in Figure 2.54) such that it operates between $R_S=50\ \Omega$ and $R_L=2000\ \Omega$ with a Q=20 at the center frequency $f_o=100$ MHz. The inductor has a $Q_C = 100$ at 100 MHz. You may use a tapped-C transformer to achieve the desired Q.

Solution:
We will use a tapped-C transformer to step up $R_S=50\ \Omega$ to 2000 Ω in order to match the load resistance for maximum power transfer.
$$R_S'=R_S(1+C_1/C_2)^2 \Rightarrow 2000=50(1+C_1/C_2)^2 \Rightarrow C_1/C_2= 5.3$$
Inductor: $Q_C=R_P/X_P=100 \Rightarrow R_P=100\ X_P \qquad (1)$
$Q= R_{tot}/X_P = (R_S' \parallel R_L \parallel R_P)/X_P \qquad (2)$

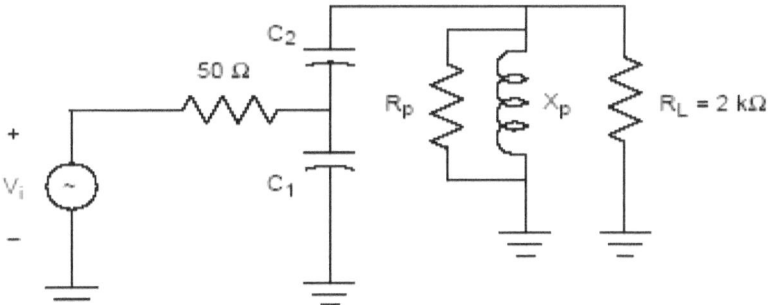

Figure 2.54 Circuit for example 2.11.

Using (1) and (2), we solve for R_P and X_P:
$R_P=4\ k\Omega$
$X_P=40\ \Omega \Rightarrow L=X_P/\omega_o = 63.6$ nH
$C_T= 1/ X_P\omega_o = 39.8$ pF
$$C_T=39.8=\frac{C_1 C_2}{C_1 + C_2} \Rightarrow C_1/C_2=5.3$$

If we select C_1=250 pF, then:
C_2=47 pF
The final design is shown in Figure 2.55.

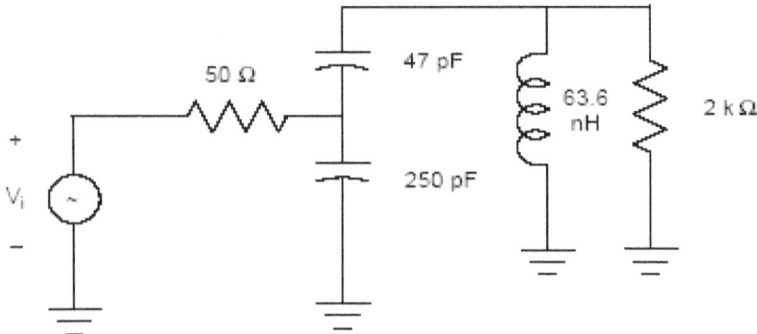

Figure 2.55 Circuit for example 6.11.

2.9 RF IMPEDANCE MATCHING

Impedance matching is often necessary in the design of RF circuitry to provide maximum possible transfer of power between a source and its load as shown in Figure 2.56.

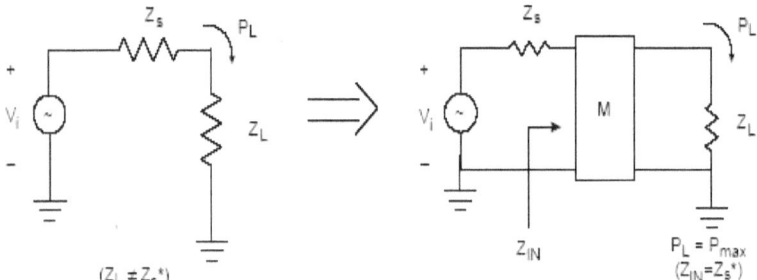

Figure 2.56 The concept of using an RF impedance matching network.

One of the applications of an RF impedance matching network is in the front end of any sensitive receiver where the signal is extremely weak and none of it can be wasted due to mismatch. Therefore the use of an appropriate matching network becomes crucial to the overall performance of the circuit.

The maximum power transfer theorem states that:

For DC: $Z_S = R_S = R_L = Z_L$ (2.27)
(i.e. There is no reactance)

For AC: $Z_L = Z_S{}^* \Rightarrow R_S = R_L$ (2.28a)

                         $X_S = -X_L$ (2.28b)

NOTE: $X_S = -X_L$ *is valid only at one frequency (the frequency of resonance). Therefore, a perfect match can occur only at the resonant frequency which poses a problem in the broadband matching of circuits.*

At all other frequencies removed from the matching center frequency, the impedance match becomes progressively worse and eventually non-existent as shown in Figure 2.57.

There is an infinite number of possible networks that could be used to perform the impedance matching function. For Example a circuit as simple as a 2-element LC network or as elaborate as a 7-element matching network would work equally well. But first we will analyze a simple matching network as illustrated in the following example.

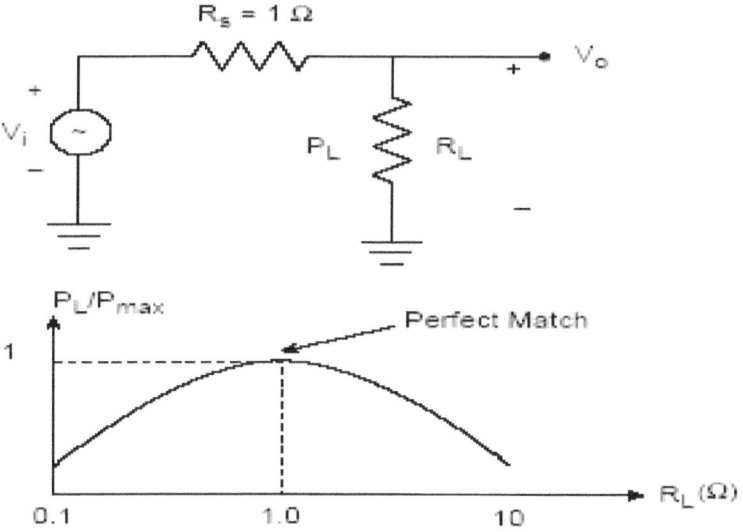

Figure 2.57 Power transfer as a function of load impedance.

EXAMPLE 2.12
Analyze the LC matching network as shown in Figure 2.58, which transforms a source resistance $R_S = 100\ \Omega$ to a load $R_L = 1000\ \Omega$..

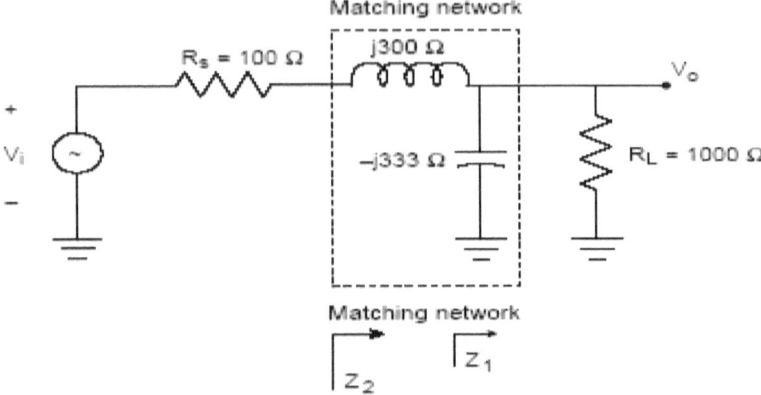

Figure 2.58 Circuit for example 2.12.

Solution:
From Figure 2.59 we have:
$Z_1 = 1000||-j333 = 100-j300$ Ω
$Z_2 = j300 + (100-j300) = 100$ Ω

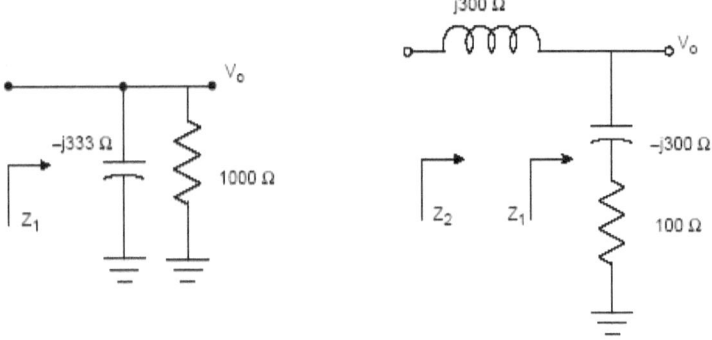

FIGURE 6.59   Circuit analysis for Example 6.12.

Figure 2.59 Circuit analysis for example 2.12.

Therefore the source sees a 100 Ω as the total load which creates a matched condition with the source resistance ($R_S=100$ Ω) and therefore maximum power transfer (for $V_i=1$ V) occurs as follows:
$V_o=0.5V_i$
$(P_L)_{max} = 1/2$ (V_o^2/R_{in})
$= 1/2(0.5^2/100) = 1.25$ mW$= 0.97$ dBm
These calculations bring us to an important question:

How much power would have been transferred if matching were not placed between R_S and R_L?
From the diagram shown in Figure 2.60, we can write:
$V_i = 1$ V
$V_o = (1000/1100)V_i = 0.91$ V_i
$P_L = (V_o^2/R_L)/2 = (0.91)^2 /2 \times 1000 = 0.41$ mW $= -3.83$ dBm

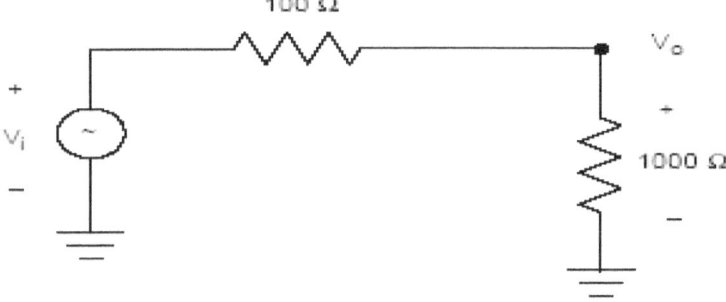

Figure 2.60 Simplified equivalent circuit.

Therefore compared to the matched case, the power loss is now:
Power loss$= 0.41/1.25 = 0.328 = -4.89$ dB

This shows that only 1/3 of the available power is transferred and the 2/3 remaining is wasted (or reflected back to the source) due to mismatch.

NOTE : *The function of the shunt component is to transform a larger impedance down to a smaller value with a real part equal to the real part of the source impedance and the reactive component capable of resonating(or canceling out) with the reactive part of the source impedance (see Figure 2.61).*

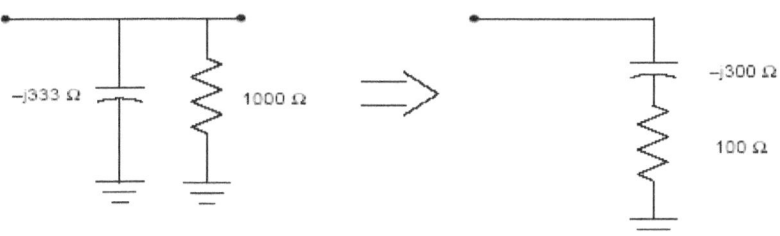

Figure 2.61 The result of adding a shunt element.

2.9.1 The L-Network

The simplest and most widely used matching circuit for lumped elements is the L- network as shown in Figures 2.62 a, b, c, and d.

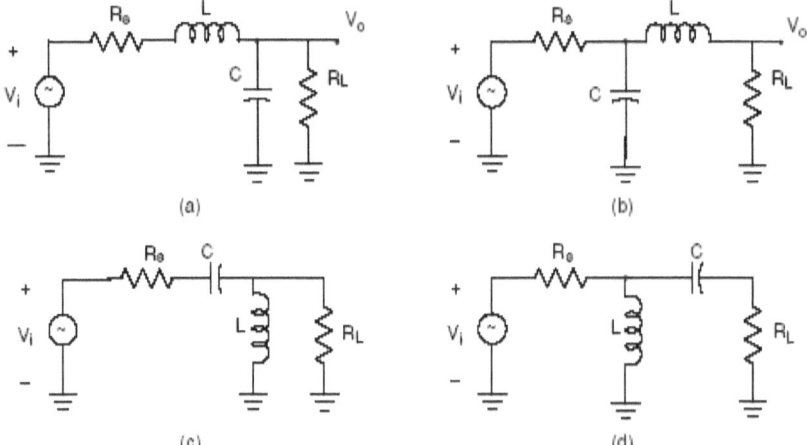

(a)   (b)

(c)   (d)

Figure 2.62 Several L-Network configurations.

The generalized configuration is shown in Figure 2.63 where we can write:

$$R_P = (Q^2 +1)R_S \qquad\qquad (2.29)$$
$$Q = Q_S = Q_P \qquad\qquad (2.30)$$
$$Q_S = Q_P = \sqrt{(R_P / R_S)-1} \qquad\qquad (2.31)$$
$$Q_S = X_S/R_S \qquad\qquad (2.32)$$
$$Q_P = R_P/X_P \qquad\qquad (2.33)$$

where Q_S, R_S and X_S are for the series leg and Q_P, R_P and X_P are for the parallel leg.

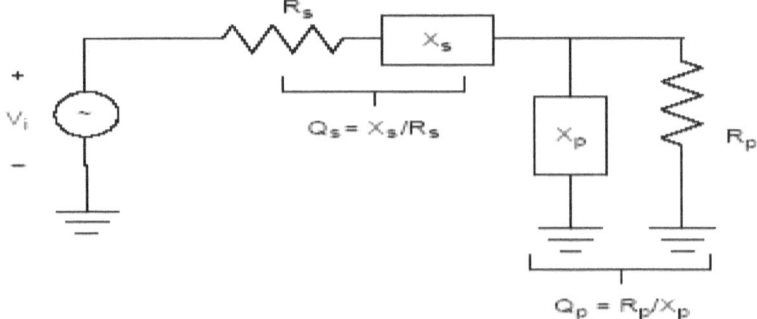

Figure 2.63 Generalized L-network configuration.

NOTE: *X_S and X_P may be either capacitive or inductive reactances but must be of opposite types. The following example illustrates the concept of L-networks more clearly.*

EXAMPLE 2.13

Design a circuit to match a 100 Ω source resistance to a 1000 Ω load resistance at f_o = 100 MHz. Assume that a DC voltage must also be transferred from the source to the load and all elements are perfect.

Solution:
The need for DC at the output dictates the need for an inductor in the series leg as shown in Figure 2.64.

R_S=100 Ω, R_P=1000 Ω

Q_S=Q_P=$(1000/100 - 1)^{1/2}$ = 3

Q_S=X_S/R_S $\Rightarrow$ X_S=3R_S=300 Ω

Q_P=R_P/X_P $\Rightarrow$ X_P=$R_P/3$=1000/3=333 Ω

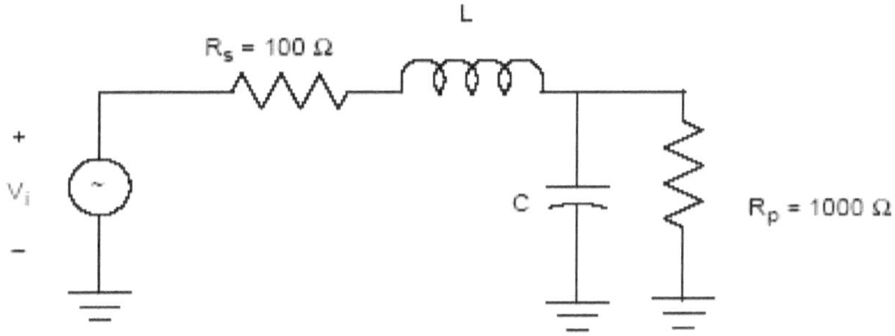

Figure 2.64 Circuit design for example 2.13.

X_S=ωL $\Rightarrow$ L=X_S/ω = 477 nH

X_P=$1/\omega C$ $\Rightarrow$ C=$1/\omega X_P$= 4.8 pF

NOTE: *This circuit was analyzed earlier in example 2.12.*

The previous examples dealt with matching two real impedances. However in actual practice we deal with transistors, transmission lines, antennas, etc., which all present complex input and output impedances as shown in Figure 2.65.

There are two basic approaches to handling complex impedances :

a. **The Absorption Method;**

b. **The Resonance Method.**

These two methods are explained below:

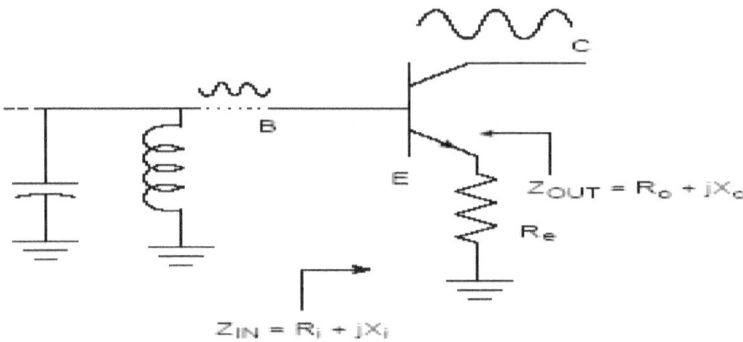

Figure 2.65a A transistor circuit.

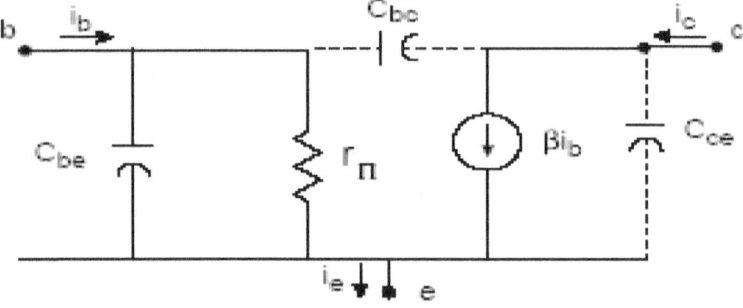

Figure 2.65b The AC equivalent circuit.

2.9.2 The Absorption Method

This is a method in which any stray reactances can be absorbed into the impedance-matching network by prudent placement of each matching element such that the following occurs:

1. Element capacitors (C) are placed in parallel with stray capacitors (C_p),

2. Element inductors (L) are placed in series with any stray inductors (L_p).

3. Next, the stray component values are then subtracted from the calculated element values to arrive at the final matching network. The new element values C' and L', are given by:

$$C = C_p + C' \Rightarrow C' = C\text{-}C_p \qquad\qquad (2.34)$$
$$L = L_p + L' \Rightarrow L' = L\text{-}L_p \qquad\qquad (2.35)$$

The following example further illustrates this point.

EXAMPLE 2.14

Use the absorption method to match the source (100+j126 Ω) to a load (1000 -j795.8Ω) at 100 MHz as shown in Figure 2.66.

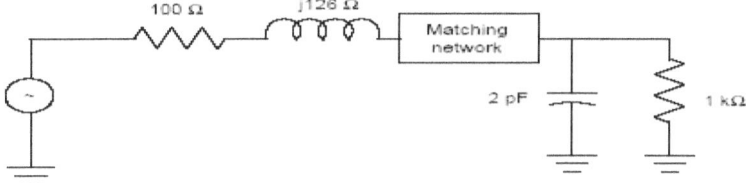

Figure 2.66 Circuit for example 2.14.

Solution:

Step 1: Totally ignore the reactances and simply match 100 Ω to 1000 Ω such that the inductors are in series and capacitors are in parallel. This step has already been done in example 2.13 and we use the results for the matching network directly (see Figure 2.67), i.e.

L=477 nH,
C=4.8 pF.

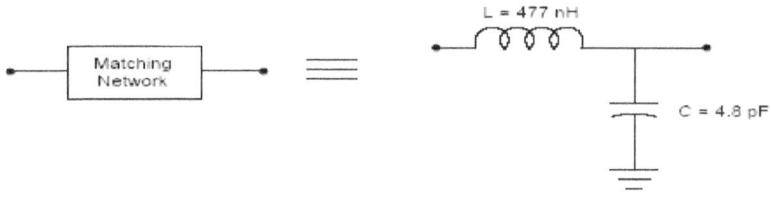

Figure 2.67 Matching network realization.

Step 2: The new elements are given by:

$$L' = L\text{-}L_p = 477\text{-}200 = 277 \text{ nH}$$
$$C' = C\text{-}C_p = 4.8\ \text{-}2 = 2.8 \text{ pF}$$

Based on these values, the final design is shown in Figure 2.68.

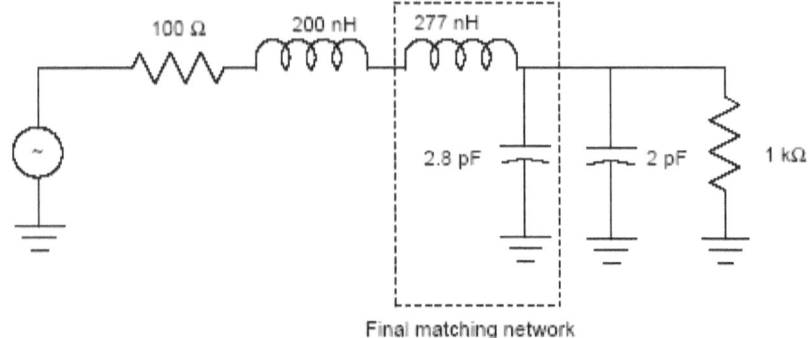

Figure 2.68 Final design for example 2.14.

NOTE: *This method is workable only if the calculated element values (L,C) are higher than the stray values (L_p, C_p).*

2.9.3 The Resonance Method

This is a method in which any stray reactances are resonated with an equal and opposite reactances at the frequency of interest. Once this is done, the design proceeds the same way as two pure resistances: one at the source and the other at load. The following example illustrates this point.

EXAMPLE 2.15

Design a matching network that will match a source resistance of 50 Ω to a capacitive load (at f_o= 75 MHz) as shown in Figure 2.69. The matching circuit should block the DC to the output. Use the resonance method.

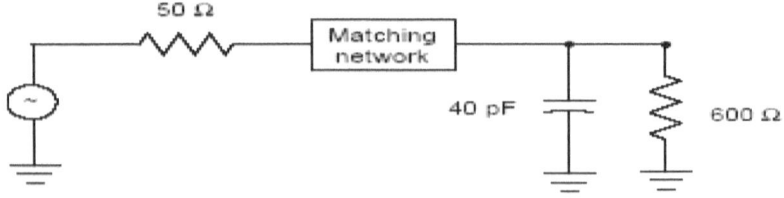

Figure 2.69 Circuit for example 2.15.

Solution:
Step 1: Resonate 40 pF with a shunt inductor (L_1) with the following value (see Figure 2.70):

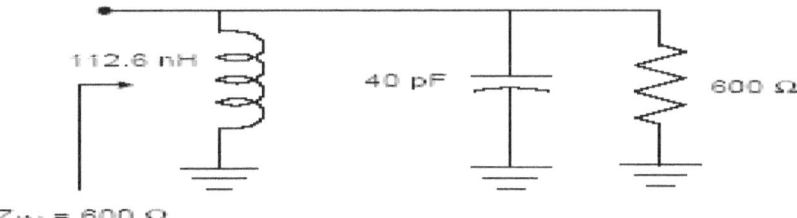

Figure 2.70 Resonating shunt inductor with the capacitor.

$\omega L_1 = 1/\omega C \Rightarrow L_1 = 1/C\omega^2 \Rightarrow L_1 = 112.6$ nH

Step 2: Now we match 50 Ω to 600 Ω by the same technique as before (see Figure 2.71):

$$Q_S = Q_P == \sqrt{(600/50) - 1} = 3.32$$

$X_S = Q_S R_S = 50 \times 3.32 = 166 \ \Omega$

$X_P = R_P/Q_P = 600/3.32 = 181 \ \Omega$

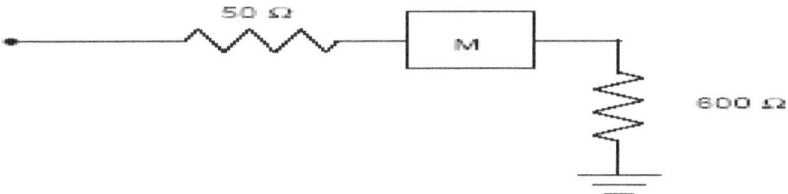

Figure 2.71 Matching 50 Ω to 600 Ω.

Using a series cap to block the DC: $C = 1/\omega X_S = 12.8$ pF

$L_2 = X_P/\omega = 384$ nH

Combining L_1 and L_2 in parallel (as shown in Figure 2.72) we obtain:

$$L_{tot} = \frac{L_1 L_2}{L_1 + L_2} = 87 \text{ nH}$$

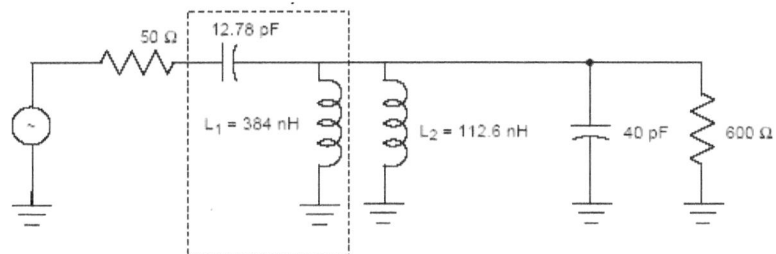

Figure 2.72 Using L1 and L2 to match a capacitive load to 50 Ω.

The final circuit is shown in Figure 2.73.

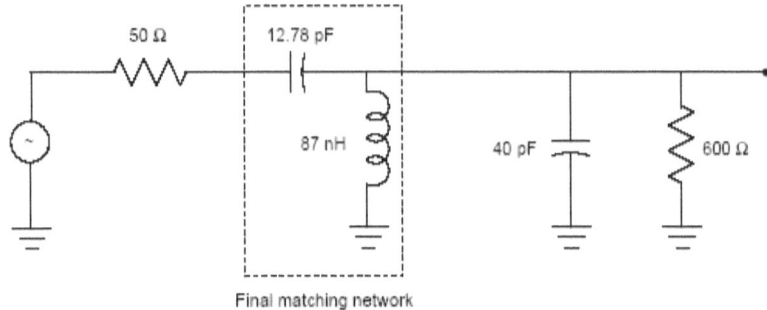

Final matching network

Figure 2.73 The final circuit.

2.10 THREE-ELEMENT MATCHING

We can observe with the L-networks that once Rs and Rp are given, the Q of the network is defined and the designer no longer has a choice over its value because:

$$Q = \sqrt{(R_p / R_s) - 1} \qquad (2.36)$$

If a narrow bandwidth is desired, this will cause a design problem. An example of a circuit with a low Q is shown in Figure 2.74.

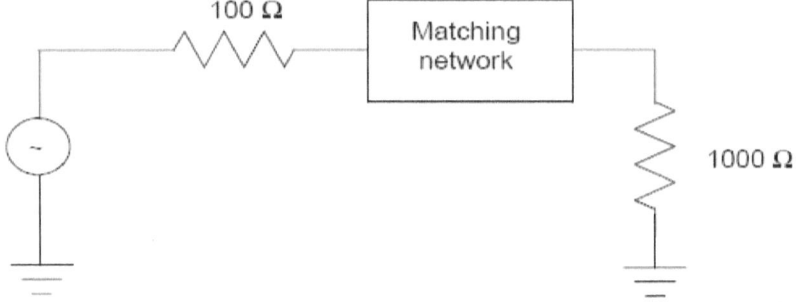

Figure 2.74 An example of a circuit with a low Q.

The use of 3-element networks overcomes this disadvantage and can be used for narrow-band high-Q applications. The minimum Q available is the circuit Q established with an L-matching network. There are two types of three-element networks: Pi-network and T-network, as shown in Figure 2.75.

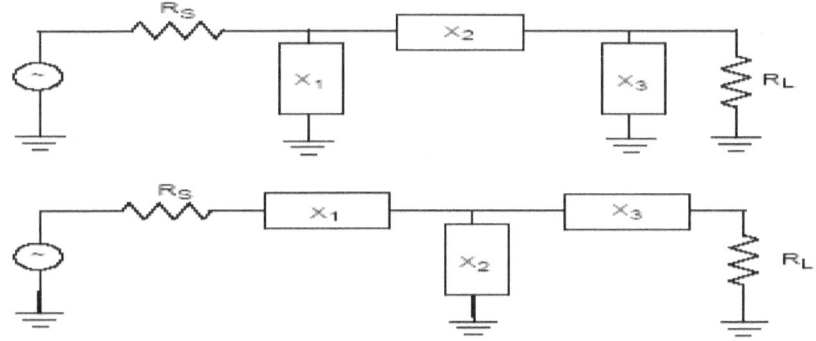

Figure 2.75 Use of three-element networks for matching.

Using more than 3 elements in the matching network design would bring about a greater amount of design flexibility which can lead to tedious mathematical equations. For that reason mathematical calculations of the reactive elements X_1, X_2, and X_3, which are too complicated, are omitted here. Instead we will show a simpler way to calculate these values later using a graphical tool called a Smith chart. As will be seen shortly, the use of a smith chart greatly simplifies a very complex design process.

Chapter 2- Symbol List

A symbol will not be repeated again, once it has been identified and defined in an earlier chapter, with its definition remaining unchanged.

AWG – American Wire Gauge

BW – Bandwidth

C_d- Distributed capacitance

C_T – Total capacitance of a center tapped capacitor

f_0 – Frequency where Q is maximum; Center of the passband.

f_r – Resonance frequency

H – Voltage Gain

IL – Insertion Loss

PF – Power Factor

μ - Permeability

Q – Quality Factor

Q_0 – Maximum Quality Factor where f= f_0

R_P – Insulation resistance

R_S – Series resistance
SF – Shape Factor
V_P – Phase velocity
σ - Conductivity
δ - Skin depth or depth of penetration
ϕ - Phase angle
ε – Dielectric constant of a material
ε_r – Relative dielectric constant of a material ($=\varepsilon/\varepsilon_o$)
ρ - Resistivity

CHAPTER-2 PROBLEMS

2.1) Design a resonant circuit with a loaded Q of 50 that operates between a source resistance of 100 Ω and a load of 2000 Ω at a frequency of 100 MHz.

2.2) Transform a series configuration of a 250 nH inductor into an equivalent parallel configuration at 50 MHz (Q=10).

2.3) Using the tapped-C method, design a resonant circuit with a loaded "Q" of 40 at a center frequency of 100 MHz that operates between a source resistance of 100 Ω and a load resistance of 3000 Ω. The capacitors are all lossless and the inductor has a Q of 100 at 100 MHz.

2.4) Design a simple parallel LC resonant circuit to provide a bandwidth of 10 MHz at a center frequency of 100 MHz. The resonant circuit is operating between a source and a load impedance of 2000 Ω each. The capacitor is lossless and the Q of the inductor is 85. Calculate the insertion loss of the resonant circuit in operation.

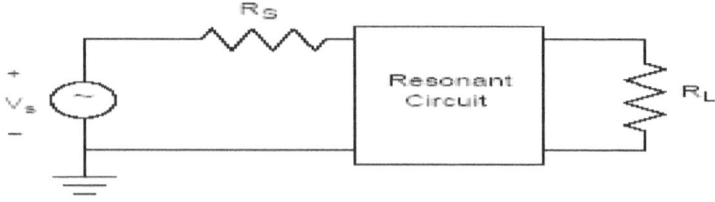

Figure P2.4

2.5) Using an L-network, design a circuit to match a 50 Ω source resistance to an 850 Ω load at 50 MHz. Assume that the DC must also be transferred from the source to the load.

2.6) Using the absorption method, design a matching network to match the source and the load at 50 MHz as shown in Figure P2.6.

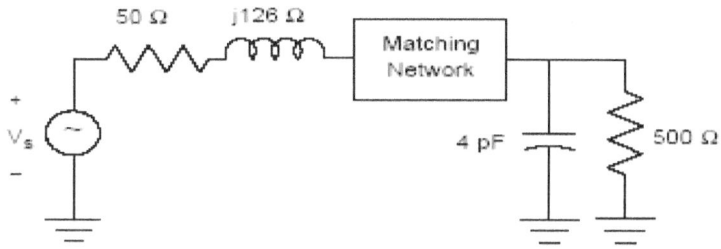

Figure P2.6

2.7) Using the resonance method, design an impedance matching network that will block the flow of DC from the load as shown in Figure P2.7. Assume f=100 MHz.

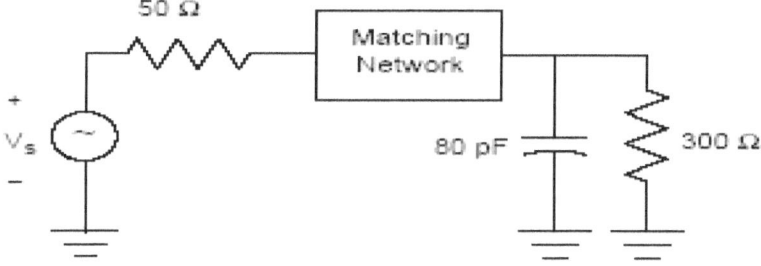

Figure P2.7

REFERENCES
[2.1] Bowick, C. *RF Circuit Design*. Carmel: SAMS-Prentice Hall, 1982.
[2.2] Carr, J. J. *Secrets of RF Circuit Design*. New York: McGraw-Hill, 1991.
[2.3] Gottlieb, I. W. *Practical RF Power Design Techniques*. New York: McGraw-Hill, 1993.
[2.4] Harsany, S. C. *Principles of Microwave Technology*.

Upper Saddle River: Prentice Hall, 1997.

[2.5] Krauss, H. L., C. W. Bostian, and F. H. Raab. *Solid State Radio Engineering*. New York: John Wiley & Sons, 1980.

[2.6] Lenk, J. D. *L enk's RF Handbook*. New York: McGraw-Hill, 1992.

[2.7] Matthaei, G., L. Young, and E. M. *Jones Microwave Filters, Impedance-Matching Networks, and Coupling Structures*. Dedham: Artech House, 1980.

[2.8] Radmanesh, M. M. *Radio Frequency and Microwave Electronics Illustrated*. Prentice Hall, 2001.

[2.9] Scott, A. W. *Understanding Microwaves*. New York: John Wiley & Sons, 1993.

[2.10] Vizmuller, P. *RF Design Guide*. Norwood: Artech House, 1995.

CHAPTER 3

The Wave Fundamentals

3.1 INTRODUCTION

The subject of "RF/Microwaves" primarily deals with electrical energy at high frequencies. Therefore to know microwaves, one needs to know the three qualities of energy in general.

3.2 Qualities of Energy

The following qualities apply to any and all types of energy forms whether electrical, mechanical, chemical, etc. at high or low frequencies. However, since we are dealing with electronics, we will narrow the following discussion to electrical energy and waves only.

A. QUALITY #1: EXISTING CHARACTERISTICS

These characteristics can be divided into three classes:

1. A Flow: is the transfer of energy from one point to another. The energy in a flow can have any type of waveform. So a flow is a transfer of energy. This is shown in Figure 3.1.

Figure 3.1 A flow.

2. A divergence (also referred to as a "dispersal"): is a generalized case of a "flow" where a number of flows travel from or to a common center as shown in Figures 3.2 (a) and (b).

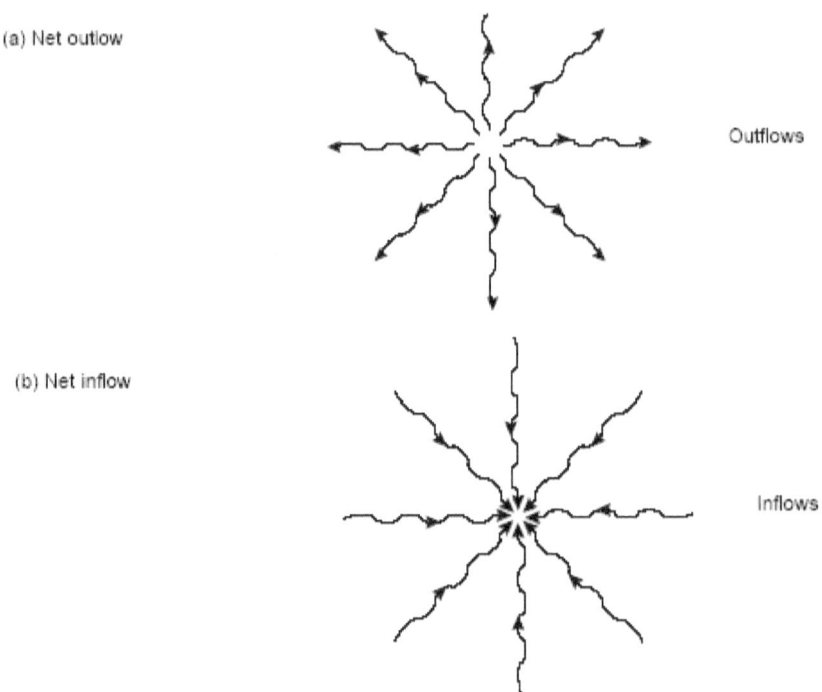

Figure 3.2 A divergence.

NOTE: *"A divergence" is similar in concept but different (in definition) from "divergence of a vector quantity" which is an exact mathematical operation measuring the net outflux (or influx) of a vector quantity.*

3. A standing wave (also called a ridge of energy): is energy suspended in space and comes about when two flows or divergences of approximately equal magnitude and exact frequency impinge against one another with sufficient amplitude to cause an enduring state of energy, which may last after the flow itself has ceased. For example, a resonator or a cavity oscillator falls into the category of devices that generate this type of wave characteristic. A few examples are shown in Figures 3.3 (a) and (b).

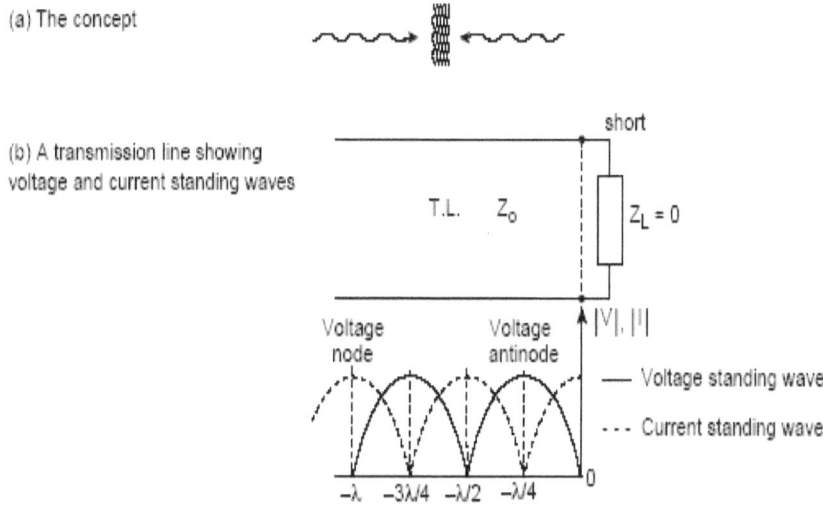

Figure 3.3 A standing wave.

B. QUALITY #2: WAVELENGTH

Wavelength is a characteristic of an orderly flow of motion and describes its regular and repeated pattern by the distance between its peaks. Many motions are too random and too chaotic to have an orderly flow and thus have no wavelength.

DEFINITION- WAVELENGTH: *Is defined to be the physical distance between two points having the same phase in two consecutive cycles of a periodic wave along a line in the direction of propagation as shown in Figure 3.4.*

As frequency increases, the wavelength (λ) decreases as can be observed. Thus higher frequency waves have shorter wavelengths as already discussed in Chapter 5 (see Figure 3.4).

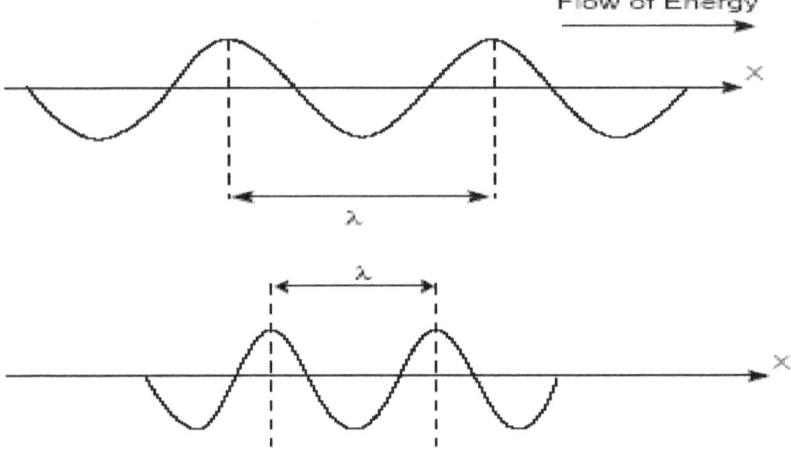

Figure 3.4 Wavelength for low and high frequencies.

Wavelength has no bearing on the wave characteristics (quality #1) but applies to the repetition property of the wave flow. A standing wave has a potential flow when released, therefore it may be considered to have a wavelength even though it is not a flow or a wave in the truest sense of the word.

If a random wave is periodic, it can be considered to have a wavelength using "the Fourier theorem". It can be proven mathematically that through the use of Fourier analysis, any wave can be decomposed into its Fourier harmonics, provided that the wave is continuously flowing and periodic as shown in Figure 3.5.

Figure 3.5 Examples of wave patterns.

C. QUALITY #3: A FLOW'S DIRECTION (OR ABSENCE THEREOF)
This quality describes the direction or the absence of direction of flow. A few examples are shown in Figure 3.6. This quality is an important one, since energy can have a flow with no net transfer of energy i.e. absence of the direction of flow.

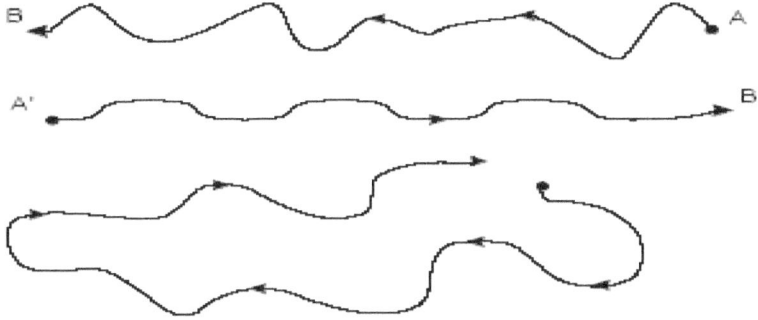

Figure 3.6 Flows with and without directions.

For example, a wave traveling from a transmitter to a receiver, or electrons moving in a wire under the influence of an electric field is said to have a "direction of flow".

Examples of "absence of direction" include a) a free electron moving in the lattice of a solid at equilibrium (i.e. when no external field is applied) which is a flow with an absence of direction or b) an electron in an atom moving in an orbit around the nucleus. Both are flows without a net transfer of energy.

3.3 DEFINITION OF A WAVE

So far we have loosely used "a wave" to mean a special case of a flow of energy. Now we need to define it exactly:

DEFINITION- A WAVE: *Is a disturbance that propagates from one point in a medium to other points without giving the medium, as a whole, any permanent displacement.*

This general definition of a wave includes any and all disturbances that could be of electrical or non-electrical origins. However, now we further restrict our definition to a special class of waves which are of electrical origin. These waves are called electromagnetic (EM) waves. Now, we need to define an important term:

DEFINITION-AN ELECTROMAGNETIC (EM) WAVE: *Is a radiant energy flow produced by the oscillation of an electric charge. In free space and away from the source (which is moving electric charges), EM rays of waves consist of vibrating electric and* *magnetic fields that move at the speed of light (in vacuum), are at right angles to each other and to the direction of motion.*

The propagation of a simple electromagnetic wave in free space is shown in Figure 3.7. EM waves propagate with no actual transport of matter and grow weaker in amplitude as they travel farther in space.

Oscillation

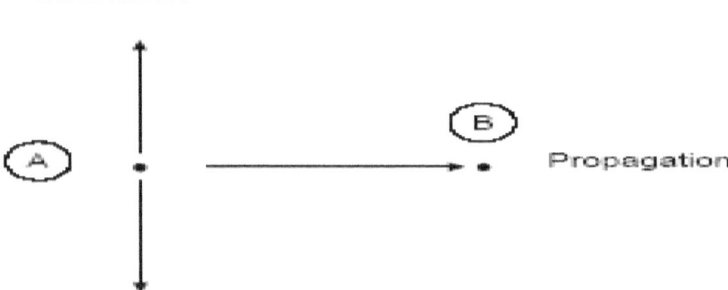

Figure 3.7 Propagation of a wave in space.

EM waves include Radio, microwaves, infrared, visible/ultraviolet light waves, X-, gamma- and cosmic- rays. (See Electromagnetic spectrum in Chapter 6).

These are all different types of electrical energy and all follow the same principles that we have discussed so far in this chapter (see Figures 3.8 and 3.9).

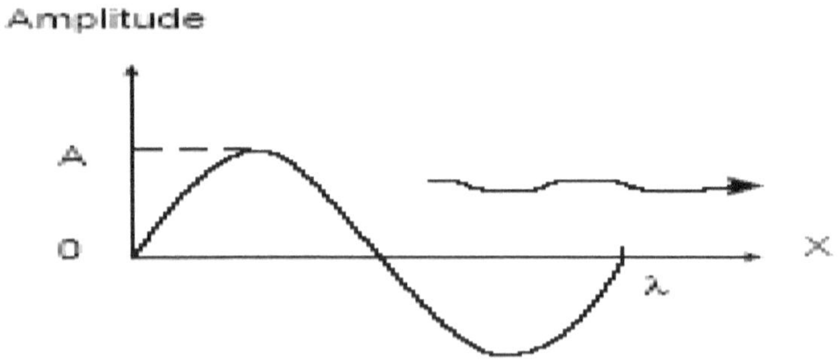

Amplitude

Figure 3.8 Wave amplitude in space.

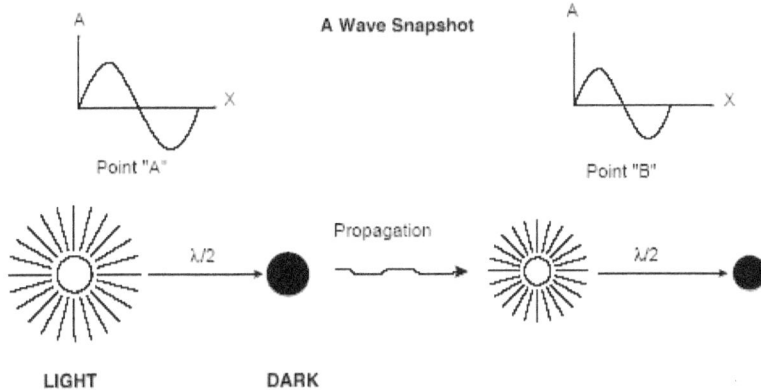

Figure 3.9 A reduction of wave amplitude as it propagates.

On a larger view of things, we can observe that RF and Microwaves are a special case of EM waves, which itself is a subset of a larger field of study, i.e., waves. Of course, this last itself is a subset of a much larger sphere of existence known as "energy", as shown in Figure 3.10.

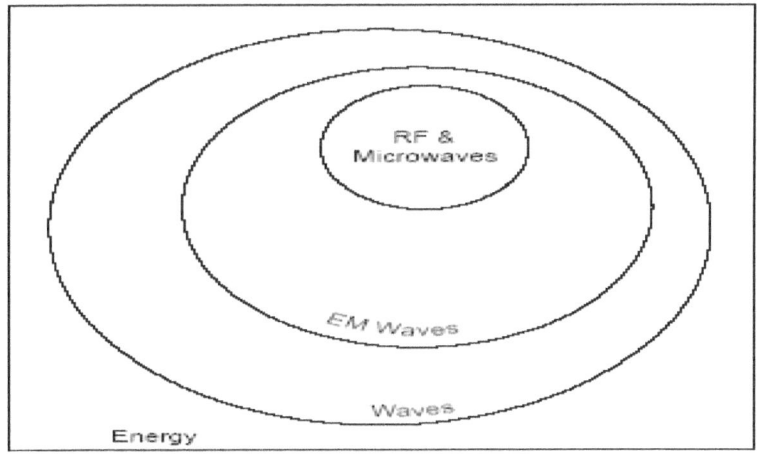

Figure 3.10 Bird's-eye view of EM energy forms.

3.4 MATHEMATICAL FORM OF PROPAGATING WAVES

We know that $f(x-x_0)$ is the same function as $f(x)$ except shifted to the right a distance "x_0" along +x axis. If instead, we consider $f(x-vt)$

then the function f(x) is shifted to the right a distance x_o=vt, where "v" and "t" can be considered to be the velocity of motion and the elapsed time, respectively. The distance (x_o) increases as time elapses, therefore the function is displaced continuously farther out along the +x axis as time elapses.

3.4.1 An Important Special Case: Sinusoidal Waves

Assume f(x) is a sinusoidal function:

f(x)=A cosβx $\qquad\qquad\qquad\qquad\qquad\qquad$ (3.1)

Where A is the amplitude and β is the phase constant. Then a sinusoidal wave propagating in +x direction would be represented in time and phasor domain by:

a. Time domain form:

$\qquad$ **f(x,t)= Acosβ(x-vt)**

$\qquad\qquad$ **= Acos(βx-ωt)** $\qquad\qquad\qquad\qquad$ (3.2)

Or,

b. in Phasor form:

$\qquad$ **F=Ae$^{-jβx}$** $\qquad\qquad\qquad\qquad\qquad\qquad$ (3.3)

Where ω (=βv) is the angular frequency.

To find the wavelength (λ), we know that it is defined to be the physical distance between two peaks (or valleys). We note that at t=0, the wave's peak is at x = 0. The next peak is at x = λ and the sinusoidal wave has a phase of 2π, thus:

βλ = 2π ⇒ λ = 2π/β $\qquad\qquad\qquad\qquad\qquad$ (3.4)

For the wave propagating in the "-x" direction, the following can be written:

c. Time domain: f(x,t)=Acos(βx+ωt) $\qquad\qquad$ (3.5)

Or,

d. Phasor domain: F=Ae^{jβx} $\qquad\qquad\qquad\qquad$ (3.6)

The phase velocity (V_P), which is defined to be the velocity at which the plane of the constant phase propagates, can be obtained from:

βx-ωt = k, $\qquad\qquad\qquad\qquad\qquad\qquad\qquad$ (3.7)

Where k is an arbitrary constant.

Differentiating Equation (3.7) with respect to time gives the phase velocity:

$$\beta dx/dt - \omega = 0 \Rightarrow V_p = dx/dt = \omega/\beta \qquad (3.8)$$

In an unrestricted or "free" space, a plane wave travels at velocity V_P which is given by:

$$V_p = 1/\sqrt{\mu_0 \varepsilon_0}$$

Where μ_0 and ε_0 are the permeability and permittivity of free space.

Equations (3.2) and (3.5) show a simple wave that keeps its size and shape while propagating at a constant velocity V_P. This type of propagation is said to be undistorted and unattenuated since it is propagating in free space (or vacuum) which is a non-dispersive medium.

DEFINITION-A DISPERSIVE MEDIUM: *is a medium in which the phase velocity (V_P) of a wave is a function of its frequency.*

This means that a complex wave, consisting of several frequencies, travels through a dispersing medium at different velocities i.e. each frequency component travels at $V_p = \omega/\beta$ with different time delays. This would cause the wave to be distorted at the exit point.

For example a square-pulse waveform entering and traveling through a dispersive medium will lose its shape and will appear rounded at both of its edges when exiting the medium.

EM waves can have two types of oscillation: a "rise and fall" for transverse energy fields and an "advance and retreat" for longitudinal energy fields as they propagate (see Figure 3.11).

3.4.2 Types of Waves

Waves are like fluids and propagate according to the medium in which they find themselves. If the medium is unrestricted, then it would be called "Free space wave propagation". When the source is a point and the medium of propagation is free space, waves have spherical wavefronts as shown in Figure 3.12.

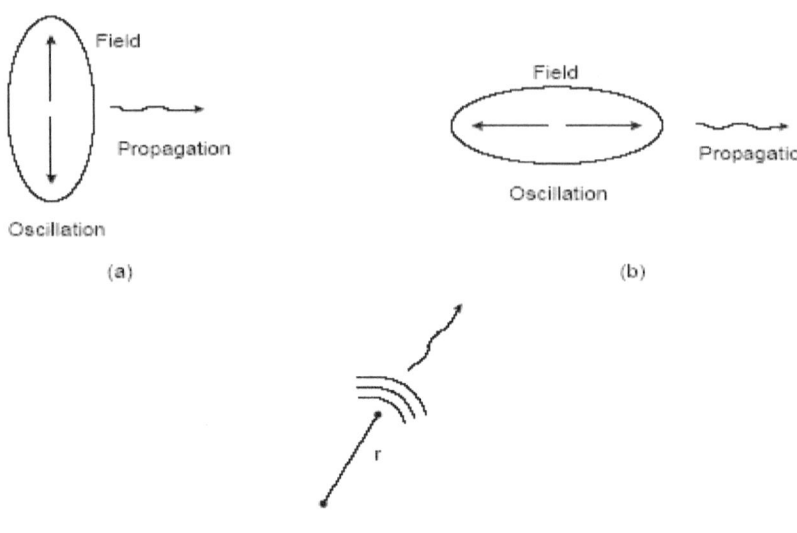

Figure 3.11 Propagation of EM waves.

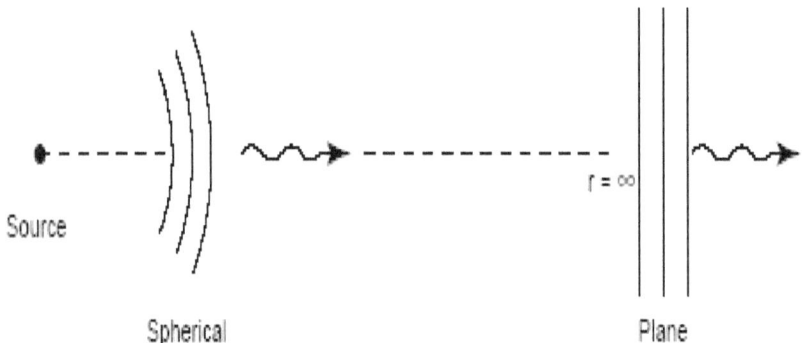

Figure 3.12 Relationship of spherical waves to plane waves.

3.4.3 A Special Case: Plane Waves

When waves under consideration are at an infinite distance away from the source of disturbance, then the "wavefront" of each wave is a plane surface and it is called a plane wave (see Figure 3.13). These plane waves are in the TEM mode of propagation.

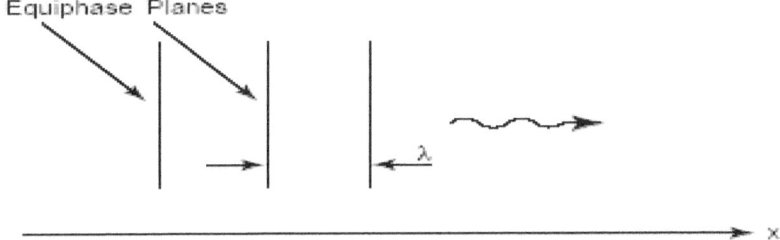

Figure 3.13 Wave fronts in a plane wave propagation.

DEFINITION-TEM (TRANSVERSE ELECTRO-MAGNETIC) MODE: *Is defined to be waves having the electric and magnetic fields perpendicular to each other and to the direction of propagation. These waves have no field components in the direction of propagation.*

A typical TEM wave in free space is shown in Figure 3.14.

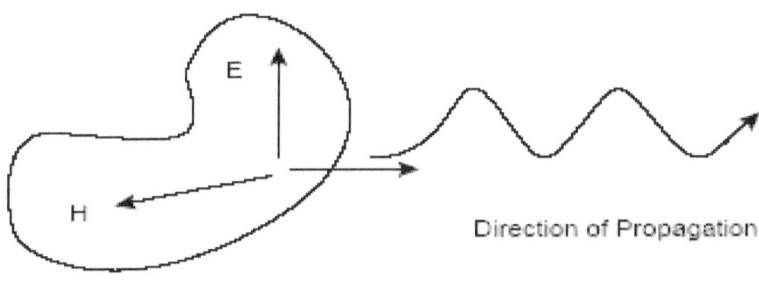

Figure 3.14 A typical TEM wave in free space.

The mathematical expression, **a(x,t),** for the plane wave propagation is defined below.

a. General time domain form is given by:

$$a(x,t) = A_o f(\omega t - \beta x) \tag{3.9}$$

For a time harmonic wave, we can write:

$$a(x,t) = A_o \cos(\omega t - \beta x)$$
$$= Re(A_o e^{-j\beta x} e^{j\omega t}) \tag{3.10}$$

b. In phasor domain, we have:

$$A(x) = A_o e^{-j\beta x}, \tag{3.11}$$

Which is a plane wave propagating in +x direction as shown in Figure 3.12 and 3.13.

3.5 PROPERTIES OF WAVES

There are several properties of waves that are worthy of consideration at the outset of this section:

PROPERTY #1: FLOW PROPERTY

This property is in common with quality #1 for energy. A wave is a flow. It goes from point "A" to point "B" and in doing so a transfer of energy takes place, of course with a reduced amplitude at the destination, as shown in Figure 3.15.

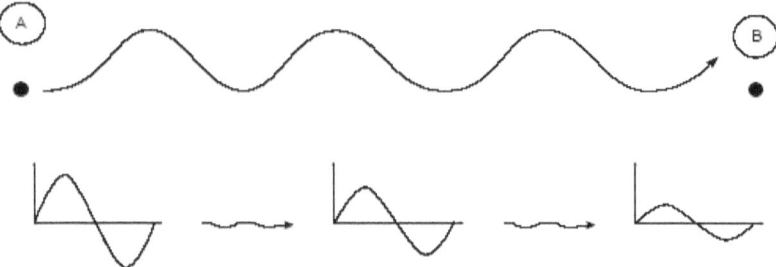

Figure 3.15 An EM wave as a flow of energy.

PROPERTY #2: WAVELENGTH PROPERTY

This property was discussed as quality #2 of energy. A wave with a regular and periodic (or repeating) waveform has a wavelength which is the physical distance between two peaks (or valleys) in two consecutive cycles as defined earlier in a more precise way. This concept is shown in Figure 3. 16.

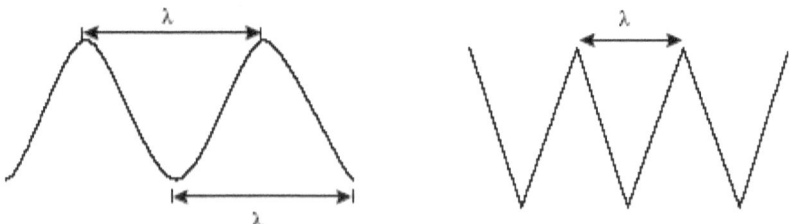

Figure 3.16 Concept of wavelength.

Wavelength (λ) is the distance that the wave travels in one wave period (T=1/f) at the speed of propagation (v), thus we have:

$$\lambda = vT = v/f \tag{3.12}$$

Where $v=c_o$ in air. At high RF and microwaves the wavelength ranges from one meter to one millimeter corresponding to a frequency of 300 MHz to 300 GHz.

PROPERTY #3: REFLECTION AND TRANSMISSION PROPERTY
When a wave encounters an obstacle or a different medium, some of it reflects back (called a reflected wave) and the rest of it transmits through (called a transmitted wave). This is true for any and all types of waves.

EXAMPLE 3.1a: PERFECT REFLECTION
How does a perfect mirror behave for an incident wave?
Solution:
For a perfect mirror we have a perfect reflection, i.e. 100% of the incident wave reflects back and zero transmission takes place as shown in Figure 3.17.

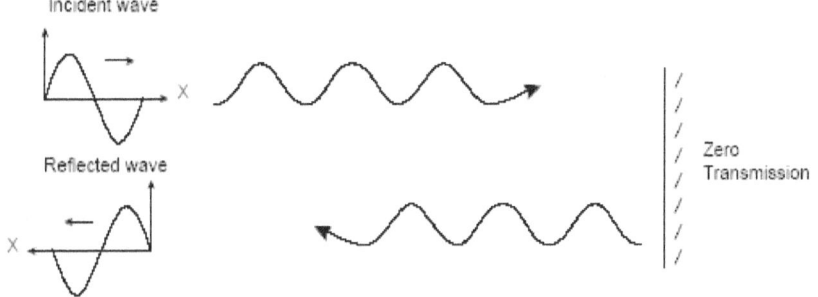

Figure 3.17 A perfect mirror.

Example 3.1b: Perfect transmission
What would constitute a perfect transmission condition?

Solution:
For a perfect transmission, the two media have to be identical in their electrical properties (such as permittivity, permeability, etc.) as shown in Figure 3.18. This means that for this condition to occur, the second medium has to continue to behave electrically the same as the first.

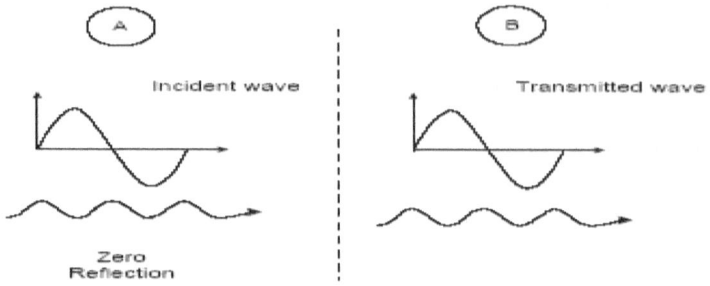

Figure 3.18 A perfect transmission.

Property #4) Standing-wave property

When two waves of exactly the same magnitude and frequency travel opposite to each other, the result is not a wave but an "Oscillation with no propagation" called a "Standing wave" which has a fixed location, as shown in Figure 3.19.

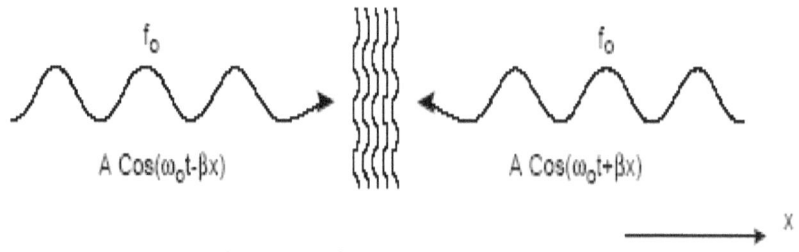

Figure 3.19 A standing wave.

The standing wave can be written mathematically in:

a. Phasor domain:

$$Ae^{-j\beta x} + Ae^{+j\beta x} = 2A\cos\beta x \qquad (3.13a)$$

b. Time domain:

$$2A\cos(\beta x)\cos(\omega t) \qquad (3.13b)$$

Since Equation (3.13b) is not of the form $f(\beta x - \omega t)$, thus it is not a wave but a pure oscillation at a fixed location!

NOTE: *A definite pre-requisite for a standing wave is two opposite waves of exact frequency. However, their amplitudes should be comparable, if not equal. The result would be a standing wave plus a traveling wave and not a pure standing wave as described above.*

3.6 TRANSMISSION MEDIA

When waves are constricted to a limited transmission space (also called a line, guide, channel, etc.), then the waves take on different forms and patterns according to the shape of the guide, just like fluid flow in a pipe.

3.6.1 Types of Transmission Media

A few examples of the wave patterns in different transmission media are: a) Coaxial line, b) two-wire transmission line, c) a waveguide, d) a microstrip line, e) a parallel plate waveguide, and f) a stripline, as shown in Figure 3.20.

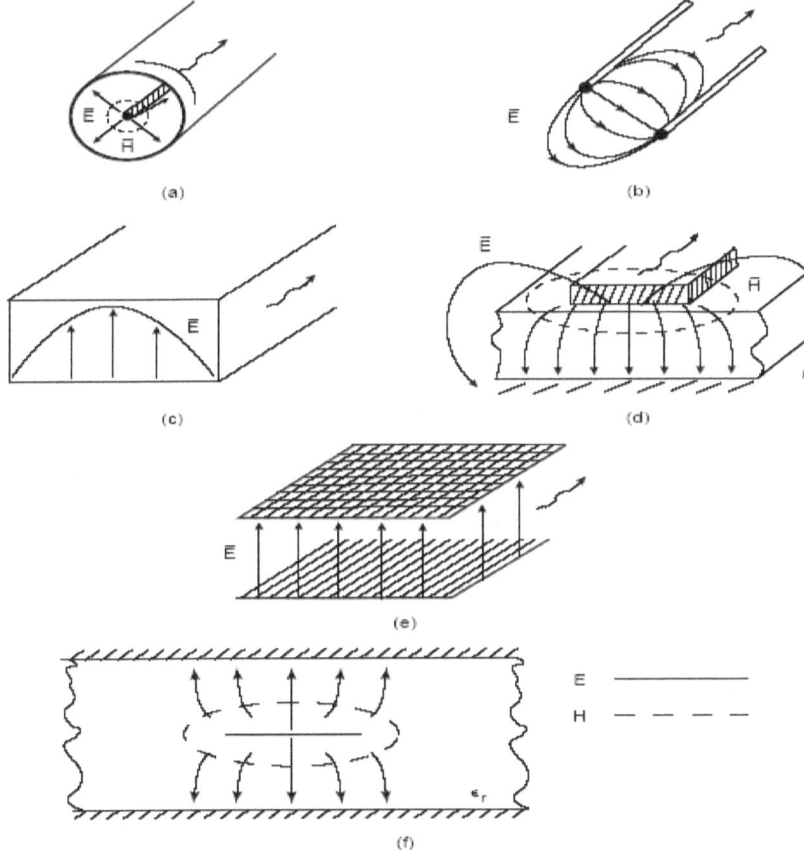

Figure 3.20 The electric and magnetic field patterns for: a) Coaxial line, b) two-wire transmission line, c) a waveguide, d) a microstrip line, e) a parallel plate waveguide, and f) a stripline.

Generally, any and all of these five transmission media could be called "transmission lines", but the terminology has been made more specific to convey more exact concepts, thus:

1. (a), (b), (e) and (f) are generally labeled as transmission lines (TL's).

2. (c) Is labeled a waveguide and,

3. (d) Is labeled a Microstrip line.

(a), (b), (e) and (f) all will support propagation of Transverse Electromagnetic (TEM) waves and will be used specifically in this book. An example of a TEM wave was shown earlier in Figure 3.14, where the direction of propagation is perpendicular to the oscillating electric and magnetic fields.

NOTE 1: *Structure (d), a microstrip line, supports a quasi-TEM wave which is a wave with a small axial field. This type of transmission line has gained tremendous popularity in Microwave integrated circuits due to its planar structure and ease of fabrication using printed circuit technology. Microstrip lines will be discussed in detail in a later section.*

NOTE 2: *Structures (a) and (c), a coaxial line and a waveguide, are closed structures and are preferred since they have much less radiation losses than the other three open structures.*

NOTE 3: *Structure (f), a stripline transmission line, can be thought of as a "flattened out" coaxial line, where both have a center conductor that is enclosed by an outer ground conductor with a uniform dielectric material filling the space between the two.*

NOTE 4: *Other types of transmission media include: Slotline, Dielectric waveguide, Coplanar waveguide, and Ridge waveguide. These transmission media have non-TEM modes of propagation and are beyond the scope of this book and can be found in advanced texts.*

A summary of Transmission media and their different characteristics is shown in table 3.1. The comparison made in this table can be roughly divided into two important general areas:

a. Electrical considerations: mode of propagation, dispersion, Bandwidth, Power loss and power capacity (items 1 through 5), and

b. Mechanical considerations: Physical size, ease of fabrication and ease of integration with other elements and components (items 6 through 8).

Feature	Coaxial	Stripline	Microstrip	Waveguide
1.) Propagating Mode	Main: TEM Other: TM, TE	Main: TEM Other: TM, TE	Main: Quasi-TEM Other: TM, TE	Main: TE_{10} Other: TM, TE
2.) Dispersion	None	None	Low	Medium
3.) Bandwidth	High	High	High	Low
4.) Power loss	Medium	High	High	Low
5.) Power capacity	Medium	Low	Low	High
6.) Size	Large	Medium	Small	Large
7.) Ease of fabrication	Medium	Easy	Easy	Medium
8.) Ease of integration	Hard	Fair	Easy	Hard

Table 3.1 Comparison of various transmission lines.

3.6.2 A Short History of Transmission Media

Waveguides were used for most microwave systems during the 1930s and 1940s but they have a limited bandwidth, are bulky and expensive, even though they have the advantage of being able to handle high powers much needed for Radar applications. During this same period, coaxial lines were also developed as a broadband and medium power transmission line, but they are difficult to integrate into or fabricate in the integrated circuit technology which is suited to planar type transmission lines.

Planar transmission lines received attention in the 1950s. They are low cost compact and capable of being integrated with planar microwave integrated devices and circuits. Therefore, they play an important role in planar microwave technology for transmission of signals between devices, circuits and networks. Examples of planar transmission lines include microstrip line (developed in 1952), stripline (developed circa 1955), slotline (developed in 1969).

Other planar transmission lines (e.g. coplanar waveguides, finlines, etc.) have also been developed through time. Overall and amongst all planar transmission lines, none have proven as popular as microstrip line technology, which has gained tremendous interest in planar circuit applications. For this reason microstrip lines are discussed and analyzed in-depth in a later section in this chapter.

3.6.3 Waves on a Transmission Line (TEM Mode)

When we mention a "transmission line", it is commonly understood to be any system of conductors suitable for conducting TEM-mode electromagnetic waves efficiently between two or more terminals. Common examples of TEM-mode transmissions lines are telephone lines, power lines, coaxial lines, parallel plate lines, etc.

At lower frequencies the length of the line is much smaller than the signal wavelength and thus the transmission line can be treated as a "lumped element" with almost zero loss and no time delay for signal propagation between two points.

However, at high RF/microwave frequencies the length of the line is comparable to the signal wavelength and the time delay of propagation (and the corresponding signal phase shift) can no longer be ignored.

Under these conditions, the "distributed circuit model" is used to analyze a transmission line. Such a model provides the governing differential equations for voltage and current waves propagating along a transmission line without a need to resort to Maxwell's equations to solve for the electromagnetic field quantities.

3.6.4 The Governing Equations

At high frequencies, an infinitesimal length of a transmission line can be modeled by two series elements (R, L) in conjunction with two shunt elements (G, C) as shown in Figure 3.21a.

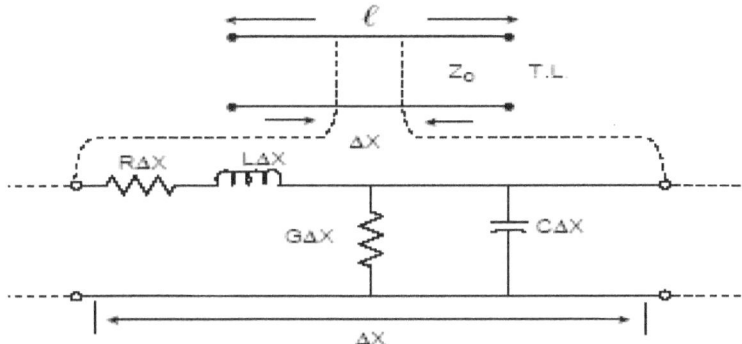

FIGURE 3.21a An infinitesimal portion of a transmission line (TL).

Juxtaposing an infinite number of this infinitesimal model into a long chain, will create a workable model for a transmission line as shown in Figure 3.21b.

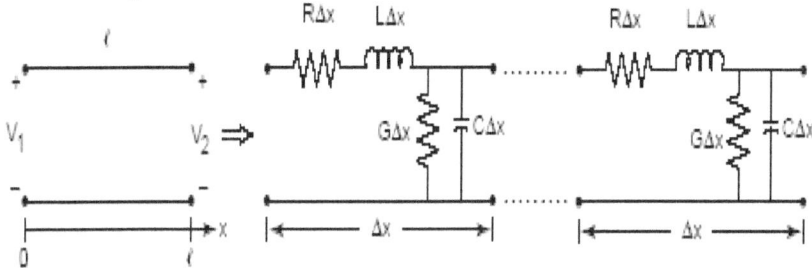

FIGURE 3.21b   The equivalent circuit of a TL at high frequencies.

To develop the governing differential equations, we will examine one Δx section of a transmission line as shown in Figure 3.21c.

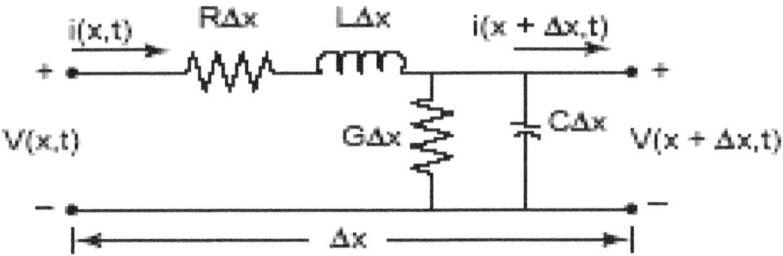

FIGURE 3.21c Voltage and current in an infinitesimal length of TL.

Using KVL for the Δx section, we can write:

v(x,t)= i(x,t) RΔx+ LΔx ∂i(x,t)/∂t +v(x+Δx,t)

Upon rearranging terms and dividing both sides by Δx, we obtain:

$$-\frac{v(x+\Delta x)-v(x,t)}{\Delta x} = Ri(x,t)+L\frac{\partial i(x,t)}{\partial t}$$

Letting $\Delta x \to 0$, yields:

$$-\frac{\partial v(x,t)}{\partial x} = Ri(x,t)+L\frac{\partial i(x,t)}{\partial t} \tag{3.14a}$$

Similarly, using KCL we can write:

$$i(x,t)= v(x+\Delta x,t)\ G\Delta x+ C\Delta x\frac{\partial v(x+\Delta x,t)}{\partial t} +i(x+\Delta x,t)$$

Upon rearranging terms, dividing by Δx and letting $\Delta x \to 0$, we have:

$$-\frac{\partial i(x,t)}{\partial x} = Gv(x,t) + C\frac{\partial v(x,t)}{\partial t} \tag{3.14b}$$

Equations (3.14a) and (3.14b) are two cross-coupled equations in terms of v and i. These two equations can be separated by first differentiating both equations with respect to "x" and then properly substituting for the terms, which leads to:

$$-\frac{\partial^2 v(x,t)}{\partial x^2} = R\frac{\partial i(x,t)}{\partial x} +L\frac{\partial^2 i(x,t)}{\partial x \partial t}$$

$$=-R\left(Gv(x,t)+C\frac{\partial v(x,t)}{\partial t}\right)-L\left(G\frac{\partial v(x,t)}{\partial t}+C\frac{\partial^2 v(x,t)}{\partial t^2}\right)$$

Or,

$$\frac{\partial^2 v(x,t)}{\partial x^2} = LC\frac{\partial^2 v(x,t)}{\partial t^2}+(RC+LG)\frac{\partial v(x,t)}{\partial x}+RGv(x,t)$$

$$\tag{3.14c}$$

Similarly for "i," we can write:

$$\frac{\partial^2 i(x,t)}{\partial x^2} = LC\frac{\partial^2 i(x,t)}{\partial t^2}+(RC+LG)\frac{\partial i(x,t)}{\partial x}+RGi(x,t)$$

$$\tag{3.14d}$$

For sinusoidal signal variation for "v" and "i", we can write the corresponding Phasors as follows:

$$v(x,t)=Re[V(x)e^{j\omega t}]$$
$$i(x,t)=Re[I(x)e^{j\omega t}]$$

Using phasor differentiation results from Chapter 3, Equations (3.14c) and (3.14d) can be written as:

$$\frac{d^2V(x)}{dx^2}-\gamma^2V(x)=0 \tag{3.14e}$$

$$\frac{d^2I(x)}{dx^2}-\gamma^2I(x)=0 \tag{3.14f}$$

Where

$$\gamma=\alpha+j\beta=\sqrt{(R+j\omega L)(G+j\omega C)}$$

γ is the **propagation constant**, with real part (α) and imaginary part (β), called the **attenuation constant** (Np/m) and **phase constant** (rad/m), respectively.

The solution to the second order differential equations as given by Equations (3.14e) and (3.14f), can be observed to be of exponential type format ($e^{\pm\gamma x}$). Thus we can write the general solutions for $V(x)$ as follows:

$$V(x)=V_0^+ e^{-\gamma x}+V_0^- e^{\gamma x} \tag{3.14g}$$

Where the complex constants V_0^+ and V_0^- are determined from the boundary conditions imposed by the source voltage and the load value.

Similarly, $I(x)$ can be obtained from $V(x)$ (see Equation 3.14a) as:

$$I(x)=\left(\frac{-1}{R+j\omega L}\right)\frac{dV(x)}{dx}=\frac{V_0^+ e^{-\gamma x}-V_0^- e^{\gamma x}}{Z_0} \tag{3.14h}$$

Where $Z_0=\sqrt{\dfrac{R+j\omega L}{G+j\omega C}}$ (3.14i)

is the characteristic impedance of the transmission line.

SPECIAL CASE: A LOSSLESS TRANSMISSION LINE
For this case, we have R=G=0. This yields the following simplifications:

$$\gamma = j\omega\sqrt{LC} = j\beta,$$

$$Z_0 = \sqrt{\frac{L}{C}}$$

Where $\beta = \omega\sqrt{LC}$ is the phase constant.   In this case   Equations 3.14e-f can be written as:

$$\frac{d^2V(x)}{dx^2} + \beta^2 V(x) = 0 \tag{3.14j}$$

$$\frac{d^2I(x)}{dx^2} + \beta^2 I(x) = 0 \tag{3.14k}$$

Similar to Equations 3.14g-h, the solutions to Equations 3.14j-k are given by:

$$V(x) = V_0^+ e^{-j\beta x} + V_0^- e^{j\beta x} \tag{3.14ℓ}$$

$$I(x) = \frac{V_0^+ e^{-j\beta x} - V_0^- e^{j\beta x}}{Z_0} \tag{3.14m}$$

NOTE 1: *Transmission line Equations 3.14e-f and 3.14j-k could have all been derived using the Maxwell's equations directly from the field quantities E and H as delineated in an appendix presented at the end of the book entitled "Laws of electricity and Magnetism", under items 14 and 19 .*

It will be seen shortly in Chapter 7, that the term $e^{-\gamma x}$ [or $e^{-j\beta x}$] represents a propagating wave in "+x" direction while $e^{\gamma x}$ [or $e^{j\beta x}$] represents a propagating wave in "-x" direction on a transmission line. The combination of the two waves propagating in opposite directions to each other forms a standing wave on the transmission.

NOTE 2: *Based on a given set of boundary conditions for the source and the load, we can find the constants in the equations 3.14ℓ-m. For example, if the source voltage (at x=0) is known to be V=V_g and the load voltage (at x= ℓ) is V=V_L, then the constants V_0^+ and V_0^- can easily be found from the following two equations:*

$$x=0, \quad V_g = V_0^+ + V_0^- \tag{3.14n}$$

$$x=\ell, \quad V_L = V_0^+ e^{-j\beta\ell} + V_0^- e^{j\beta\ell} \tag{3.14o}$$

EXERCISE 3.1

a. Derive expressions for Vo^+ and Vo^- from equations 3.14n-o in terms of Vg and V_L.

b. Given the load value as $Z=Z_L$, find Vo^+ and Vo^- in terms of V_g and Z_L [as in part (a)].

Hint: Use $V_2 = Z_L [Vo^+ e^{-j\beta\ell} - Vo^- e^{j\beta\ell}]/Z_o$

3.6.5 Sinusoidal Waves on a Transmission Line

Consider a transmission line as shown in Figure 3.21d. Assuming a sinusoidal signal excitation, the propagating voltage and current waves on a transmission line are also sinusoidal and can be expressed as:

$$v(x,t)=Re[V(x)e^{j\omega t}] \qquad (3.14p)$$
$$i(x,t)=Re[I(x)e^{j\omega t}] \qquad (3.14q)$$

Where complex quantities $V(x)$ and $I(x)$ are phasor quantities.

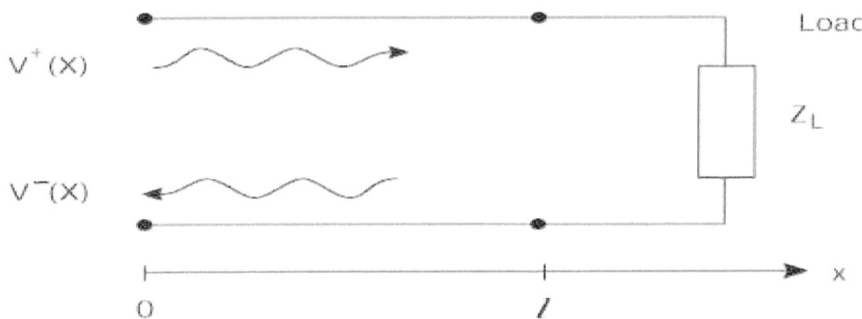

FIGURE 3.21d Incident and reflected waves on a transmission line.

Using the distributed circuit model of a transmission line and its corresponding equivalent circuit, the following differential equations for $I(x)$ and $V(X)$ can be derived (see Chapter 5, example 5.5) for an infinitesimal length of a transmission line:

$$\frac{d^2V(x)}{dx^2} - \gamma^2 V(x)= 0 \qquad (3.15a)$$

$$\frac{d^2I(x)}{dx^2} - \gamma^2 I(x) =0 \qquad (3.15b)$$

Where γ is the complex propagation constants given by:

$$\gamma = \alpha + j\beta = [(R+j\omega L)(G+j\omega C)]^{1/2} \qquad (3.16)$$

Where:

α = attenuation constant (in Nepers/m)
β = phase constant (in radian/m)
R = resistance per unit length in Ω/m
L = inductance per unit length in H/m
G = conductance per unit length in S/m
C = capacitance per unit length in F/m

By observation, we notice that the general solution to the problem is of exponential form, therefore we can write:

$$V_1(x) = V_o^+ e^{-\gamma x} \quad \Rightarrow \quad I_1(x) = \frac{V_o^+}{Z_0} e^{-\gamma x} \qquad (3.17a)$$

$$V_2(x) = V_o^- e^{\gamma x}, \quad \Rightarrow \quad I_2(x) = -\frac{V_o^-}{Z_0} e^{\gamma x} \qquad (3.17b)$$

Where Z_0 is the characteristic impedance of the transmission line; V_o^+ and V_o^- are complex constants in general, whose values depend upon the source and the transmission line characteristics, as will be seen shortly.

Since we are dealing with a linear system, the general solution for voltage and current is obtained using the superposition theorem as follows:

$$V(x) = V_1(x) + V_2(x) = V_o^+ e^{-\gamma x} + V_o^- e^{\gamma x} \qquad (3.17c)$$

$$I(x) = I_1(x) + I_2(x) = \frac{V_o^+}{Z_0} e^{-\gamma x} - \frac{V_o^-}{Z_0} e^{\gamma x} \qquad (3.17d)$$

From Equation (3.17), we observe that voltage and current are a pair of waves co-existing and are inseparable for a distributed circuit. Each solution for voltage or current consists of two waves which will be labeled as follows:

a. An incident wave: $e^{-\gamma x} = e^{-\alpha x} e^{-j\beta x}$ \qquad (3.18a)
b. A reflected wave: $e^{\gamma x} = e^{\alpha x} e^{j\beta x}$ \qquad (3.18b)
Where "βx" is referred to as the electrical length.

Each wave travels at the phase velocity (V_P) given by:

$V_P = \omega/\beta = c_0$ (air), $\qquad\qquad\qquad$ (3.19)

Where c is the speed of light in vacuum given by:

$c_0 = 1/(\mu_0\varepsilon_0)^{1/2} = 2.9988 \times 10^8 \approx 3 \times 10^8$ m/s.

Note: *In general, V_P in a medium is given by*:

$V_P = \omega/\beta = 1/(\mu\varepsilon)^{1/2} = c = c_0/(\mu_r\,\varepsilon_r)^{1/2}$,

where

$\varepsilon = \varepsilon_r\varepsilon_0$, and

$\mu = \mu_r\,\mu_0$.

The time-average incident power propagating along a transmission line is given by (assuming Z_O is a real number):

$$P^+(x) = \frac{1}{2}\,Re\left[V^+(x)I^+(x)\,*\right] = \frac{V_0^{+2}}{2Z_O}\,e^{-2\alpha x} \qquad\qquad (3.20)$$

The same can be written for the reflected power propagating back to the source.

The law of conservation of energy requires that the rate of decrease of propagating power $P(x)$ along the line should equal the average power loss per unit length (P_{loss}). Thus we can write:

$$-\frac{\partial P(x)}{\partial x} = P_{loss} = 2\alpha P(x)$$

$$\alpha = \frac{P_{loss}}{2P(x)} \approx \frac{-\Delta P/\Delta x}{2P} = \frac{-[P(x+\Delta x)-P(x)]/\Delta x}{2P(x)} \qquad (3.21)$$

Equation (3.21) shows an interesting and yet very practical way to measure the actual attenuation constant (α). This method is particularly helpful if one is trying to establish the integrity of a faulty line, since a simple comparison of the measured (α) with the nominal (α) would reveal the needed information. The following example elucidates this point further.

EXAMPLE 3.3

The microwave power at one point (P_1) on a transmission line is measured to be 10 mW. At a distance of d=50 cm away, another power measurement (P_2) indicates a power of 7 mW. Determine the attenuation constant of the transmission line.

Solution:

$$\alpha = \frac{P_{loss}}{2P(x)} \approx \frac{-\Delta P / \Delta x}{2P} = \frac{-[P(x+\Delta x) - P(x)]/\Delta x}{2P(x)}$$

Where

$\Delta P / \Delta x = (P_2 - P_1)/d$

Thus we have:

$$\alpha = \frac{-(P_2 - P_1)/d}{2P_1} = \frac{-(7-10)/0.5}{2 \times 10} = 0.3 \; 1/m$$

$\alpha \; (Np/m) = \ln(1/0.3) = 1.20 \; Np/m$

3.6.6 The Concept of the Reflection Coefficient

Any time an incident wave encounters a second medium different than the first, it is partly reflected (creating a reflected wave) while the remaining is transmitted through (creating a transmitted wave). The reflected wave encountering the incident waveforms a standing wave as described earlier. Thus we can see that there are four possible waves in a transmission line:

a. An incident wave,
b. A reflected wave,
c. A transmitted wave, and
d. A standing wave.

Let us now define an important term:

DEFINITION- REFLECTION COEFFICIENT: *Is defined to be the ratio of the reflected wave phasor to the incident wave phasor.*

In the special case of a uniform transmission line when the incident wave encounters a second medium such as a termination (load) or a discontinuity, then under these conditions, the ratio of the reflected wave phasor to the incident wave phasor is "The reflection coefficient".

To illustrate this concept, consider a transmission line circuit with a load (Z_L) located at $x=\ell$, as shown in Figure 3. 21d.

$V^{+}(x) = V_0^{+} e^{-\gamma x}$ (3.22a)

$V^{-}(x) = V_0^{-} e^{\gamma x}$ (3.22b)

At the load end ($\mathbf{x} = \ell$), $V^+(\ell)$ is given by:

$$V^+(\ell) = V_0^+ e^{-\gamma \ell}$$

However, to find $V^-(\ell)$, we need to realize that the reflected wave reflects from the load by a factor of Γ_L (i.e., Γ_L is the load reflection coefficient at $x=\ell$):

$$V^-(\ell) = \Gamma_L V_0^+ e^{-\gamma \ell} \tag{3.23d}$$

The reflected wave travels back a distance of $x'' = \ell - x$ towards the source as:

$$V^-(x'') = V^-(\ell-x) = \Gamma_L V_0^+ e^{-\gamma \ell} e^{-\gamma x''} = \Gamma_L V_0^+ e^{-\gamma \ell} e^{-\gamma(\ell-x)} \tag{3.23b}$$

where x'' is an imaginary reference frame set up at the load ($x''=0$) and is directed toward the source ($x'' = \ell$). Thus $V^-(x)$ can be written as:

$$V^-(x) = \Gamma_L V_0^+ e^{-2\gamma \ell} e^{\gamma x} = V_0^- e^{\gamma x} \tag{3.23c}$$

The reflection coefficient can now be defined as:

$$\Gamma(x) = \frac{V^-(x)}{V^+(x)} = \frac{V_0^- e^{\gamma x}}{V_0^+ e^{-\gamma x}} = \frac{V_0^-}{V_0^+} e^{2\gamma x} = \Gamma_L e^{-2\gamma \ell} e^{2\gamma x} \tag{3.23e}$$

Thus the total voltage and current phasors [$V(x)$, $I(x)$] along the transmission line can now be written as:

$$V(x) = V^+(x) + V^-(x) = V_0^+ e^{-\gamma x} + V_0^- e^{\gamma x}$$

Or,

$$V(x) = V_0^+ (e^{-\gamma x} + \frac{V_0^-}{V_0^+} e^{\gamma x}) = V_0^+ (e^{-\gamma x} + \Gamma_L e^{-2\gamma \ell} e^{\gamma x}) \tag{3.24}$$

Similarly,

$$I(x) = I^+(x) - I^-(x) = V^+(x)/Z_0 - V^-(x)/Z_0$$

Or,

$$I(x) = \frac{V_0^+}{Z_0} (e^{-\gamma x} - \Gamma_L e^{-2\gamma \ell} e^{\gamma x}). \tag{3.25}$$

The input impedance, $Z_{IN}(x)$, at any point along the transmission line is obtained through dividing equation 3.24 over 3.25 and is given by:

$$Z_{IN}(x) = \frac{V(x)}{I(x)} = Z_0 \frac{e^{-\gamma x} + \Gamma_L e^{-2\gamma \ell} e^{\gamma x}}{e^{-\gamma x} - \Gamma_L e^{-2\gamma \ell} e^{\gamma x}} \tag{3.26}$$

A SPECIAL CASE:

At the load end (where x = ℓ), the following is observed:

$$Z_{in}(\ell) = Z_L = Z_0 \frac{1 + \Gamma_L}{1 - \Gamma_L}$$

$$\Rightarrow \Gamma_L = \frac{Z_L - Z_0}{Z_L + Z_0} \qquad (3.27a)$$

We can generalize equation (3.27a) for any arbitrary point along the transmission line with an input impedance (Z_{in}), and write the reflection coefficient (Γ_{IN}) at that point as:

$$\Gamma_{IN} = \frac{Z_{IN} - Z_0}{Z_{IN} + Z_0} \qquad (3.27b)$$

Using Equation (3.27a) and letting the distance from the load as d=ℓ-x, Equation (3.26) can be written as:

$$Z_{IN}(d) = Z_0 \frac{Z_L + Z_0 \tanh \gamma d}{Z_0 + Z_L \tanh \gamma d} \qquad (3.28)$$

3.6.7 Lossless Transmission Lines

Since most of the transmission lines at RF/microwave frequencies have negligible losses, we will focus exclusively on lossless transmission lines.

In a lossless transmission line, there is no series resistance (R) or shunt leakage conductance (G). Thus the energy propagating on the line does not get attenuated in strength (or power). Considering Figure 3.22a, the following simplifications can be made:

$\alpha = 0$,

$\gamma = j\beta$,

$$Z_0 = \sqrt{\frac{L}{C}} \qquad (3.29a)$$

$V_P = \omega/\beta = 1/(LC)^{1/2}$ \hfill (3.29b)

$\lambda = V_P/f = 2\pi/\beta$ \hfill (3.29c)

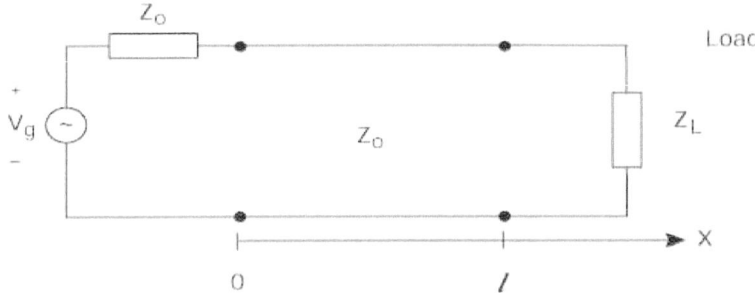

FIGURE 3.22a   A lossless Transmission line.

Using Equation (3.23e), we can write:

$$\Gamma(x) = = \frac{V^-(x)}{V^+(x)} = \frac{V_0^- e^{j\beta x}}{V_0^+ e^{-j\beta x}} = \frac{V_0^-}{V_0^+} e^{j2\beta x} = \Gamma_L e^{-j2\beta\ell} e^{j2\beta x} \qquad (3.30a)$$

Using (3.24) and (3.25), we can write the voltage and current on a lossless transmission line as:

$$V(x) = V^+(x) + V^-(x) = V_0^+ e^{-j\beta x} + V_0^- e^{j\beta x} \qquad (3.30b)$$

$$V(x) = V_0^+ e^{-j\beta x} \left(1 + \frac{V_0^-}{V_0^+} e^{j2\beta x} \right)$$

Or,

$$V(x) = V_0^+ e^{-j\beta x} [1 + \Gamma(x)] \qquad (3.31a)$$

Similarly, $I(x)$ can be written as:

$$I(x) = I^+(x) - I^-(x) = \frac{V_0^+}{Z_0} e^{-j\beta x} - \frac{V_0^-}{Z_0} e^{j\beta x}$$

$$I(x) = \frac{V_0^+}{Z_0} e^{-j\beta x} [1 - \Gamma(x)] \qquad (3.31b)$$

The input impedance at any point (x) on the transmission line from Equations 3.31 can now be written as:

$$Z_{IN}(x) = Z_0 \frac{1 + \Gamma(x)}{1 - \Gamma(x)} \qquad (3.32)$$

Using Equation (3.30) and letting the distance from the load as $d = \ell - x$, Equation (3.32) can be written as:

$$Z_{IN}(d) = Z_O \frac{Z_L + jZ_O \tan\beta d}{Z_O + jZ_L \tan\beta d} \tag{3.33}$$

The time-average incident (P_i or P^+), reflected (P_r or P^-) and transmitted (P_t or P_L) powers propagating along a transmission line are given by (assuming Z_O is a real number, $\alpha=0$):

$$P^+(x) = \frac{1}{2} Re\left[V^+(x)I^+(x)^*\right] = \frac{\left|V_o^+\right|^2}{2Z_O} \tag{3.34a}$$

$$P^-(x) = \frac{1}{2} Re\left[V^-(x)I^-(x)^*\right] = \left|\Gamma\right|^2 \frac{\left|V_o^+\right|^2}{2Z_O}, \tag{3.34b}$$

$$P_t(x) = P^+(x) - P^-(x) = (1 - \left|\Gamma\right|^2)\frac{\left|V_o^+\right|^2}{2Z_O} = P_L, \tag{3.34c}$$

Where P_L is given by:

$$P_L = \frac{1}{2} Re(V_L I_L^*) = \frac{1}{2} Re(V_L V_L^* / Z_L^*) = \left|V_L\right|^2 \frac{Re(Z_L)}{2\left|Z_L\right|^2} \tag{3.34d}$$

3.6.8 Determination of V_o^+ and V_o^-

From the earlier discussion, we know that the incident wave is given by:

$$V^+(x) = V_o^+ e^{-j\beta x} \tag{3.34e}$$

To find V_o^+ we consider the source end ($x=0$) as follows:

$$V(0) = V_g - Z_o I(0) \tag{3.34f}$$

From (3.31a) for x=0, we can write

$$V(0) = V_o^+ [1 + \Gamma(0)]$$

$$I(0) = \frac{V_o^+}{Z_O} [1 - \Gamma(0)]$$

Upon substitution for V(0) and I(0) in Equation 3.34f, we have:

$$V_o^+ = (V_g/2) \quad \text{(for x=0 at source)} \tag{3.34g}$$

Equation (3.34g) is simply stating a voltage division of the source voltage between the source impedance (Z_g) and the characteristic impedance of the line (Z_o), where $Z_g=Z_o$. This can be easily visualized by noting that the incident wave does not see the load at first but only Z_o, thus the voltage division!

Moreover, we know that the reflected wave is given by:
$$V^-(x)= V_o^- e^{j\beta x} \tag{3.34h}$$
To find V_o^- we need to visualize the incident wave traveling toward the load with a magnitude of $V_g/2$ given by:
$$V^+(x)= (V_g/2)e^{-j\beta x}$$

At the load ($x=\ell$), the incident wave reflects back to the source by a factor of Γ_L. This is called the reflected wave and is given by:
$$V^+(\ell)= (V_g/2)e^{-j\beta \ell}$$
$$V^-(\ell)= \Gamma_L (V_g/2) e^{-j\beta \ell}$$

The reflected wave travels back a distance of $x"= \ell-x$ and arrives at the source as:
$$V^-(\ell)= \Gamma_L (V_g/2) e^{-j\beta \ell} e^{-j\beta x"}= \Gamma_L (V_g/2) e^{-j\beta \ell} e^{-j\beta(\ell-x)}= (V_g/2) \Gamma_L e^{-j2\beta \ell} e^{j\beta x} \tag{3.34i}$$

where x" is an imaginary reference frame set up at the load ($x"=0$) and is directed toward the source ($x"= \ell$). Comparing 3.34i with 3.34h, we obtain:
$$V_o^- = (V_g/2) \Gamma_L e^{-j2\beta \ell} \quad \text{(for x=0 at source)} \tag{3.34j}$$

Therefore using Equations (3.34g) and (3.34j) in Equation (3.30b), the total voltage at each point (x) along the transmission line can be written as:
$$V(x)= V_o^+ e^{-j\beta x} + V_o^- e^{j\beta x} =(V_g/2) e^{-j\beta x} + (V_g/2)\Gamma_L e^{-j2\beta \ell} e^{j\beta x}$$
Or by using (3.30a), we can write:
$$V(x)= (V_g/2) e^{-j\beta x}[1 + \Gamma_L e^{-j2\beta \ell} e^{j2\beta x}]= (V_g/2)e^{-j\beta x}[1+\Gamma(x)] \tag{3.34k}$$

It should be noted that Equation (3.34k) is the same as Equation (3.31a) derived earlier.

3.6.9 A Summary of Analysis
Let us now recapitulate what we have developed for the incident and reflected waves mathematically and write:
$$V^+(x)= (V_g/2)e^{-j\beta x} =(V_g/2)\angle-\beta x \tag{3.34m}$$

$$V^-(x)= V_o^- e^{j\beta x} =(V_g/2)\Gamma_L \ e^{-j2\beta\ell} \ e^{j\beta x}=(V_g/2)\Gamma_L \angle -2\beta\ell+\beta x \qquad (3.34n)$$

At the source end (x=0) we have:
$$V^+(0)= (V_g/2) =(V_g/2)\angle 0° \qquad (3.34o)$$
$$V^-(x)= (V_g/2)\Gamma_L \angle -2\beta\ell \qquad (3.34p)$$

At the load end (x=ℓ) we have:
$$V^+(x)= (V_g/2)e^{-j\beta\ell} =(V_g/2)\angle -\beta \ \ell \qquad (3.34q)$$
$$V^-(x)= (V_g/2)\Gamma_L \ e^{-j2\beta\ell} \ e^{j\beta x} = (V_g/2)\Gamma_L \angle -\beta\ell \qquad (3.34r)$$

What we have presented so far can be summarized in one diagram, which reveals all of the complexities of transmission line analysis and makes them into great simplicities. Figure 3.22b shows these simplicities clearly and makes the analysis or design of any complex transmission line an expedient task!

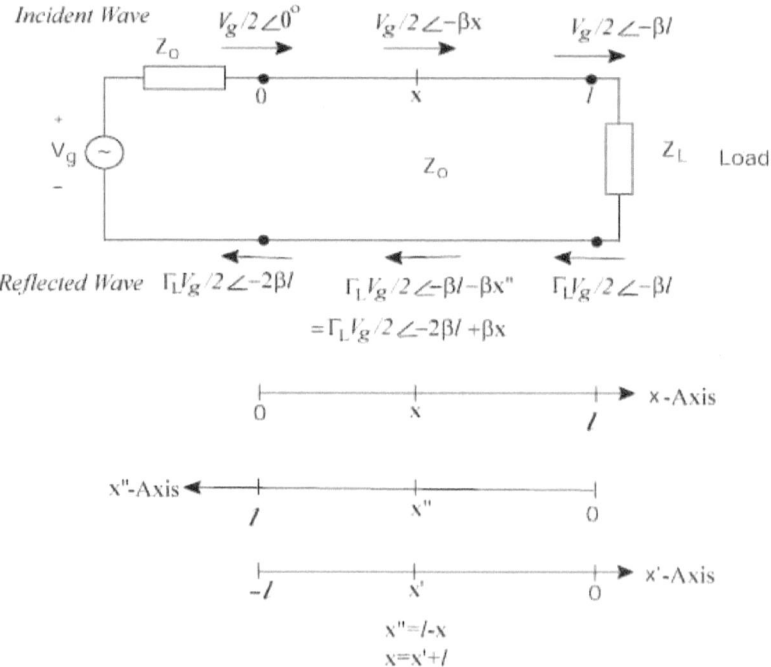

FIGURE 3.22b   A summary of transmission line key points.

3.6.10 Voltage Standing Wave Ratio (VSWR)

As described earlier, a standing wave results from two waves having the same frequency traveling in opposite directions on a transmission line. The meeting of these two waves produces a standing wave pattern of voltage and current on a transmission line.

The maximum value of voltage anywhere along the transmission line is given by:

$$V_{max}=|V(x)|_{max}=|\ V_o^+|+|\ V_o^-|=|\ V_o^+|(1+|\Gamma_L|) \qquad (3.35)$$

The minimum value of the voltage is given by:

$$V_{min}=|V(x)|_{min}=|\ V_o^+|-|\ V_o^-|=|\ V_o^+|(1-|\Gamma_L|) \qquad (3.36)$$

Similarly, for the current standing wave we have:

$$I_{max}=|I(x)|_{max}=|I^+|+|I^-|=\frac{\left|V_o^+\right|}{Z_0}+\frac{\left|V_o^-\right|}{Z_0}$$

$$=\frac{\left|V_o^+\right|}{Z_0}\left(1+|\Gamma_L|\right) \qquad (3.37)$$

$$I_{min}=|I(x)|_{min}=|I^+|-|I^-|=\frac{\left|V_o^+\right|}{Z_0}-\frac{\left|V_o^-\right|}{Z_0}$$

$$=\frac{\left|V_o^+\right|}{Z_0}\left(1-|\Gamma_L|\right) \qquad (3.38)$$

Equations (3.35) to (3.38) are used to define the standing wave ratio (often referred to as voltage standing wave ratio "VSWR") as follows:

$$VSWR=\frac{V_{max}}{V_{min}}=\frac{I_{min}}{I_{max}}=\frac{1+|\Gamma_L|}{1-|\Gamma_L|} \qquad (3.39)$$

Or,

$$|\Gamma_L|=\frac{VSWR-1}{VSWR+1} \qquad (3.40)$$

EXAMPLE 3.4

What is the VSWR for a matched transmission line ($Z_o=50\Omega$)?
Solution:

$Z_L=Z_o \Rightarrow \Gamma_L=0$
VSWR=$(1+0)/(1-0)=1$

EXAMPLE 3.5
What is the VSWR for:
a. An open load ($Z_L=\infty$),
b. A short load ($Z_L=0$)? Assume $Z_o=50\ \Omega$.
Solution:
a) $Z_L=\infty \Rightarrow \Gamma_L=\lim_{Z_L\to\infty}(Z_L-50)/(Z_L+50)=1$
VSWR=$(1+1)/(1-1)=\infty$
b) $Z_L=0$
$\Rightarrow \Gamma_L=(0-50)/(0+50)=-1 \Rightarrow |\Gamma_L|=1$
VSWR=$(1+1)/(1-1)=\infty$

Conclusion: From examples 3.2 and 3.3, we can see that:
$1\leq$ VSWR $\leq \infty$ (3.41)

EXAMPLE 3.6
What is the Z_{IN} of a TL at x=0 (d=ℓ-x = ℓ) for an open circuit load?
Solution:
For $Z_L=\infty$, (3.34) can be written as:

$$\mathbf{Z_{OC} = Z_{IN}(\ell) = \lim_{Z_L\to\infty} Z_O \frac{Z_L + jZ_O \tan\beta\ell}{Z_O + jZ_L \tan\beta\ell}}$$

$\mathbf{Z_{OC} = -jZ_o \cot\beta\ell}$ (3.42)

EXAMPLE 3.7
What is the Z_{IN} of a TL at x=0 (d=ℓ-x = ℓ) for an short circuit load?

Solution:
For $Z_L=0$, (3.34) can be written as:
$$\mathbf{Z_{SC} = Z_{IN}(\ell) = Z_O \frac{0 + jZ_O \tan\beta\ell}{Z_O + 0}}$$

$\mathbf{Z_{SC} = jZ_o \tan\beta\ell}$ (3.43)

3.6.11 Quarter-Wave Transformers

The two main functions of any transmission line at any frequency, are two-folded as follows:

a. Transmission of power, and/or
b. Transmission of information.

At RF/microwave frequencies, it becomes essential that all lines be matched to each other, to the source and finally to the load. This is due to the obvious fact that reflections due to mismatch or discontinuities (e.g. at a connection, at a junction, etc.) will result in echoes and will reduce the transmitted power and will distort the information-carrying signal.

A simple method for matching a resistive load Z_L to a lossless feed line (having a real characteristic impedance Z_O) is the use of a quarter-wave transformer which is a piece of a transmission line having a $\lambda/4$ length and a characteristic impedance of $(Z_O')_{\lambda/4}$.

The characteristic impedance of the quarter-wave transformer $(Z_O')_{\lambda/4}$ terminated in a real load Z_L can be derived as follows:

DERIVATION OF $(Z_O')_{\lambda/4}$

$x=0 \Rightarrow d=\ell-x = \ell$

$d=\lambda/4 \Rightarrow \beta d=(2\pi/\lambda)(\lambda/4)= \pi/2 \Rightarrow \tan\beta d=\infty$

Thus the input impedance of the quarter-wave transformer (Equation 3.33b) terminated in a real load Z_L can be written as:

$d=\lambda/4$

$$Z_{in}(\lambda/4) = \lim_{\ell \to \lambda/4}(Z_O')\frac{Z_L + jZ_O' \tan\beta\ell}{Z_O' + jZ_L \tan\beta\ell} = (Z_O')\frac{jZ_O' \tan\beta\ell}{jZ_L \tan\beta\ell}$$

$$\Rightarrow Z_{in} = \frac{Z_O'^{\,2}}{Z_L} \qquad\qquad (3.44)$$

Or,

$$(Z_O')_{\lambda/4} = \sqrt{Z_{in}Z_L} \qquad\qquad (3.45)$$

POINT OF CAUTION: *This simple method of matching is applicable to when both of the following conditions are met:*

a) *The feed transmission line is lossless (this leads to a characteristic impedance value which is a real number), and*
b) *The load is resistive*

There are cases where the load is a complex number, and thus at first glance a quarter-wave transformer does not seem to lend itself for matching purposes.

However, in such a case the load should first be converted into a real number by adding a reactance having the same value as the load's reactance but with the opposite sign. The resultant load is resistive and can then be transformed to the feed line's characteristic impedance through the use of a quarter-wave transformer as described above (For further details on matching techniques, please see Chapter 10). The following example further elucidates this-simple method of matching.

EXAMPLE 3.8
What is $(Z_T)_{\lambda/4}$ of a quarter-wave transformer to transform a load of 100 Ω to a 50 Ω feed line as shown in Figure 3.23?

Solution:
To create a match, we require that Z_{in} to be the same as the characteristic impedance of the feed line, i.e., $Z_{in}=50\ \Omega$. Using (3.45), we can write:

$$(Z_T)_{\lambda/4}=(Z_L Z_{in})^{1/2}$$

Thus we obtain:

$$(Z_T)_{\lambda/4}=(100 \times 50)^{1/2}=70.7\ \Omega$$

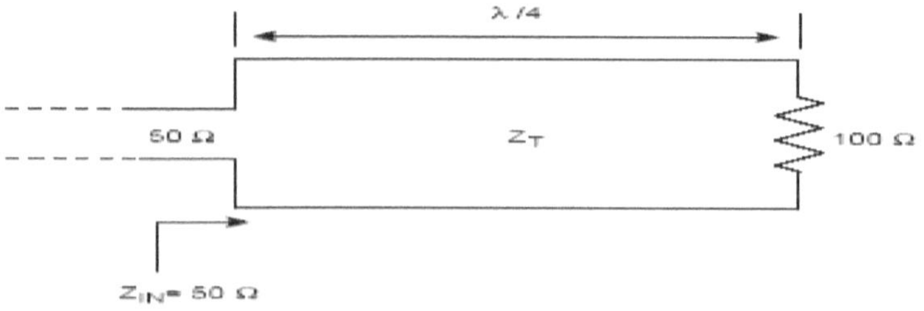

Figure 3.23 A quarter-wave transformer.

NOTE: *Example 3.8 clearly shows why these types of shorted transformers are ideal for electrically isolating the RF circuitry from the DC bias source in an amplifier circuit as will be discussed later in Chapter 15, RF/Microwave Amplifiers. This is because the RF circuitry is connected at the input side of the transformer while DC bias source is at the short-circuited end of the transformer (of course, the short circuit is created by the use of a high-value capacitor to ground). In this fashion, the RF signals "see" an open circuit at the RF side and would not be able to travel to the DC bias source, while the DC bias "sees" a direct connection (i.e. a short circuit) to the RF circuitry.*

3.6.12 A Generalized Transmission Line Circuit
In the previous examples, the main focus has been on the effects of a load on the current and voltage waves traveling on a transmission line. However the source of the waves, which is the generator located at the other end, plays an important role in the propagation of the waves along the transmission line.

Up to this point in our discussion, the generator's internal impedance has been a real number equal to the characteristic of the transmission line. In effect, the generator was matched to the line and only the effects of the mismatch of the load were studied so far. Obviously, this is a special case. The most general case is having mismatches at both ends (i.e., at the generator and at the load ends), which will now be discussed in detail.

3.6.13 Analysis
Consider a finite lossless transmission line (T.L.) of length (ℓ) with a characteristic impedance (Z_O) driven by a generator (V_g) with an internal impedance (Z_g) at x=0 and terminated in a load (Z_L) at x=ℓ as shown in Figure 3.24.

The boundary condition (B.C.) at each end can be written as:
a. B.C. #1-Voltage and current at x=0 is given by: $V_i=V_g-Z_gI_i$

b. B.C. #2-Voltage and current at x=ℓ is given by: $V_L=Z_LI_L$

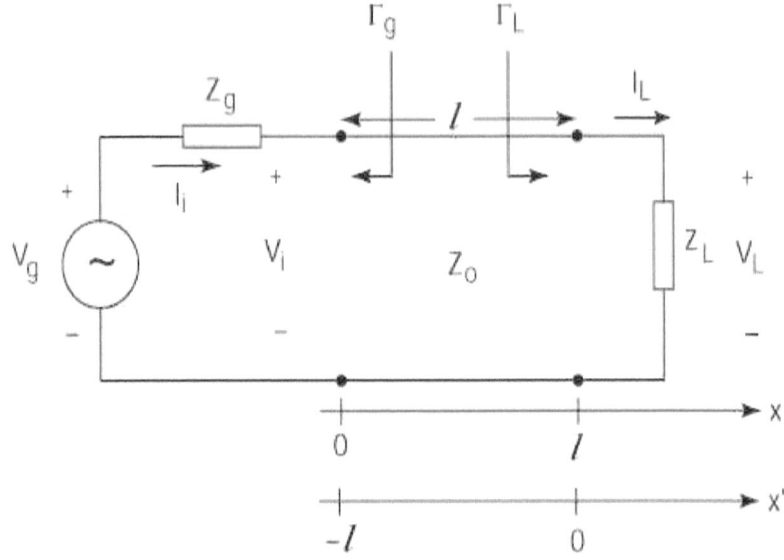

FIGURE 3.24   A general transmission line circuit.

c. Voltage and current on the T.L. for $0 \leq x \leq \ell$, from (3.31) and (3.32) is given by:

$V(x)=V^+(x)+V^-(x)$

$V(x)= V_0^+ e^{-j\beta x}[1+\Gamma(x)]$ (3.46a)

$I(x)= I^+(x)-I^-(x)= \dfrac{V_0^+}{Z_0} e^{-j\beta x}[1-\Gamma(x)]$ (3.46b)

Where, from (3.30), $\Gamma(x)$ is given by:

$\Gamma(x)= \dfrac{V^-(x)}{V^+(x)} = \dfrac{V_0^- e^{j\beta x}}{V_0^+ e^{-j\beta x}} = \dfrac{V_0^-}{V_0^+} e^{j2\beta x} = \Gamma_L \ e^{j2\beta(x-\ell)}$ (3.46c)

Applying the boundary condition given by (a), we can solve for V_0^+ as follows:

$V_i=V(0)=V_0^+ e^{-j\beta x}[1+\Gamma(0)]= Vg- Z_g \dfrac{V_0^+}{Z_0} [1-\Gamma(0)]$ (3.47)

Where

$\Gamma(0)=\Gamma_L \ e^{-j2\beta\ell}$

From Equation 3.47, we can solve for V_0^+ in terms of V_g to obtain:

$$V_0^+ = \frac{Z_0 V_g}{Z_0 + Z_g}\left(\frac{e^{j\beta\ell}}{1 - \Gamma_L\Gamma_g e^{-j2\beta\ell}}\right) \tag{3.48a}$$

where

$$\Gamma_g = \frac{Z_g - Z_0}{Z_g + Z_0} \tag{3.48b}$$

Thus V(x) and I(x) under this general condition may be obtained by substituting for V_0^+ from Equation (3.48) in (3.46) as follows:

$$V(x) = \frac{Z_0 V_g}{Z_0 + Z_g} e^{-j\beta x}\left(\frac{1 + \Gamma_L e^{j2\beta(x-\ell)}}{1 - \Gamma_L\Gamma_g e^{-j2\beta\ell}}\right) \tag{3.49a}$$

$$I(x) = \frac{V_g}{Z_0 + Z_g} e^{-j\beta x}\left(\frac{1 - \Gamma_L e^{j2\beta(x-\ell)}}{1 - \Gamma_L\Gamma_g e^{-j2\beta\ell}}\right) \tag{3.49b}$$

NOTE: *Equations 3.49a, b show phasor expressions for the voltage and current due to a sinusoidal voltage source (V_g) feeding a finite transmission line, which is terminated in a general load(Z_L).*

These equations represent the summation of an infinite number of reflections from both ends of the transmission line, i.e.,

$$V(x) = V_1^+ + V_1^- + V_2^+ + V_2^- + ... = \sum_{i=1}^{\infty}(V_i^+ + V_i^-) \tag{3.50a}$$

where
$|V_1^+| = |V_0'^+|,$
$|V_1^-| = |\Gamma_L|\,|V_0'^+|,$
$|V_2^+| = |\Gamma_g||\Gamma_L|\,|V_0'^+|,$
$|V_2^-| = |\Gamma_g||\Gamma_L|^2\,|V_0'^+|,$ etc.
and,

$$V_0'^+ = V_g[Z_0/(Z_0 + Z_g)]$$

EXERCISE 3.1
Prove that the summation of infinite number of voltage reflections as given by (3.50a) converges to (3.49a).

HINT: Note that:

$$\sum_{n=0}^{\infty} x^n = 1 + x + x^2 + \ldots + x^n + \ldots = 1/(1-x), \quad |x| < 1 \qquad (3.50b)$$

SPECIAL CASES
From Equation (3.49) we can derive several special cases as shown in Figures 3.25a, b,c.

Case I. Matched at Both Ends
$Z_g = Z_L = Z_O$ (see Figure 3.25a)
$\Gamma_g = \Gamma_L = 0$
$V_0^+ = (V_g/2)$ (for x=0 at source)

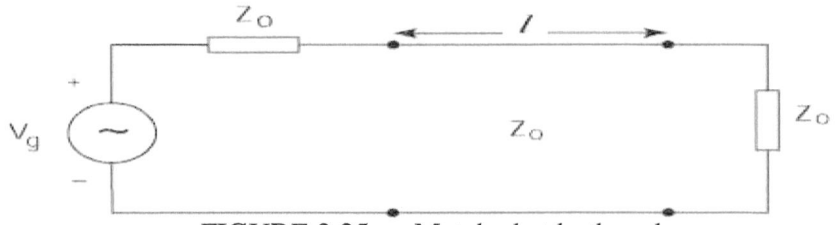

FIGURE 3.25a   Matched at both ends.

Since $\Gamma_L = 0$, there is no reflected wave from the load, and thus Equation 3.49 can be written as:

$$V(x) = \frac{V_g}{2} e^{-j\beta x} \qquad (3.51a)$$

$$I(x) = \frac{V_g}{2Z_O} e^{-j\beta x} \qquad (3.51b)$$

At the source end (x=0), we have:

$$V(0) = V_1 = \frac{V_g}{2} \qquad (3.51c)$$

$$I(0) = I_i = \frac{V_g}{2Z_O} \qquad (3.51d)$$

and at the load end ($x=\ell$), we can write:

$$V(\ell) = V_L = \frac{V_g}{2} e^{-j\beta\ell} \qquad (3.51e)$$

$$I(\ell) = I_L = \frac{V_g}{2Z_O} e^{-j\beta\ell} \qquad (3.51f)$$

In this case, there are no standing waves on the transmission line and magnitude of the voltage and current is the same everywhere on the line, that is,

$$|V_i|=|V_L|=|V(x)|=\frac{V_g}{2} \qquad (3.51g)$$

$$|I_i|=|I_L|=|I(x)|=\frac{V_g}{2Z_O} \qquad (3.51h)$$

NOTE: *In some texts, the reference for length is located at the load end rather than the generator end. This axis is designated by the x'-axis in figure 3.24. This means that there is a shift in the x-axis (by +ℓ), i.e.,*

$$x=x'+\ell \qquad (3.51i)$$

Thus Equations (3.51a,b) can now be written in terms x' as:
$V(x')= (Vg/2)e^{-j\beta(x'+\ell)}$ (for x'=0 at load)

Or,

$$V(x') = \frac{V_g}{2} e^{-j\beta\ell} e^{-j\beta x'} \qquad (3.51j)$$

$$I(x') = \frac{V_g}{2Z_O} e^{-j\beta\ell} e^{-j\beta x'} \qquad (3.51k)$$

Case II. Matched at the Source End
$Z_g= Z_O$, $Z_L \neq Z_O$
$\Gamma_g =0$ (see Figure 3.25b)

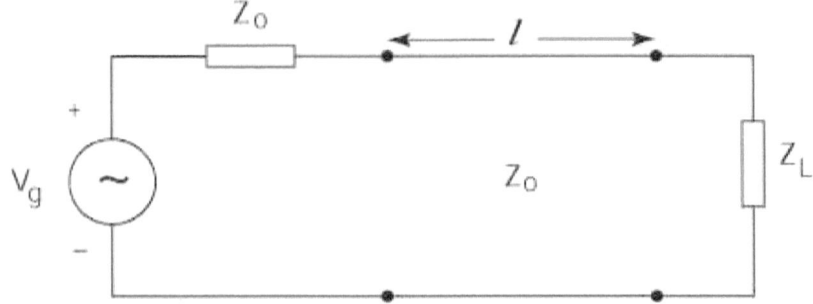

FIGURE 3.25b   Matched at the source end only.

Since $\Gamma_g=0$, there is no reflected wave from the source, but there is a reflected wave from the load. Thus Equation 3.49 can be written as:

$$V(x) = \frac{V_g}{2} e^{-j\beta x} \left(1 + \Gamma_L e^{2j\beta(x-\ell)}\right) \tag{3.52a}$$

$$I(x) = \frac{V_g}{2Z_0} e^{-j\beta x} \left(1 - \Gamma_L e^{j2\beta(x-\ell)}\right) \tag{3.52b}$$

In terms of the shifted axis $(x=x'+\ell)$, we can write Equations (3.52) as:

$$V(x') = \frac{V_g}{2} e^{-j\beta(x'+\ell)} \left(1 + \Gamma_L e^{j2\beta x'}\right) \tag{3.52c}$$

$$I(x') = \frac{V_g}{2Z_0} e^{-j\beta(x'+\ell)} \left(1 - \Gamma_L e^{j2\beta x'}\right) \tag{3.52d}$$

At the generator end $(x'=-\ell)$, we have:

$$V(-\ell) = V_i = \frac{V_g}{2} \left(1 + \Gamma_L e^{-j2\beta\ell}\right) \tag{3.52e}$$

$$I(-\ell) = I_i = \frac{V_g}{2Z_0} \left(1 - \Gamma_L e^{-j2\beta\ell}\right) \tag{3.52f}$$

At the load end $(x'=0)$ we have:

$$V(0) = V_L = \frac{V_g}{2} e^{-j\beta\ell} \left(1 + \Gamma_L\right) \tag{3.52g}$$

$$I(0) = I_L = \frac{V_g}{2Z_0} e^{-j\beta\ell} \left(1 - \Gamma_L\right) \tag{3.52h}$$

It should be noted that Equation (3.52g) is the same as Equation (3.34k) when x=ℓ.

Case III. Matched at the Load End

$Z_g \neq Z_O$, $Z_L = Z_O$ (see Figure 3.25c)

$\Gamma_L = 0$

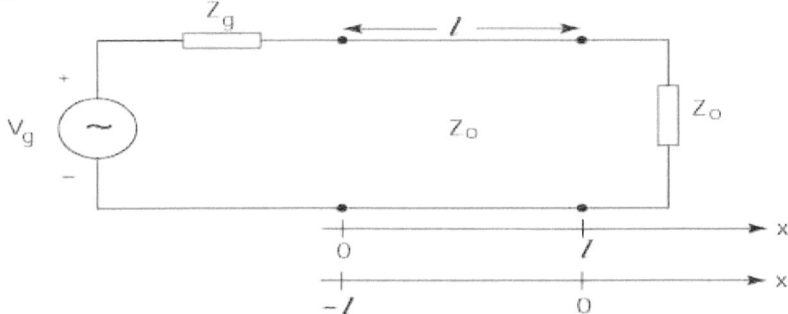

FIGURE 3.25c   Matched at load end only.

Using Equation 3.49, we can write:

$$V(x) = \frac{Z_0 V_g}{Z_0 + Z_g} e^{-j\beta x} \qquad (3.53a)$$

$$I(x) = \frac{V_g}{Z_0 + Z_g} e^{-j\beta x} \qquad (3.53b)$$

At the generator end (x=0), we have:

$$V(0) = V_i = \frac{Z_0 V_g}{Z_0 + Z_g} \qquad (3.53c)$$

This result could have easily be written using the voltage division principle, without using Equation (3.49). In other words, this case reduces to a simple voltage division between the source internal impedance (Z_g) and the transmission line presenting a constant input impedance of (Z_o).

$$I(0) = I_i = \frac{V_g}{Z_0 + Z_g} \qquad (3.53d)$$

And at the load end (x=ℓ) we have:

$$V(\ell) = V_L = \frac{Z_0 V_g}{Z_0 + Z_g} e^{-j\beta \ell} \qquad (3.53e)$$

$$I(\ell) = I_L = \frac{V_g}{Z_0 + Z_g} e^{-j\beta\ell} \tag{3.53f}$$

The voltage and current on the transmission line have the same magnitude except for a phase shift with length:

$$V_L = V_i e^{-j\beta\ell} \quad \text{for } x=\ell \tag{3.53g}$$

Or in general,

$$V(x) = V_i e^{-j\beta x} \tag{3.53h}$$

Similarly, we can write for current:

$$I_L = I_i e^{-j\beta\ell} \quad \text{for } x=\ell \tag{3.53i}$$

Or in general,

$$I(x) = I_i e^{-j\beta x} \tag{3.53j}$$

Equations (3.53a) and (3.53b) can be written in terms of the shifted axis $(x=x'+\ell)$ as:

$$V(x') = V_i e^{-j\beta(x'+\ell)} \tag{3.53k}$$

$$I(x') = I_i e^{-j\beta(x'+\ell)} \tag{3.53\ell}$$

EXAMPLE 3.9

Consider a 50 Ω lossless transmission line of length $\ell=1$ m, connected to a generator operating at $f=1$ GHz and having $V_g=10$ V and $Z_g=50$ Ω at one end and to a load $Z_L=100$ Ω at the other (see Figure 3.26).

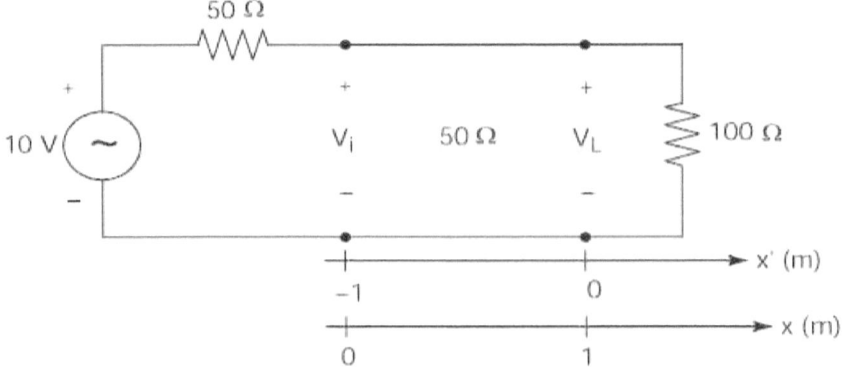

FIGURE 3.26 Circuit for Example 3.10.

Determine:

a. The voltage and current at any point on the transmission line.

b. The voltage at the generator (V_i) and load (V_L) ends.
c. The reflection coefficient and VSWR at any point on the line.
d. The average power delivered to the load.

Solution:
a. Since $Z_g=Z_O=50\ \Omega \Rightarrow \Gamma_g=0$
Since $\Gamma_g=0$, special case II applies here. Thus we can write:
$\beta=\omega/c=2\pi\times10^9/3\times10^8=20\pi/3$

$$\Gamma_L = \frac{Z_L - Z_O}{Z_L + Z_O} = \frac{100-50}{100+50} = \frac{1}{3}$$

$$V(x) = \frac{V_g}{2}e^{-j\beta x}\left(1+\Gamma_L e^{j2\beta(x-\ell)}\right)=5e^{-j20\pi x/3}\left(1+\frac{1}{3}e^{j40\pi(x-1)/3}\right)$$

$$I(x) = \frac{V_g}{2Z_O}e^{-j\beta x}\left(1-\Gamma_L e^{j2\beta(x-\ell)}\right)=0.1e^{-j20\pi x/3}\left(1-\frac{1}{3}e^{j40\pi(x-1)/3}\right)$$

b. At the generator end (x=0 m), we have:

$$V_i = V(-1) = 5\left(1+\frac{1}{3}e^{-j40\pi/3}\right)=-4.16+j1.44\ \text{V}$$

At the load end (x=1 m), we have:

$$V_L = V(1) = 5e^{-j20\pi/3}\left(1+\frac{1}{3}\right)=\frac{20}{3}e^{-j20\pi/3}$$

c. The reflection coefficient and VSWR are as follows:

$$\Gamma(x) =\Gamma_L\, e^{j2\beta x}=\frac{1}{3}e^{j40\pi x/3}$$

$$\text{VSWR} = \frac{1+|\Gamma_L|}{1-|\Gamma_L|} = \frac{1+1/3}{1-1/3} = 2$$

d. The average power delivered to the load is:
$\alpha=0$

$$P(x) = \frac{1}{2}\text{Re}\left[V_L(x)I_L^{*}(x)\right]=\frac{|V_L|^2}{2Z_L}=\frac{\left|20e^{-j20\pi/3}/3\right|^2}{2x100}=\frac{2}{9}=0.22\ \text{W}$$

NOTE: *If the load was completely matched to the line the power delivered to the load would have been:*
$Z_L=50\ \Omega$
$|V_i|=|V_L|=V_g/2=5\ \text{V}$

$$\left(P_{av}\right)_{max} = \frac{\left|V_L\right|^2}{2Z_L} = \frac{5^2}{2 \times 50} = 0.25 \text{ W}$$

Since there is no reflected power, $(p_{av})_{max}$ is also the incident power (P_i) which is higher than the (P_{av}) calculated earlier under unmatched conditions. The difference in the two powers is due to the reflected power back to the source:
$P_r = |\Gamma_L|^2 P_i = (1/9)(0.25) = 0.03$ W

3.7 MICROSTRIP LINE

Amongst all planar transmission lines, microstrip line has gained much popularity and importance in microwave planar circuit technology, and thus will be considered and analyzed in this section.

A microstrip line is a transmission line consisting of a strip of conductor of thickness (t), width (w), and a ground plane separated by a dielectric medium of thickness (h) as shown in Figure 3. 27. Since it is an open conduit for wave transmission, not all of the electric or magnetic fields will be confined in the structure. This fact along with the existence of a small axial E-field, leads to a not-purely TEM wave propagation, but a quasi-TEM mode of propagation.

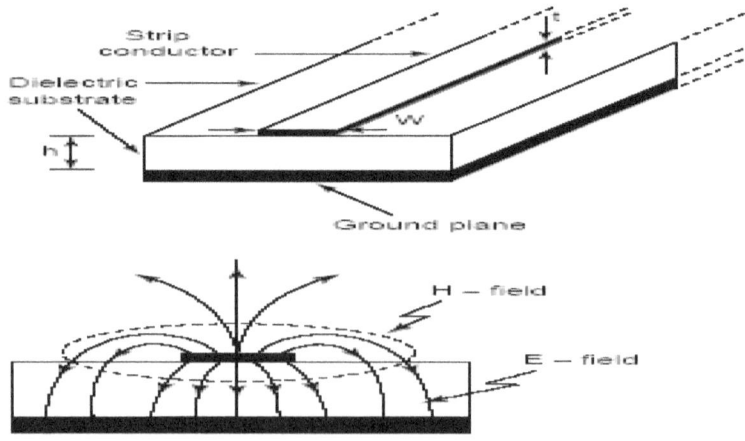

Figure 3.27 A microstrip line: geometry and field configurations.

These types of transmission lines are very popular and are used extensively in microwave planar circuit design and microwave integrated circuit (MIC) technology. Use of printed circuit board technology and its simplicity of fabrication, along with ease of placement and interconnection of lumped elements and components has made this type of transmission line very popular and much superior to other types of planar transmission lines

3.7.1 Wave Propagation in Microstrip Lines

The dielectrics used in the fabrication of the Microstrip line are characterized by a dielectric constant (ε_r) defined by:

$\varepsilon_r = \varepsilon / \varepsilon_0,$ (3.54a)

$\varepsilon_0 = 8.854 \times 10^{-12}$ F/m

where ε and ε_0 are the dielectric's and vacuum's permittivity, respectively. Some of the most popular dielectrics are: Duroid ($\varepsilon_r = 2.23$, 6, 10.5), alumina ($\varepsilon_r = 9.5-10$), Quartz ($\varepsilon_r = 3.7$), silicon ($\varepsilon_r = 11.9$), etc.

The EM-wave propagation in a microstrip line is approximately non-dispersive below a cut-off frequency (f_0), which is given by:

$$f_0 \, (\text{GHz}) = 0.3 \sqrt{\frac{Z_0}{h\sqrt{\varepsilon_r - 1}}}$$ (3.54b)

Where "h" is in centimeters.

The phase velocity for a quasi-TEM is given by:

$V_P = c/\sqrt{\varepsilon_{ff}}$

Where c is the speed of light and ε_{ff} is the effective relative dielectric constant.

Since the field lines are not contained in the structure and some exist in the air (see Figure 3.27b), the effective dielectric constant satisfies the following relation:

$1 < \varepsilon_{ff} < \varepsilon_r$

In general, the effective dielectric constant (ε_{ff}) is a function of not only the substrate material (ε_r) but also of the dielectric thickness (h) and conductor width (W).

The characteristic impedance (Z_0) is given by:

$$Z_0 = \frac{1}{V_p C_0} \tag{3.55}$$

Where C_0 is the capacitance per unit length.

The wavelength (λ) of a propagating wave in the microstrip line is given by:

$$\lambda = V_P/f = \lambda_o/\sqrt{\varepsilon_{ff}} \tag{3.56a}$$

where $\lambda_o = c/f$ is the wavelength in free space. **NOTE:** *The wavelength of a TEM wave (λ_{TEM}) propagating in the dielectric material is different than the wavelength (λ_o) of a propagating wave in free space as follows:*

$$\lambda_{TEM} = \lambda_o/\sqrt{\varepsilon_r} \tag{3.56b}$$

As can be seen from these equations the characteristic impedance (Z_o) and the wavelength (λ) both are functions of the geometry (w, h) of the microstrip line. This variation is shown in Figures 3.28 and 3.29.

3.7.2 Empirical Formulas

The essential empirical formulas for a microstrip line can be categorized as follows (assuming zero or negligible thickness of the strip of metal on top of the dielectric, i.e., t/h<0.005):

a. ε_{ff} FORMULA

The effective dielectric constant (ε_{ff}) is given by [assuming that the dimensions of the microstrip line (W, h) are known]:

For W/h ≤ 1:

$$\varepsilon_{ff} = \frac{\varepsilon_r + 1}{2} + \frac{\varepsilon_r - 1}{2}\left[\left(1 + 12\frac{h}{W}\right)^{-1/2} + 0.04\left(1 - \frac{W}{h}\right)^2\right], \tag{3.57}$$

For W/h ≥ 1:

$$\varepsilon_{ff} = \frac{\varepsilon_r + 1}{2} + \frac{\varepsilon_r - 1}{2}\left(1 + 12\frac{h}{W}\right)^{-1/2} \tag{3.58}$$

The effective dielectric constant (ε_{ff}) can be thought of as the dielectric constant of a homogeneous medium that would fill the entire space, replacing air and dielectric regions.

b. Z_0 FORMULA

The Characteristic impedance is given by [assuming that the dimensions of the microstrip line (W, h) are given or known]:

For W/h ≤ 1:

$$Z_O = \frac{60}{\sqrt{\varepsilon_{ff}}} \ln\left(\frac{8h}{W} + \frac{W}{4h}\right) \tag{3.59}$$

For W/h ≥ 1:

$$Z_O = \frac{120\pi}{\sqrt{\varepsilon_{ff}}\left[W/h + 1.393 + 0.667\ln(W/h + 1.444)\right]} \tag{3.60}$$

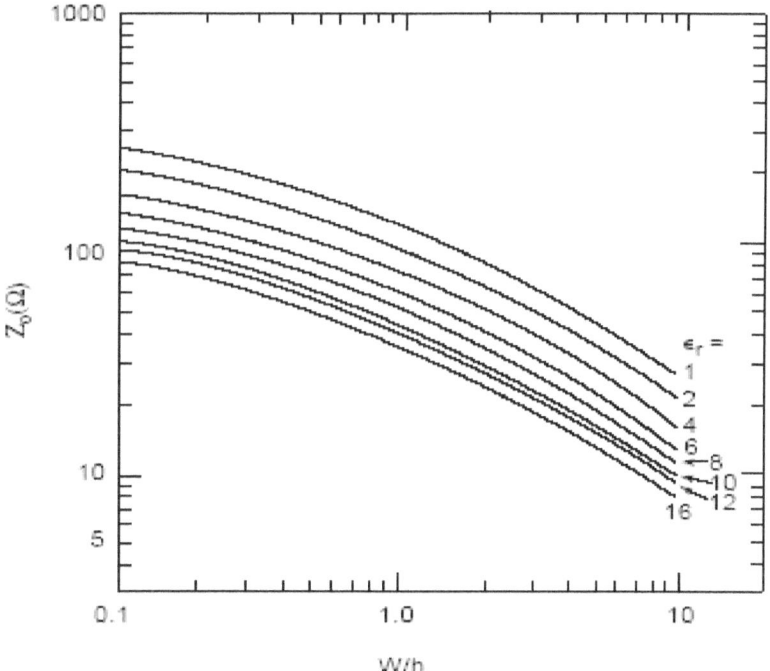

Figure 3.28 The characteristic impedance of the microstrip line.

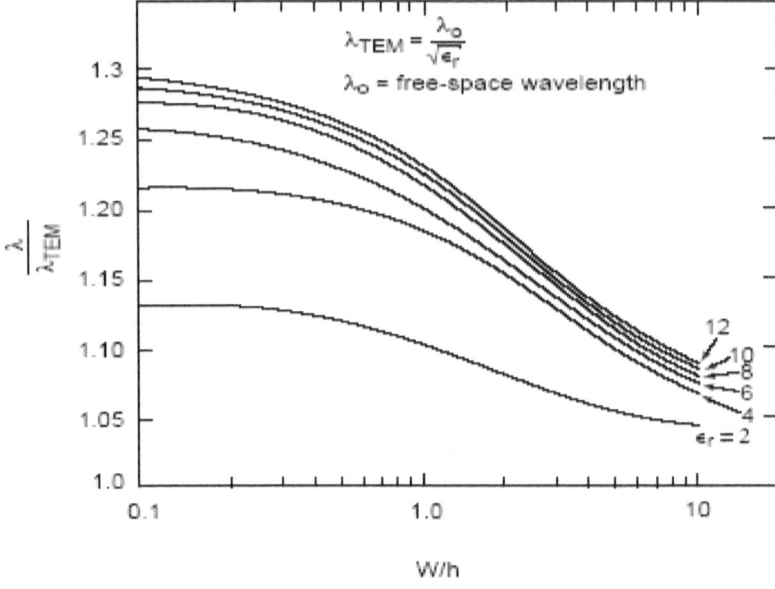

Figure 3.29 The normalized wavelength of the microstrip line.

c.   W/h FORMULA

Assuming (ε_{ff}) and Z_o are given, then the microstrip dimensions (W/h) can be found as follows (a design problem):

For W/h ≤ 2:

$$\frac{W}{h} = \frac{8e^A}{e^{2A} - 2} \tag{3.61}$$

For W/h ≥ 2:

$$\frac{W}{h} = \frac{2}{\pi}\left[B - 1 - \ln(2B - 1) + \frac{\varepsilon_r - 1}{2\varepsilon_r}\left\{\ln(B - 1) + 0.39 - \frac{0.61}{\varepsilon_r}\right\}\right] \tag{3.62}$$

Where

$$A = \frac{Z_o}{60}\sqrt{\frac{\varepsilon_r + 1}{2}} + \frac{\varepsilon_r - 1}{\varepsilon_r + 1}\left(0.23 + \frac{0.11}{\varepsilon_r}\right) \tag{3.63}$$

and

$$B = \frac{377\pi}{2Z_0\sqrt{\varepsilon_r}}$$ (3.64)

d. λ FORMULA
The wavelength in the microstrip line (λ) is given by:
For W/h < 0.6:

$$\lambda = \frac{\lambda_0}{\sqrt{\varepsilon_r}}\left[\frac{\varepsilon_r}{1+0.6(\varepsilon_r-1)(W/h)^{0.0297}}\right]^{1/2}$$ (3.65)

For W/h ≥ 0.6:

$$\lambda = \frac{\lambda_0}{\sqrt{\varepsilon_r}}\left[\frac{\varepsilon_r}{1+0.63(\varepsilon_r-1)(W/h)^{0.1255}}\right]^{1/2}$$ (3.66)

e. α FORMULAS (Attenuation factors)
Another characteristic of the microstrip line is its attenuation when signals travel on it. There are two types of losses in a microstrip line:
- Dielectric substrate loss due to dielectric conductivity
- Conductor Ohmic loss due to skin effect

The loss factor (α) can be found by noting that the power carried along a transmission line in (+x direction) in a quasi-TEM mode can be written as:

$$P^+(x) = \frac{1}{2}\left[V^+(x)I^+(x)^*\right] = \frac{\left[V^+(x)\right]^2}{Z_0}$$ (3.67)

Where
$$V^+(x) = |V^+|e^{-\alpha x}e^{-j\beta x}$$ (3.68)
Thus we have:

$$P^+(x) = \frac{\left|V^+\right|^2}{2Z_0}e^{-2\alpha x}e^{-j2\beta x} = \left|P^+\right|e^{-j2\beta x}$$ (3.69)

Where,

$$\left|P^{+}\right| = \frac{\left|V^{+}\right|^{2}}{2Z_{O}} e^{-2\alpha x}$$

and α is the total attenuation factor which is composed of two components:

$$\alpha = \alpha_{d} + \alpha_{c} \tag{3.70}$$

where

α_{d} = Dielectric loss factor, and

α_{c} = Conductor loss factor

These two loss factors are discussed next:

f. α_{d} FORMULA
Attenuation Due to Dielectric Loss

Attenuation due to dielectric loss identified by "dielectric loss factor (α_{d})" using the quasi-TEM mode of propagation, is given by:

1) For Low-loss dielectric

$$\alpha_{d} = 27.3 \frac{\tan\delta}{\lambda_{O}} \left(\frac{\varepsilon_{r}}{\varepsilon_{r}-1}\right) \left(\frac{\varepsilon_{ff}-1}{\sqrt{\varepsilon_{ff}}}\right) \quad \text{(dB/cm)} \tag{3.71}$$

Where $(\tan\delta)$ is the loss tangent given by:

$$\tan\delta = \frac{\sigma}{\omega\varepsilon}$$

2) For high-loss dielectric

$$\alpha_{d} = 4.34\sigma \left(\frac{\mu_{O}}{\varepsilon_{O}}\right)^{1/2} \left(\frac{1}{\varepsilon_{r}-1}\right) \left(\frac{\varepsilon_{ff}-1}{\sqrt{\varepsilon_{ff}}}\right) \quad \text{(dB/cm)} \tag{3.72}$$

Where σ is the conductivity of the dielectric and $\mu_{o} = 4\pi \times 10^{-7}$ (H/m) is the permittivity of the free space.

g. α_{c} FORMULA
Attenuation Due to Conductor Loss

Attenuation due to dielectric loss identified by "conductor loss factor (α_{c})" using the quasi-TEM mode of propagation (for W/h $\rightarrow\infty$), is given approximately by:

W/h $\rightarrow\infty$,

$$\alpha_c = \frac{R_S}{Z_o W} \quad \text{(Np/m)} \tag{3.73}$$

Where

$$R_S = \sqrt{\frac{\pi f \mu_o}{\sigma}} \tag{3.74}$$

is the surface resistivity of the conductor. Usually, conductor loss is more dominant than the dielectric loss in most microstrip lines, i.e.,

$$\alpha_c >> \alpha_d; \ \Rightarrow \alpha = \alpha_c + \alpha_d \approx \alpha_c$$

However, there are some cases (such as in silicon substrates) where the dielectric loss factor (α_d) is of the same order or larger than the conductor loss Factor (α_c).

EXAMPLE 3.10

A 50 Ω microstrip transmission line needs to be designed using a sheet of Epsilam-10® (ε_r=10) with h=1.02 mm. Determine W, λ and ε_{ff}.by:
a. An exact method
b. An approximate method

Solution:

a. Exact method

we will design a microstrip line with W/h≤2. Thus from (3.61) we have:

$$\frac{W}{h} = \frac{8e^A}{e^{2A} - 2}$$

Where

$$A = \frac{Z_o}{60}\sqrt{\frac{\varepsilon_r + 1}{2}} + \frac{\varepsilon_r - 1}{\varepsilon_r + 1}\left(0.23 + \frac{0.11}{\varepsilon_r}\right) = \frac{50}{60}\sqrt{\frac{10+1}{2}} + \frac{10-1}{10+1}\left(0.23 + \frac{0.11}{10}\right)$$

$$\Rightarrow A = 2.152$$

Thus (W/h) is obtained to be:

$$\frac{W}{h} = 0.96 \ \Rightarrow W = 1.02 \times 0.96 = 0.98 \text{ mm}$$

Since W/h>0.6, we use (3.66) to find λ and then use (3.56a) to find ε_{ff} as follows:

$$\lambda = \frac{\lambda_o}{\sqrt{\varepsilon_r}} \left[\frac{\varepsilon_r}{1+0.63(\varepsilon_r -1)(W/h)^{0.1255}} \right]^{1/2} = \frac{\lambda_o}{\sqrt{10}} \left[\frac{\varepsilon_r}{1+0.63(10-1)(0.96)^{0.1255}} \right]^{1/2}$$

$$\Rightarrow \lambda = 0.39\lambda_o = \lambda_o/\sqrt{\varepsilon_{ff}} \Rightarrow \varepsilon_{ff} = (1/0.39)^2 = 6.6$$

b. Approximate method
Using Figure 3.28, we obtain W/h for $Z_O=50\ \Omega$ to be:
$Z_O=50 \Rightarrow W/h\approx1 \Rightarrow W=h=1.02$ mm
From Figure 3.29 for W/h=1, we obtain:
$\lambda/\lambda_{TEM}=1.23$

From (3.56b) we have;
$\lambda_{TEM}=\lambda_o/\sqrt{\varepsilon_r}=\lambda_o/\sqrt{10}=0.316\lambda_o$

Thus λ is found to be:
$\lambda=1.23\times0.316\lambda_o=0.39\lambda_o$
and from (3.56a) we have:
$\lambda=\lambda_o/\sqrt{\varepsilon_{ff}} \Rightarrow \varepsilon_{ff}=(\lambda_o/\lambda)^2$
$\varepsilon_{ff}=(1/0.39)^2=6.6$

EXAMPLE 3.11
Design a 50 Ω transmission line that provides 90° phase shift at 2.5 GHz. Assume h=1.27 mm and $\varepsilon_r=2.2$.

Solution:
To find "W", we assume that W/h≥2 and will verify this assumption later. From Equation (3.62) and (3.64), we find:

$$B = \frac{377\pi}{2Z_O\sqrt{\varepsilon_r}}$$

B=7.985,
And

$$\frac{W}{h} = \frac{2}{\pi}\left[B-1-\ln(2B-1) + \frac{\varepsilon_r -1}{2\varepsilon_r}\left\{ \ln(B-1)+0.39 - \frac{0.61}{\varepsilon_r} \right\} \right]$$

Yielding:

W/h=3.08 $\Rightarrow$ W=3.08x1.27 =3.91 mm

The value of W/h=3.08 is obviously greater than 2, which justifies our earlier assumption.

So far we have found the width of the line, now we need to know the length of the line. Using the given phase shift of 90° yields:

$\phi=\beta\ell=\omega\ell/V_p=2\pi f\ell/(c/\sqrt{\varepsilon_{ff}})=2\pi f\ell\sqrt{\varepsilon_{ff}}/c=90°=\pi/2$

$\Rightarrow \ell=c/(4f\sqrt{\varepsilon_{ff}})$

From the above equation we can see that in order to find ℓ, we need to find ε_{ff}. Using Equation (3.58), we obtain:

For W/h $\geq$ 1:

$$\varepsilon_{ff} = \frac{\varepsilon_r +1}{2} + \frac{\varepsilon_r -1}{2}\left(1+12\frac{h}{W}\right)^{-1/2}$$

ε_{ff}=1.87

Thus the length of the transmission line is given by:

$\ell=3\times10^8/(4\times2.5\times10^9\times\sqrt{1.87})=0.0219$ m=2.19 cm

---

Chapter 3- Symbol List

A symbol will not be repeated again, once it has been identified and defined in an earlier chapter, with its definition remaining unchanged.

C_0 - Capacitance per unit length

EM – Electro-Magnetic

k – Arbitrary constant

TEM – Transverse Electro-Magnetic

TL – Transmission Line

v – Velocity of motion

V_P - Phase velocity

V^+ - Incident voltage

V^- - Reflected voltage

Z_0 – Characteristic Impedance

Z_{OC} - Open circuit impedance

Z_{SC} - Short circuit impedance

$Z_{\lambda/4}$ - Impedance at the location $\lambda/4$.

β - Phase constant
Γ - Reflection coefficient
Γ_L - Reflection coefficient at the load
$\Gamma(x)$ - Reflection coefficient at location x
ε – Dielectric permittivity
ε_{ff} – Effective relative dielectric constant
ε_o – Permittivity of vacuum (8.85×10^{-12} F/m)
ε_r – Dielectric constant of a material
γ - Propagation constant
λ_o - Wavelength in free space
ω - Angular frequency ($\omega = \beta v$)

CHAPTER -3 PROBLEMS

3.1)   In the two-port network shown in Figure P3.1, assume that $(V_S)_{RMS} = 20\angle0°$ V, and $Z_L = 50 + j50$ Ω.
a) Find $V^+(0)$, $V^+(\lambda/8)$, $V^-(0)$, $V^-(\lambda/8)$.
b) Calculate net voltages: $V(0)$, $V(\lambda/8)$, $I(0)$ and $I(\lambda/8)$.
c) Calculate the input powers at x=0, $\lambda/8$ and show: $P(0) = P(\lambda/8)$.
d) Find $Z_{IN}(0)$

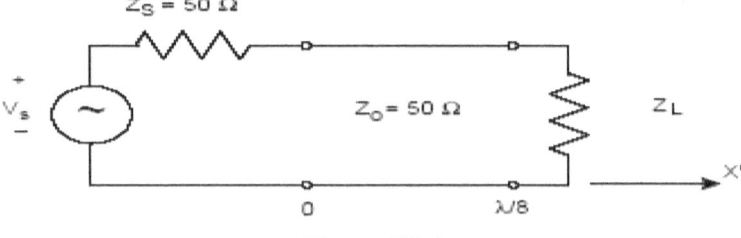

Figure P3.1

3.2)   Find the input impedance and the reflected power at Port(1) and the power delivered to the load at port(2) for the circuit shown in Figure P3.2. Assume $V_s = \cos 2\pi \times 10^9 t$ Volts.

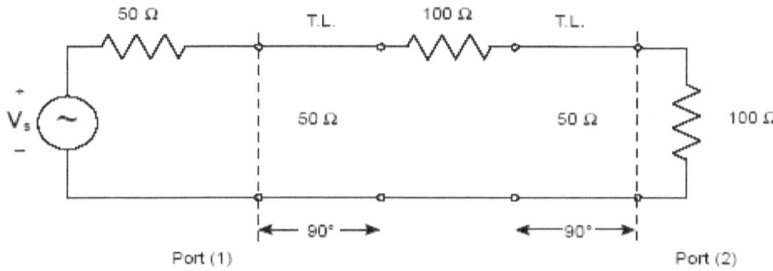

Figure P3.2

3.3) In the lossless transmission line circuit shown in Figure P3.3, calculate the incident power, the reflected power and the power transmitted into the 75 Ω line.   Show that: $P_{INC}=P_{REF}+P_{TRANS}$

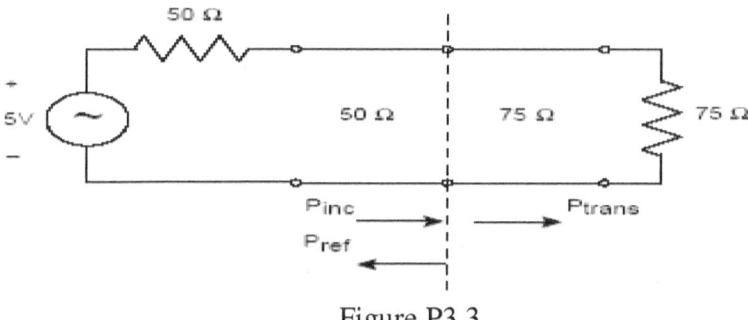

Figure P3.3

3.4) A lossless transmission line (l=0.6λ) is terminated in a load impedance (Z_L=40+j20 Ω). Find the reflection coefficient at the load, the input impedance of the line and the VSWR on the line.

3.5) In the circuit shown in Figure P3.5, calculate the reflection coefficient at the load, the VSWR on the line and the power to load. **Hint:** Use the Thevenin Theorem for the source and two 100 Ohm resistors before solving the problem.

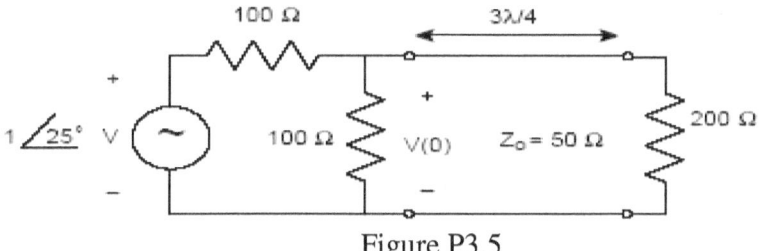

Figure P3.5

3.6)   Consider the lossless transmission line circuit shown in Figure P3.6. Calculate:
a) The load impedance (Z_L)?
b) The reflection coefficient at the input of the line.
c) The VSWR on the line.

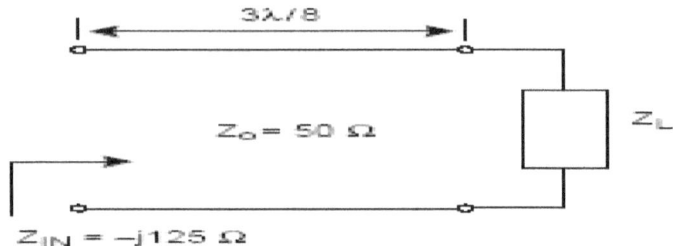

Figure P3.6

3.7)   A lossless transmission line is terminated with a 200 Ω load if the VSWR on the line is 2.0, find the possible values for the line's characteristic impedance.

3.8)   For a lossless transmission line, terminated in a reactive load ($Z_L=jX$), find the reflection coefficient and the VSWR. What is |Γ|?

REFERENCES

[3.1] Cheng, D. K. *Fundamentals of Engineering Electromagnetics.* Reading: Addison Wesley, 1993.

[3.2] Cheung, W. S. and F. H. Levien. *Microwave Made Simple, Principles and Applications.* Dedham: Artech House, 1985.

[3.3] Collin, R. E. *Foundation For Microwave Engineering,* 2nd Ed., New York: McGraw-Hill, 1992.

[3.4] Edwards, T. C. *Foundations for Microstrip Circuit Design.* New York: John Wiley & Sons, 1981.

[3.5] Gardiol, F. *Microstrip Circuits.* New York: John Wiley & Sons, 1994.

[3.6] Gonzalez, G. *Microwave Transistor Amplifiers, Analysis and Design,* 2nd ed. Upper Saddle River: Prentice Hall, 1997.

[3.7] Kraus, J. D. *Electromagnetics,* 3rd Ed., New York: McGraw-Hill, 1984.

[3.8] Plonsey, R. and R. E. Collin. *Principles and*

Applications of Electromagnetic Fields, $_2$nd Ed., New York: McGraw-Hill, 1982.

[**3.9**] Radmanesh, M. M. and B. W. Arnold, *Generalized Microstrip-Slotline Transitions, Theory and Simulation Vs. Experiment,* Microwave Journal, Vol. 36, No. 6, pp. 88–95, June 1993.

[**3.10**] Radmanesh, M. M. and B. W. Arnold, *Microstrip-Slotline Transitions: Simulation Versus Experiment*, EESof User's Group, IEEE MTT-S International Microwave Symposium, Albuquerque, New Mexico, June 1992.

[**3.11**] Radmanesh, M. M. *The Gateway to Understanding: Electrons to Waves and Beyond,* AuthorHouse, 2005.

[**3.12**] Radmanesh, M. M. *Cracking the Code of Our Physical Universe,* AuthorHouse, 2006.

CHAPTER 4

Two-Port Network Representations

4.1 INTRODUCTION

RF/microwaves devices, circuits, and components can be classified as one-, two-, three- or N-port networks. However, a majority of circuits under analysis are two-port networks. Therefore, we will focus primarily on two-port network characterization and will study its representation in terms of a set of parameters that can be cast into a matrix format.

DEFINITION- A TWO PORT NETWORK: *is a network that has only two access ports, one for input or excitation and one for output or response.*

The description of two-port networks from a circuit viewpoint can best be achieved both at low and high frequencies through the use of network parameters. These parameters are discussed in the upcoming sections.

4.2 LOW-FREQUENCY PARAMETERS

To characterize a linear network at low frequencies, several different sets of parameters are available, where one may be selected to fit the application to obtain the most optimum results.

Voltages and currents at each port provide us with four variables of interest: v_1, v_2, i_1, and i_2. There are six ways of picking two out of a set of four variables, but only four combinations (or sets) will yield non-trivial and unique parameters. These are called Z-, Y-, h-, and ABCD-parameters. A two-port network with four voltage and current parameters is shown in Figure 4.1.

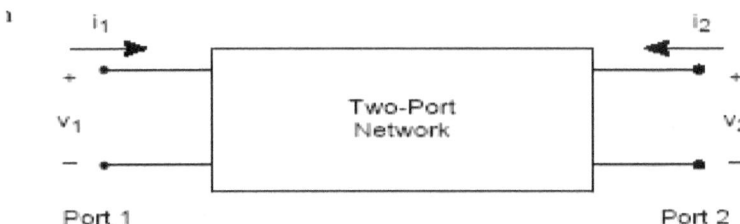

Figure 4.1 A block diagram of a two-port network.

These parameters are defined next.

A. IMPEDANCE OR Z-PARAMETERS

$v_1 = z_{11}i_1 + z_{12}i_2$ (4.1a)

$v_2 = z_{21}i_1 + z_{22}i_2$ (4.1b)

Or, in matrix form:

$[V] = [Z][I]$ (4.2)

$$[V] = \begin{bmatrix} v_1 \\ v_2 \end{bmatrix}$$ (4.3a)

$$[I] = \begin{bmatrix} i_1 \\ i_2 \end{bmatrix}$$ (4.3b)

$$[Z] = \begin{bmatrix} Z_{11} & Z_{12} \\ Z_{21} & Z_{22} \end{bmatrix}$$ (4.4)

B. ADMITTANCE OR Y-PARAMETERS

Similarly, we can write the Y-parameters in matrix form as:

$[I] = [Y][V]$ (4.5)

where [I] and [V] are defined as before and [Y] as follows:

$$[Y] = \begin{bmatrix} Y_{11} & Y_{12} \\ Y_{21} & Y_{22} \end{bmatrix}$$

(4.6)

C. HYBRID OR H-PARAMETERS

$$\begin{bmatrix} v_1 \\ i_2 \end{bmatrix} = \begin{bmatrix} h_{11} & h_{12} \\ h_{21} & h_{22} \end{bmatrix} \cdot \begin{bmatrix} i_1 \\ v_2 \end{bmatrix}$$

(4.7)

D. TRANSMISSION OR ABCD PARAMETERS

$$\begin{bmatrix} v_1 \\ i_1 \end{bmatrix} = \begin{bmatrix} A & B \\ C & D \end{bmatrix} \cdot \begin{bmatrix} v_2 \\ -i_2 \end{bmatrix}$$

(4.8)

EXAMPLE 4.1

Find the [ABCD] matrix for a series impedance element (Z) as shown in Figure 4.2.

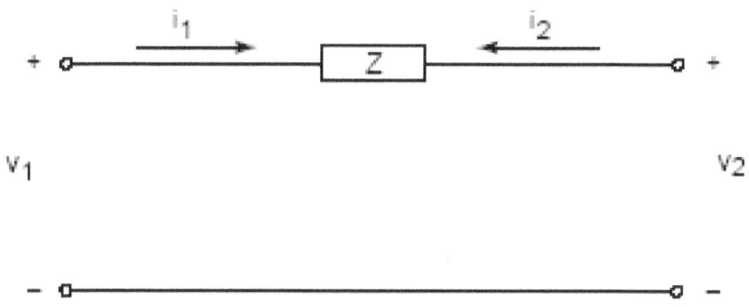

Figure 4.2 A series element.

Solution:

Using KVL and KCL, the following can be written:

$v_1 = v_2 - Z i_2 = A v_2 - B i_2$

$i_1 = -i_2 = 0 - i_2 = C v_2 - D i_2$

Thus the [ABCD] matrix is given by:

$$\begin{bmatrix} A & B \\ C & D \end{bmatrix} = \begin{bmatrix} 1 & Z \\ 0 & 1 \end{bmatrix}$$

EXAMPLE 4.2
Find the [ABCD] matrix for a shunt element (Y) as shown in Figure 4.3.

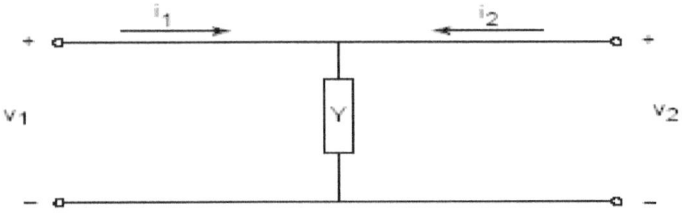

Figure 4.3 A shunt element.

Solution:
Using KVL and KCL, the following can be written:
$$v_1 = v_2 = v_2 + 0 = Av_2 - Bi_2$$
$$i_1 = Yv_2 - i_2 = Cv_2 - Di_2$$
Thus the [ABCD] matrix is given by:
$$\begin{bmatrix} A & B \\ C & D \end{bmatrix} = \begin{bmatrix} 1 & 0 \\ Y & 1 \end{bmatrix}$$

EXAMPLE 4.3
Find the [ABCD] matrix for a circuit consisting of a series element (Z) and a shunt element (Y) as shown in Figure 4.4.

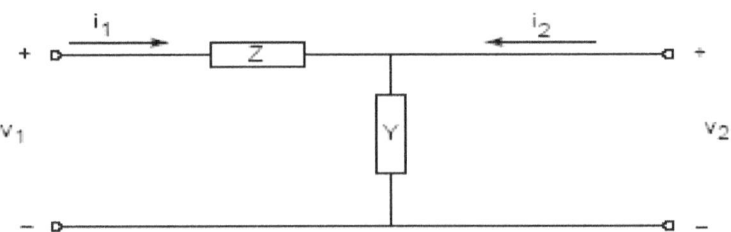

Figure 4.4 A series-shunt element.

Solution:
The [ABCD] matrix for the whole circuit which is a cascade of a series and
a shunt element, is a multiplication of the two matrices as follows:

$$\begin{bmatrix} A & B \\ C & D \end{bmatrix} = \begin{bmatrix} A_1 & B_1 \\ C_1 & D_1 \end{bmatrix} \cdot \begin{bmatrix} A_2 & B_2 \\ C_2 & D_2 \end{bmatrix}$$

$$= \begin{bmatrix} 1 & Z \\ 0 & 1 \end{bmatrix} \cdot \begin{bmatrix} 1 & 0 \\ Y & 0 \end{bmatrix}$$

Thus the [ABCD] matrix is given by:

$$\begin{bmatrix} A & B \\ C & D \end{bmatrix} = \begin{bmatrix} 1+ZY & Z \\ Y & 1 \end{bmatrix}$$

Example 4.4

Find the [ABCD] matrix for a transformer as shown in Figure 4.5.

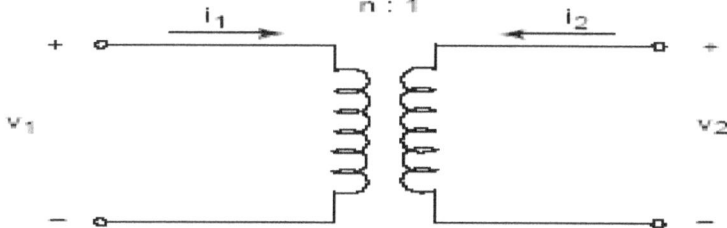

Figure 4.5 A transformer.

Solution:

Using the transformer voltage and current rule, which states that if the voltage is stepped down then in order to preserve the power flow the current must be proportionately stepped up, we have:

$v_1 = nv_2 = Av_2 - Bi_2$

$i_1 = -\dfrac{1}{n} i_2 = Cv_2 - Di_2$

Thus the [ABCD] matrix is given by:

$$\begin{bmatrix} A & B \\ C & D \end{bmatrix} = \begin{bmatrix} n & 0 \\ 0 & \dfrac{1}{n} \end{bmatrix}$$

Example 4.5

Find the [ABCD] matrix for a lossless transmission line of length (ℓ) and characteristic impedance (Z_O) as shown in Figure 4.6.

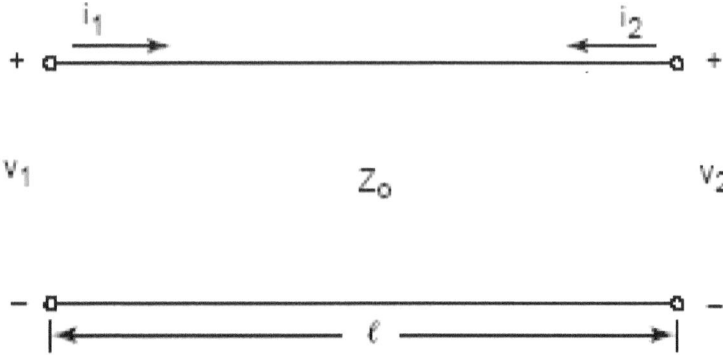

Figure 4.6 A series element.

Solution:

Using results obtained in Chapter 7, we know that the output voltage or current have the same magnitude as the input but lag behind in phase by $e^{-j\beta\ell}$. Thus the following can be written:

$v_2 = v_1 e^{-j\beta\ell} \Rightarrow v_1 = v_2 e^{+j\beta\ell} = v_2 \cos\beta\ell + j v_2 \sin\beta\ell$

$i_2 = -i_1 e^{-j\beta\ell} \Rightarrow i_1 = -i_2 e^{+j\beta\ell} = -i_2 \cos\beta\ell - j i_2 \sin\beta\ell$

Since the load end is considered to be matched to the transmission line, we can write:

$v_2 = -Z_o i_2$

$v_1 = (\cos\beta\ell)v_2 - (jZ_o \sin\beta\ell)i_2 = A v_2 - B i_2$

$i_1 = (jY_o \sin\beta\ell)v_2 - (\cos\beta\ell)i_2 = C v_2 - D i_2$

Thus the [ABCD] matrix can be written as:

$$\begin{bmatrix} A & B \\ C & D \end{bmatrix} = \begin{bmatrix} \cos\beta\ell & jZ_o \sin\beta\ell \\ jY_o \sin\beta\ell & \cos\beta\ell \end{bmatrix}$$

4.3 HIGH-FREQUENCY PARAMETERS

We note that Z-, Y-, h- and ABCD-parameters are based upon the following considerations at each of the network ports:

- Net voltage (v) and net current (i)
- Short and open circuit terminations

Simple observations at high RF/microwave frequencies reveal that:

a.   Shorts and open circuit terminations are difficult to implement over a broad range of frequencies and thus cannot be used to characterize networks, and

b.   At high RF/microwave frequencies, the net voltage (or net current) is a combination of two or more voltage (or current) traveling waves.

Based on these observations, the Z-, Y-, h- and ABCD-parameters cannot be accurately measured at these higher frequencies and therefore we have to use the concept of propagating or traveling waves to define the network parameters.

The network representation of a two-port network at high RF/microwave frequencies is called "scattering parameters" (or "S-parameters" for short).

When cascading networks, a variation of S-parameters called chain scattering parameters (or T-parameters) are used to simplify the analysis.

These two types of high frequency parameters are very popular and are primarily used at the high RF/microwave frequencies.

4.4 FORMULATION OF THE S-PARAMETERS

The high frequency S- and T-parameters are used to characterize high RF/microwave two-port networks (or N-Port networks, in general). These parameters are based on the concept of traveling waves and provide a complete characterization of any two-port network under analysis or test at high RF/microwave frequencies.

In view of the linearity of the Electromagnetic field equations and the linearity displayed by most microwave components and networks, the "scattered waves" (i.e. the reflected and transmitted wave amplitudes) are linearly related to the incident wave amplitude. The matrix describing this linear relationship is called the "scattering matrix," or [S].

While the lower frequency network parameters (such as Z-or Y-matrices, etc.) are defined in terms of net (or total) voltage and currents at the ports, these concepts are not practical at high RF/microwaves frequencies where it is found that any set of parameters to be meaningful, must be defined in terms of a combination of traveling waves.

To characterize a two-Port network, which has identical characteristic impedances at both the input and output ports, let us consider the incident and reflected voltage waves at each port as shown in Figure 4.7.

To accurately define the S-parameters, we will consider a voltage $[V_i^+]$ incident on and a voltage $[V_i^-]$ reflected from the terminals of a two-port network (i=1,2) as shown in Figure 4.7.

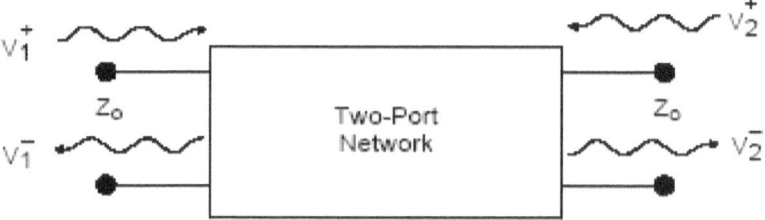

Figure 4.7 A two-port with the incident and reflected waves.

The scattering matrix, [S], is now defined to describe the linear relationship between the incident voltage wave phasor matrix $[V_i^+]$ and the reflected or transmitted wave phasor matrix $[V_i^-]$ at any of the two ports as follows:

$$V_1^- = S_{11}V_1^+ + S_{12}V_2^+$$
$$V_2^- = S_{21}V_1^+ + S_{22}V_2^+$$

Or, in matrix form we can write:

$$\begin{bmatrix} V_1^- \\ V_2^- \end{bmatrix} = \begin{bmatrix} S_{11} & S_{12} \\ S_{21} & S_{22} \end{bmatrix} \cdot \begin{bmatrix} V_1^+ \\ V_2^+ \end{bmatrix}$$

(4.9)

Or,

$[V^-]=[S][V^+]$

Where,

$$[V^-] = \begin{bmatrix} V_1^- \\ V_2^- \end{bmatrix},$$

$$[V^+] = \begin{bmatrix} V_1^+ \\ V_2^+ \end{bmatrix}$$

and,

$$[S] = \begin{bmatrix} S_{11} & S_{12} \\ S_{21} & S_{22} \end{bmatrix} \qquad (4.10)$$

This linear relationship is expressed in terms of a ratio of two phasors which are complex numbers with the magnitude of the ratio always less than or equal to 1. Each specific element of the [S] matrix is defined as:

$$S_{11} = \frac{V_1^-}{V_1^+} \Big|_{V_2^+ = 0} = \Gamma_{in} \qquad \text{(Input reflection coefficient when the output}$$

$$\text{port is terminated in a matched load.)} \quad (4.11)$$

$$S_{21} = \frac{V_2^-}{V_1^+} \Big|_{V_2^+ = 0} \qquad \text{(Forward transmission coefficient when the}$$

$$\text{output port is terminated in a matched load.)} \quad (4.12)$$

$$S_{12} = \frac{V_1^-}{V_2^+} \Big|_{V_1^+ = 0} \qquad \text{(Reverse transmission coefficient when the}$$

$$\text{input port is terminated in a matched load.)} \quad (4.13)$$

$$S_{22} = \frac{V_2^-}{V_2^+} \Big|_{V_1^+ = 0} = \Gamma_{out} \qquad \text{(output reflection coefficient when the input}$$

$$\text{port is terminated in a matched load.)} \quad (4.14)$$

S-parameters as defined above, have many advantages at high RF/microwave frequencies which can be briefly stated as:

a. S-parameters provide a complete characterization of a network, as seen at its two ports.

b. S-parameters make the use of short or open (as prescribed at lower frequencies) completely unnecessary at higher frequencies. It is a known fact that the impedance of a short or an open varies with frequency which is one reason why they are not useful for device characterization at high RF/Microwave frequencies. Furthermore, the presence of a short or open in a circuit can cause strong reflections (since $|\Gamma_L|=1$), which usually leads to oscillations or damages to the transistor circuitry.

c. S-parameters require the use of matched loads for termination and since the loads absorb all the incident energy, the possibility of serious reflections back to the device or source is eliminated.

EXAMPLE 4.6
Given the [ABCD] matrix for a two-port network, derive its [S] matrix (see Figure 4.8).

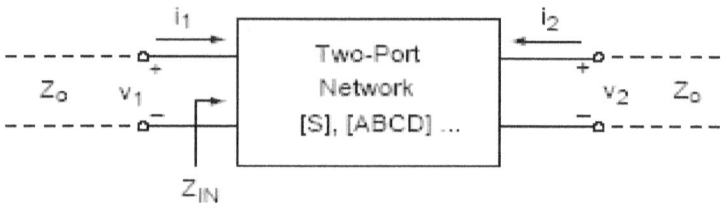

Figure 4.8 Network for example 4.6.

Solution:
To obtain S_{11}, we terminate port 2 in a matched load and find the input reflection coefficient (Γ_{in}) as follows:

$$S_{11} = \Gamma_{in} = \frac{Z_{in} - Z_O}{Z_{in} + Z_O}$$

Where $Z_{in}=v_1/i_1$. Substituting for v_1 and i_1 in terms of [ABCD] we have:

$$v_1 = Av_2 - Bi_2$$
$$i_1 = Cv_2 - Di_2$$
$$v_2 = -Z_o i_2$$

$$Z_{in} = \frac{v_1}{i_1} = \frac{Av_2 - Bi_2}{Cv_2 - Di_2} = \frac{AZ_O + B}{CZ_O + D}$$

Now, Substituting for Z_{in}, we can write S_{11} as:

$$S_{11} = \Gamma_{in} = \frac{A + BY_O - CZ_O - D}{\Delta}$$

where Δ is given by:

$$\Delta = A + BY_O + CZ_O + D$$

Similarly, S_{12}, S_{21} and S_{22} can be found as follows:

$$S_{12} = \frac{2(AD - BC)}{\Delta}$$

$$S_{21} = \frac{2}{\Delta}$$

$$S_{22} = \Gamma_{out} = \frac{-A + BY_O - CZ_O + D}{\Delta}$$

In general, using the same technique as demonstrated in example 4.6, any set of network parameters can be converted into another set of parameters. Appendix H shows the conversion relation between the z-, y-, h-, ABCD- and the S-parameters.

The conversion among the three transistor configurations is an important relation which becomes useful in many practical design situations. Appendix I shows the conversion relation between the y-parameters of a transistor in common-emitter, common-base and common-collector configurations. If parameters other than Y-parameters (e.g. S-parameters) are needed, then appendix H can be used effectively to convert Y- to S-parameters.

4.5 PROPERTIES OF S-PARAMETERS
The S-parameters of an N-port network, in general, have certain properties and inter-relationships amongst the parameters themselves which are worth considering. In the following discussion, due to the popularity and frequent use, we limit our discussion solely to two-port networks. Depending upon whether the network is reciprocal or lossless the S-parameters will have different properties which are discussed below.

4.5.1 Reciprocal Networks
A reciprocal network is defined to be a network that satisfies the reciprocity theorem which is defined as:

RECIPROCITY THEOREM: *Is a theorem stating that the interchange of electromotive force at one point (e.g. in branch k, v_k) in a passive linear network, with the current produced at any other point (e.g. branch m, i_m) results in the same current (in branch k, i_k) when the same electromotive force is applied in the new location (branch m, v_m), that is,*

$$v_k/i_m = v_m/i_k \qquad (4.15a)$$

or,

$$Z_{km} = Z_{mk} \qquad (4.15b)$$

As observed, this theorem only applies to passive networks having linear bilateral impedances. Networks that satisfy this condition include all passive networks that contain linear passive elements including resistors, capacitors, inductors, and transformers except:
a. Independent or dependent sources,
b. Non-linear elements and/or active solid state devices such as diodes, transistors, etc.

It can be shown that for all **reciprocal networks**, the [S] matrix is symmetrical, i.e.

$$S_{12} = S_{21} \qquad (4.16a)$$

Generalizing the above, it can be shown that for an N-Port network:

$$S_{ij} = S_{ji} \text{ for } i \neq j \qquad (4.16b)$$

Where,
$i = 1, \quad N$
$j = 1, \quad N.$

A SPECIAL CASE: A SYMMETRICAL RECIPROCAL NETWORK
A special case of a reciprocal network is a symmetrical network. **Symmetrical networks** are defined as *"Networks that have identical size, element values and arrangement for corresponding electrical elements in reference to a plane or line of symmetry."*

Due to the symmetry of a network topology and by observation, the input impedance obtained by looking into the input port is equal to the impedance looking into the output port. The equality of input and output impedances leads to the equality of input and output reflection coefficients in addition to the equality of S_{12} and S_{21} as required by the reciprocity theorem stated earlier. Therefore for **symmetrical passive** networks, we can always write:

$$S_{11} = S_{22} \tag{4.17a}$$
$$S_{12} = S_{21} \tag{4.17b}$$

Or, in general for any **symmetrical passive** N-port network we can write S-parameters as:

$$S_{ii} = S_{jj} \tag{4.18a}$$
$$S_{ij} = S_{ji} \tag{4.18b}$$

Where, $i \neq j$ and,
$$i=1,......N$$
$$j=1,......N.$$

4.5.2 Lossless Networks

For a lossless passive network (i.e. one containing no resistive elements) the power entering the circuit will always equal to the power leaving the network, i.e. the power is conserved. This condition will impose a number of restrictions on the s-parameters which give rise to the unity and zero property as follows:

A. THE UNITY PROPERTY OF [S] MATRIX
This property states that for a passive lossless two-port (or in general an N-Port network), the sum of the products of each term of any one row (or any one column) multiplied by its own complex conjugate is unity, i.e.

$$\sum_{i=1}^{N} S_{ij}S_{ij}^* = 1, \qquad j=1,2,....,N \tag{4.19}$$

Where i and j are row and column numbers respectively.

For a two-port network, Equation (4.19) yields two equations:

$$S_{11}S_{11}^* + S_{12}S_{12}^* = 1 \tag{4.20a}$$
$$S_{22}S_{22}^* + S_{21}S_{21}^* = 1 \tag{4.20b}$$

Furthermore, if the **lossless network is also reciprocal** (i.e. $S_{12} = S_{21}$), these two equations are greatly simplified as follows:

$$S_{12}=S_{21} \tag{4.21a}$$
$$|S_{11}|=|S_{22}| \tag{4.21b}$$

$$|S_{11}|^2+|S_{21}|^2=1 \qquad\qquad (4.21c)$$

B. THE ZERO PROPERTY OF [S] MATRIX
This property states that for a passive lossless N-port network, the sum of the products of each term of any row (or any column) multiplied by the complex conjugate of the corresponding terms of any other row (or column) is zero:

$$\sum_{k=1}^{N} S_{ki}S_{kj}^* = 0, \qquad \text{for } i\neq j, \ \& \ i,j=1,2,....,N \qquad (4.22)$$

where i and j are row and column numbers, respectively.

For a two-Port network this equation simplifies into two equations:
$$S_{11}S_{12}^* +S_{21}S_{22}^* =0 \qquad\qquad (4.23a)$$
$$S_{11}S_{21}^* +S_{12}S_{22}^* =0 \qquad\qquad (4.23b)$$

Furthermore, if the lossless network is also reciprocal (i.e. $S_{12} = S_{21}$), then the above two equations simplify into one equation:
$$S_{12}=S_{21}$$
$$S_{11}S_{21}^* +S_{21}S_{22}^* =0 \qquad\qquad (4.24)$$

Note: *A matrix satisfying the zero and unity property is called a unitary matrix.*

C. ANALYSIS OF RECIPROCAL LOSSLESS NETWORKS
From the zero and unity properties of the S-matrix, The S-parameters of a **reciprocal lossless network** are constrained by Equations (4.20), (4.21) and (4.24) as follows:
$$S_{21}=S_{12} \qquad\qquad (4.25a)$$
$$|S_{11}|=|S_{22}| \qquad\qquad (4.25b)$$
$$|S_{11}|^2+|S_{21}|^2=1 \qquad\qquad (4.25c)$$
$$S_{11}S_{21}^* +S_{21}S_{22}^* =0 \qquad\qquad (4.25d)$$
If we let:
$$S_{11} = |S_{11}|e^{j\theta_{11}},$$
$$S_{22} = |S_{22}|e^{j\theta_{22}}$$
and,
$$S_{21} = |S_{21}|e^{j\theta_{21}}$$

Then Equations (4.25c) and (4.25d) give:

$$|S_{21}| = (1-|S_{11}|^2)^{1/2} \tag{4.26a}$$

$$|S_{11}|(1-|S_{11}|^2)^{1/2}\left(e^{j(\theta_{11}-\theta_{21})} + e^{j(\theta_{21}-\theta_{22})}\right) = 0$$

which yields:

$$\left(e^{j(\theta_{11}-\theta_{21})} + e^{j(\theta_{21}-\theta_{22})}\right) = 0 \quad \Rightarrow e^{j(\theta_{11}-\theta_{21})} = e^{-j\pi}e^{j(\theta_{21}-\theta_{22})}$$

$$\Rightarrow \theta_{11}+\theta_{22} = 2\theta_{21} - \pi \pm 2n\pi$$

Or,

$$\theta_{21} = \frac{\theta_{11}+\theta_{22}}{2} + \pi\left(\frac{1}{2}\mp n\right) \qquad \text{For } n=0,1,2,... \tag{4.26b}$$

Equations (4.26a) and (4.26b) provide the magnitude and phase of S_{21} (or S_{12}) in terms of magnitude and phase of S_{11} and S_{22}.

Therefore from a measurement knowledge of S_{11} and S_{22}, one can completely describe and specify a reciprocal lossless two-port network. This use of S-parameters in specifying a lossless and reciprocal two-port network, shows its usefulness and versatility. The following will illustrate the concept of S-Parameters further.

EXAMPLE 4.7
What are the S-parameters of a series element (Z) as shown in Figure 4.9?

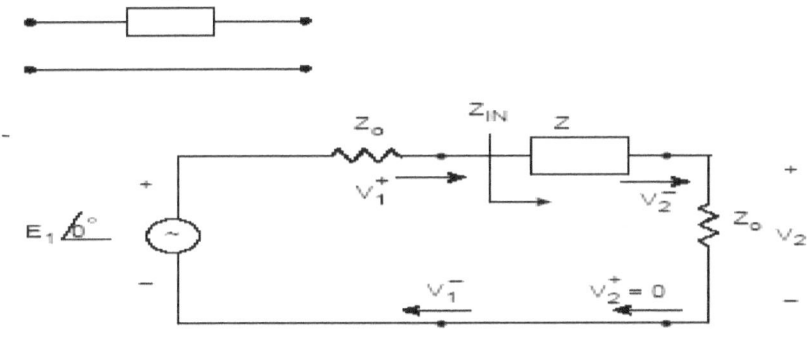

Figure 4.9 Circuit for example 4.7.

Solution:
Since this is a reciprocal and symmetrical network, we have:

$S_{11}=S_{22}$,
$S_{12}=S_{21}$
So we only need to find S_{11} and S_{21}.

NOTE: *This is not a lossless network since Z=R+jX has a lossy component!*

a. $S_{11}=\dfrac{V_1^-}{V_1^+}\Big|_{V_2^+=0}$ $\qquad\qquad\qquad$ (4.27)

According to Equation (4.27) S_{11} is the input reflection coefficient when the output is matched (see Equation 4.27); that is,
$S_{11}=\Gamma_{IN}=(Z_{IN}-Z_o)/(Z_{IN}+Z_o)$,
Where $Z_{IN}=Z+Z_o$, thus we have:
$S_{11}=Z/(Z+2Z_o)$ $\qquad\qquad\qquad\qquad\qquad$ (4.28)

b. $S_{21}=\dfrac{V_2^-}{V_1^+}\Big|_{V_2^+=0}$ $\qquad\qquad\qquad$ (4.29)

From Equation (4.29) we can see that S_{21} is the voltage gain (or loss) when the output is matched. Thus by applying a source voltage (E_1) at port 1, the voltage gain is found as follows:
$I=E_1/(Z_o+Z_{IN})$
$V_2=V_2^-+V_2^+$

Since the load is matched, $v_2^+=0$. Thus we have:
$\Rightarrow V_2^-=Z_oI$ $\qquad\qquad\qquad\qquad\qquad\qquad$ (4.30a)
$V_1=V_1^++V_1^-=V_1^+(1+S_{11})=Z_{IN}I$
$\Rightarrow V_1^+=Z_{IN}I/(1+S_{11})$ $\qquad\qquad\qquad\qquad$ (4.30b)

Dividing (4.30a) by (4.30b), we have:
$S_{21}=V_2^-/V_1^+=Z_o(1+S_{11})/Z_{IN}$
$\Rightarrow S_{21}=2Z_o/(Z+2Z_o)$ $\qquad\qquad\qquad\qquad$ (4.31)
Observation: For a "series Z" network, from Equations (4.29) and (4.31) we can see that:
$S_{21}=1-S_{11}$ $\qquad\qquad\qquad\qquad\qquad\qquad$ (4.32)

Therefore the whole S-matrix can be written as:

$$S = \begin{bmatrix} \dfrac{Z}{Z+2Z_0} & \dfrac{2Z_0}{Z+2Z_0} \\ \dfrac{2Z_0}{Z+2Z_0} & \dfrac{Z}{Z+2Z_0} \end{bmatrix}$$

EXAMPLE 4.8

What are the S-parameters of a shunt element (Y) as shown in Figure 4.10?

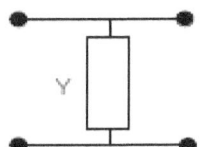

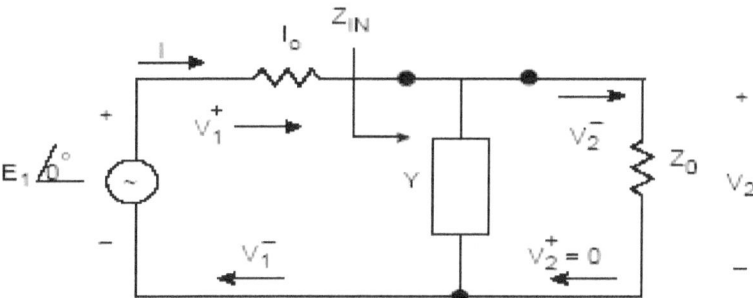

Figure 4.10 Circuit for example 4.8.

Solution:
Similar to example 4.1, this is a reciprocal and symmetrical network, thus:

$S_{11}=S_{22}$,

And,

$S_{12}=S_{21}$

So we only need to find S_{11} and S_{21}.

a. $S_{11} = \dfrac{V_1^-}{V_1^+}\Big|_{V_2^+=0}$

$S_{11}=(Z_{IN}-Z_0)/(Z_{IN}+Z_0)$

$Z_{IN}= (1/Y \| Z_0)$

Substituting for Z_{IN} in S_{11} above, we obtain:

$S_{11}=-Z_0Y/(2+Z_0Y)$ (4.33)

b. $S_{21}=\dfrac{V_2^-}{V_1^+}\Big|_{V_2^+=0}$

By applying a source voltage E1 to port 1, we obtain:

$I=E_1/(Z_o+Z_{IN})$

Since port 2 is terminated in a matched load (i.e. $V_2^+=0$), we can write:

$V_2=V_2^-=(1/Y\|Z_o)I$ (4.34a)

$V_1=V_1^++V_1^-= V_1^+(1+S_{11})=Z_{IN}I$

$\Rightarrow V_1^+= Z_{IN}I/(1+S_{11})$ (4.34b)

Dividing (4.34a) by (4.34b), we obtain:

$S_{21}= v_2^-/v_1^+=(1/Y\|Z_o)(1+S_{11})/Z_{IN}$

$\Rightarrow S_{21}=2/(2+Z_oY)$ (4.35)

Observation: *For a "shunt Y" network, from Equation (4.33) and (4.35) we can see that:*

$S_{21}=1+S_{11}$ (4.36)

Therefore the whole S-matrix can be written as:

$$S = \begin{bmatrix} \dfrac{-Z_oY}{2+Z_oY} & \dfrac{2}{2+Z_oY} \\[2ex] \dfrac{2}{2+Z_oY} & \dfrac{-Z_oY}{2+Z_oY} \end{bmatrix}$$

4.6 SHIFTING REFERENCE PLANES

The S-parameters relate amplitude and phase of traveling waves, which are incident on, transmitted through, or reflected from a network terminal. Therefore the location of the reference plane must be known precisely to calculate or measure the exact phase of the S-parameters.

Consider a two-Port network in which the reference plane at port 1 has moved a distance ℓ_1 to port 1'. Similarly, the reference plane at port 2 has moved a distance ℓ_2 to port 2' as shown in Figure 4.11.

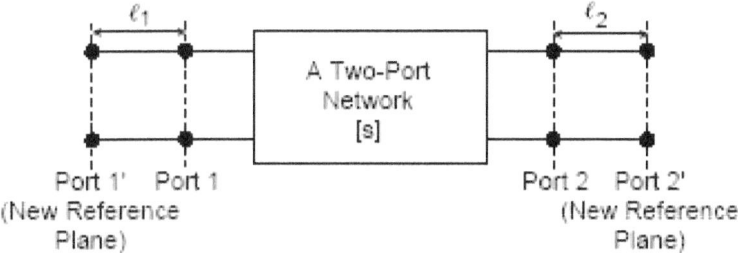

Figure 4.11 A two-port network with new reference planes.

The voltage waves at each new port (i.e. 1' and 2') can now be written as:
$$V_i^{+'}=V_i^+ e^{j\theta_i}, \quad i=1,2 \tag{4.37a}$$
$$V_i^{-'}=V_i^- e^{-j\theta_i}, \quad i=1,2 \tag{4.37b}$$
where $\theta_i=\beta\ell_i$ (i=1,2) is the electrical length corresponding to the reference plane shift at each port.

Inverting Equations in (4.37), we can write:
$$V_i^+=V_i^{+'}e^{-j\theta_i}, \quad i=1,2 \tag{4.38a}$$
$$V_i^-=V_i^{-'}e^{j\theta_i}, \quad i=1,2 \tag{4.38b}$$

Upon substitution of Equation (4.38) in:
$$[V^-]=[S][V^+]$$
and further mathematical manipulation, we obtain [S'] which is the shifted S-parameters as:

$$\left[S'\right]=\begin{bmatrix} S_{11}e^{-j2\theta_1} & S_{12}e^{-j(\theta_1+\theta_2)} \\ S_{21}e^{-j(\theta_1+\theta_2)} & S_{22}e^{-j2\theta_2} \end{bmatrix} \tag{4.39}$$

or conversely,

$$\left[S\right]=\begin{bmatrix} S_{11}'e^{j2\theta_1} & S_{12}'e^{j(\theta_1+\theta_2)} \\ S_{21}'e^{j(\theta_1+\theta_2)} & S_{22}'e^{j2\theta_2} \end{bmatrix} \tag{4.40}$$

To summarize this analysis, we note that:
$$S_{ii}'=S_{ii} e^{-j2\theta_i} \quad i=1,2 \tag{4.41}$$
$$S_{ij}'=S_{ij} e^{-j(\theta_i+\theta_j)} \quad i\neq j, i=1,2 \tag{4.42}$$

Equation (4.41) shows that the phase of S_{ii} is shifted by twice the electrical length, because the incident wave travels twice over this length upon reflection. On the other hand, at port i (i = 1,2), (4.42) shows that S_{ij} (i ≠ j) is shifted by the sum of the electrical lengths because the incident wave must pass through both lengths in order to travel from one shifted port to the other.

4.7 TRANSMISSION MATRIX

The following discussion, in general, applies to a cascade of N-port networks. However, for the sake of simplicity, we limit our analysis to two-port networks only. When cascading a number of two-port networks in series, a more useful network representation is needed in order to facilitate the calculation of the overall network parameters.

This new representation should relate the output quantities in terms of input quantities. Using such a representation will enable one to obtain a description of the complete cascade by simply multiplying together the matrices describing each network.

At low frequencies, the transmission matrix (also known as ABCD matrix) is defined in terms of the net input voltage and current as the independent variables and output net voltage and current as the dependent variables.

However, at high RF and microwave frequencies, the transmission matrix is expressed in terms of the input incident and reflected waves as the independent variables and the output incident and reflected waves as the dependent variables.

Using the latter definition at RF/microwave frequencies, the transmission matrix formulation becomes very useful when dealing with multi-stage circuits (such as filters, amplifiers, etc.) or infinitely long periodic structures such as those used in small-wave circuits for traveling wave tubes, etc.

The transmission matrix (or T-matrix) for a two port network as shown in Figure 4.12, is defined as:

$$\begin{bmatrix} V_1^+ \\ V_1^- \end{bmatrix} = \begin{bmatrix} T_{11} & T_{12} \\ T_{21} & T_{22} \end{bmatrix} \begin{bmatrix} V_2^- \\ V_2^+ \end{bmatrix} \tag{4.43}$$

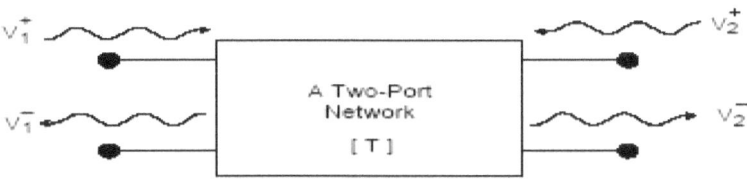

Figure 4.12 A two-port network.

The relationship between S- and T-parameters can be derived using the above basic definition as follows:

$$\begin{bmatrix} T_{11} & T_{12} \\ T_{21} & T_{22} \end{bmatrix} = \begin{bmatrix} \dfrac{1}{S_{21}} & -\dfrac{S_{22}}{S_{21}} \\ \dfrac{S_{11}}{S_{21}} & S_{12} - \dfrac{S_{11}S_{22}}{S_{21}} \end{bmatrix} \tag{4.44}$$

The reverse relationship expressing [S] in terms of [T] matrix can also be derived with the following result:

$$\begin{bmatrix} S_{11} & S_{12} \\ S_{21} & S_{22} \end{bmatrix} = \begin{bmatrix} \dfrac{T_{21}}{T_{11}} & T_{22} - \dfrac{T_{21}T_{12}}{T_{11}} \\ \dfrac{1}{T_{11}} & -\dfrac{T_{12}}{T_{11}} \end{bmatrix} \tag{4.45}$$

For a cascade connection of two port networks as shown in Figure 4.13, the overall T-matrix can be obtained as follows:

$$\begin{bmatrix} V_1^+ \\ V_1^- \end{bmatrix} = \begin{bmatrix} T_{11} & T_{12} \\ T_{21} & T_{22} \end{bmatrix} \cdot \begin{bmatrix} V_2^- \\ V_2^+ \end{bmatrix} \tag{4.46a}$$

$$\begin{bmatrix} V_1'^+ \\ V_1'^- \end{bmatrix} = \begin{bmatrix} T_{11}' & T_{12}' \\ T_{21}' & T_{22}' \end{bmatrix} \cdot \begin{bmatrix} V_2'^- \\ V_2'^+ \end{bmatrix} \tag{4.46b}$$

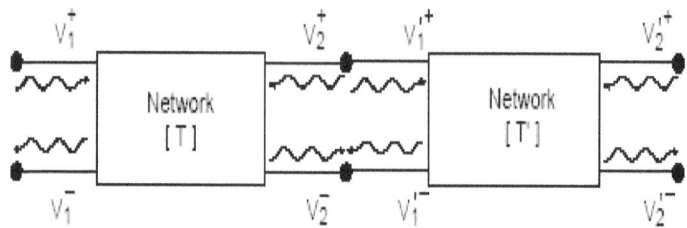

Figure 4.13 A cascade two-port networks.

But we note that:
$$V_2^+ = V_1^{'-},$$ (4.47a)
$$V_2^- = V_1^{'+}$$ (4.47b)

Therefore combining (4.46) and (4.47), yields:
$$\begin{bmatrix} V_1^+ \\ V_1^- \end{bmatrix} = \begin{bmatrix} T_{11} & T_{12} \\ T_{21} & T_{22} \end{bmatrix} \cdot \begin{bmatrix} T_{11}^{'} & T_{12}^{'} \\ T_{21}^{'} & T_{22}^{'} \end{bmatrix} \cdot \begin{bmatrix} V_2^{'-} \\ V_2^{'+} \end{bmatrix}$$ (4.48)

Thus the total T-matrix is the multiplication of the two T-matrices:
$$[T]_{tot} = [T][T^{'}]$$ (4.49)

4.8 GENERALIZED SCATTERING PARAMETERS

The scattering matrix defined earlier was based on the assumption that all ports have the same characteristic impedances (usually $Z_o = 50\ \Omega$). Even though this is the case in many practical situations, however, there are cases where this may not apply and each port has a non-identical characteristic impedance (see Figure 4.14). Thus a need to generalize the scattering parameters arises.

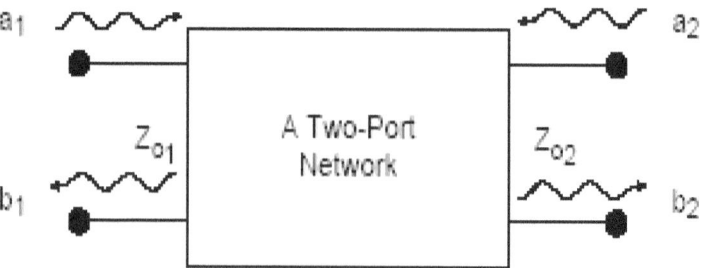

Figure 4.14 Normalized incident and reflected waves.

In this case we have to modify our ordinary definition for the [S] matrix to include the characteristic impedances at each port. Taking each port's characteristic impedances into account, we need to define two normalized voltage waves as follows:

$$a_i=V_i^+/\sqrt{Z_{oi}}, \quad i=1,2 \tag{4.50}$$
$$b_i=V_i^-/\sqrt{Z_{oi}}, \quad i=1,2 \tag{4.51}$$

where "i" is the port number, "a_i" represents the normalized incident voltage wave, "b_i" represents the normalized reflected voltage wave from the ith port and "Z_{oi}" is the characteristic impedance at the ith port (**Note**: Z_{oi} *is a real number for lossless lines*).

Thus the total voltage and current at each port can now be written as:
$$V_i=V_i^++V_i^-=\sqrt{Z_{oi}}(a_i+b_i) \tag{4.52}$$
$$I_i=I_i^+-I_i^-= V_i^+/Z_{oi} - V_i^-/Z_{oi} =(a_i-b_i)/\sqrt{Z_{oi}} \tag{4.53}$$

The average net power delivered to the ith port can now be expressed in terms of a_i and b_i with no further concern about different Z_{oi} at each port:
$$P_i=Re[V_iI_i^*]/2$$
$$=Re[|a_i|^2-|b_i|^2+(a_i^*b_i-a_ib_i^*)]/2 \tag{4.54}$$

Noticing that $(a_i^*b_i-a_ib_i^*)$ term is purely imaginary leads to the expression for the net real power:
$$P_i= (|a_i|^2-|b_i|^2)/2 \tag{4.55}$$

This equation is meaningful since it is clearly showing that the net power delivered to each port is equal to the normalized incident power less the normalized reflected power.

The generalized [s] matrix can now be defined in terms of the normalized voltage waves as follows:
$$\begin{bmatrix} b_1 \\ b_2 \end{bmatrix} = \begin{bmatrix} S_{11} & S_{12} \\ S_{21} & S_{22} \end{bmatrix} \cdot \begin{bmatrix} a_1 \\ a_2 \end{bmatrix} \tag{4.56a}$$
Or, in a simpler form:
$$[b] = [S][a] \tag{4.56b}$$
where each element of the generalized [S] matrix is now defined as:

$S_{11} = \dfrac{b_1}{a_2}\big|_{a_2=0} = \Gamma_{in}$ (reflection coefficient at port 1 with port 2

matched)   (4.57)

$S_{12} = \dfrac{b_1}{a_2}\big|_{a_1=0}$ (transmission coefficient from port 1 to port 2 with

port 2 matched)   (4.58)

$S_{21} = \dfrac{b_2}{a_1}\big|_{a_2=0}$ (transmission coefficient from port 2 to port 2

with port 1 matched)   (4.59a)

$S_{22} = \dfrac{b_2}{a_2}\big|_{a_1=0} = \Gamma_{out}$ (reflection coefficient at port 2 with port 1

matched)   (4.59b)

Clearly these definitions are very similar to the earlier ones for the [S] matrix except that "V_i^{+}" and "V_i^{-}" are replaced by "a_i" and "b_i", respectively. Alternately, each element can be expressed as a general equation by:

$S_{ij} = \dfrac{b_i}{a_j}\big|_{a_k=0}$ **for i,j,k=1,2 and k≠j**   (4.60a)

A SPECIAL CASE

Consider a network, with identical characteristic impedances at all ports, having a known [S] matrix. If transmission lines of unequal characteristic impedances (Z_{oi}) are connected to each port, the new [s] matrix for the entire network with the help of (4.60a), can now be written as :

$(S_{ij})_{new} = \dfrac{V_i^{-}\sqrt{Z_{oj}}}{V_j^{+}\sqrt{Z_{oi}}}\big|_{V_k^{+}=0}$ **for i,j,k=1,2 and k≠j**   (4.60b)

we can simplify Equation (4.60b) to yield:

$(S_{ij})_{new} = (S_{ij})_{old}\sqrt{\dfrac{Z_{oj}}{Z_{oi}}}$ **for i,j,k=1,2 and k≠j**   (4.60c)

In the next section we will discuss the subject of "signal flow graphs" whereby any complex circuit can be analyzed in terms of a simple diagram that can yield the relation between desired variables.

4.9 SIGNAL FLOW GRAPHS (SFGs)

Any linear system, or more specifically any linear electrical network, can be described by a set of simultaneous linear equations. Solutions to these equations can be obtained by the following methods:

a. The elimination theory which is a method of successive substitutions,

b. The Cramer's rule which is a method of solving by using determinants, and

c. Any of the topological techniques such as the "flow graph techniques" represented by the works of Mason.

Although the algebraic manipulation methods (a) and (b) above can be executed by a computer with relative ease and speed, however, they do not allow a pictorial analysis or perspective on the physical nature and the signal flows taking place inside the linear system.

The signal flow graphs, through the use of graphical diagrams, provide a physical insight into the cause-effect relationships between the variables of the system. This method of analysis enables the circuit analyst to gain an intuitive understanding of the system (or network) operation.

In the previous sections, we have seen how the incident, reflected, and transmitted waves are inter-related through a series of linear equations expressed concisely by the S-parameter matrix. This fact indicates that we are dealing with a linear system to which the signal flow graph can directly be applied.

In essence, a signal flow graph is an alternate and yet a simpler method to the "block diagram method" in representing a complicated linear system.

The advantage of a signal flow graph method over a block diagram method is the availability of a flow-graph gain formula that provides the relation between system variables without requiring any detailed procedure for manipulation or reduction of the flow graph.

In This section we will present a detailed discussion about the application and construction of signal flow graphs for analysis of any linear RF/MW network or system. Before we proceed into a detailed discussion of signal flow graphs, let us first define it.

DEFINITION-SIGNAL FLOW GRAPH: *is defined to be an abbreviated block diagram consisting of small circles (called nodes) representing the variables that are connected by several directed lines (called branches) representing one-way signal multipliers; an arrow on the line indicates the direction of signal flow, a letter near the arrow indicates the multiplication factor.*

Having defined what a signal flow graph is, we will now discuss the main features and its construction as well as a reduction technique to generate ratios of any set of desired variables.

4.9.1 Main Features of a Signal Flow Graph

The main components of a signal flow graph are nodes and branches which are defined as follows:

a. Node: Each port of a microwave network (e.g. the ith port) can be represented by two nodes:
- Node a_i- representing a wave entering port i (an independent variable)
- Node b_i- representing a wave reflected from port i (a dependent variable)

b. Branch: A branch is a directed path between an "a-node" (an independent variable) and a "b-node" (a dependent variable). A branch represents a signal flow from node a to b. The multiplication factor placed near the arrow is the associated S-parameter.

A signal flow graph is a convenient technique to represent and subsequently analyze the flow of waves in a Microwave network.

There are certain rules one needs to follow in constructing one:
1. *Each variable is shown as a node.*
2. *S-parameters are shown by branches.*
3. *Branches enter dependent variable nodes (reflected wave variables).*

4. Branches emanate from independent variable nodes (incident wave variables)
5. A node is equal to the sum of the branches entering it. For example consider Figure 4.15, where the dependent variable b_1 can be written as: $b_1 = S_{11}a_1 + S_{12}a_2$

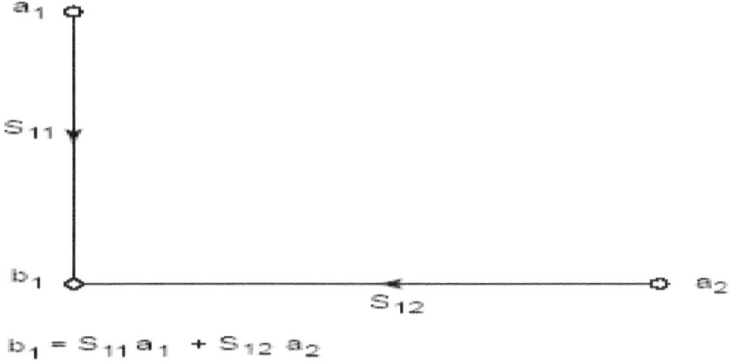

Figure 4.15 An example of nodes and branches.

Example 4.9
Draw the signal flow graph for a linear two-port microwave network as shown in Figure 4.16.

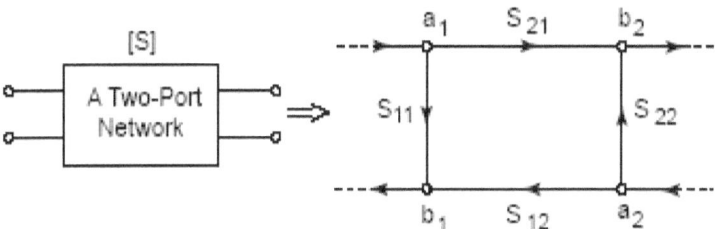

Figure 4.16 SFG for a two-port network.

Solution:
A two-port network at microwave frequencies can be characterized by S-parameters given by:
$b_1 = S_{11}a_1 + S_{12}a_2$
$b_2 = S_{21}a_1 + S_{22}a_2$
Therefore using the rules set forth earlier in (1-5), the signal flow graph for a two-port network consists of two a-nodes (a_1, a_2), two b-nodes (b_1, b_2) and four branches ($S_{11}, S_{12}, S_{21}, S_{22}$) as shown in Figure 4.16)

EXAMPLE 4.10
Find the signal flow graph (SFG) of a microwave amplifier shown in Figure 4.17.

Figure 4.11 A two-port network with new reference planes.

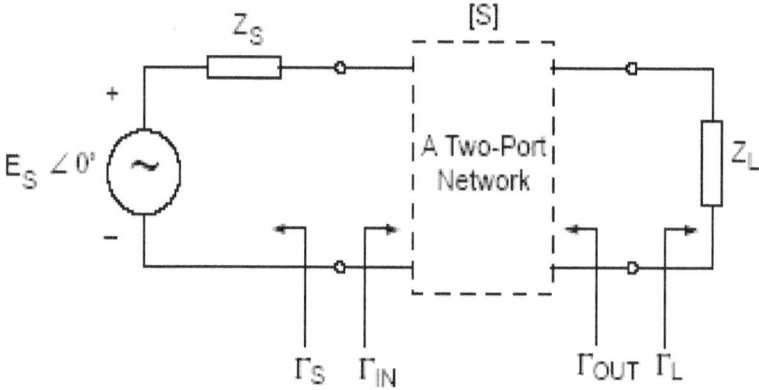

Figure 4.17 A circuit diagram of a microwave amplifier.

Solution:
In order to obtain the SFG for the microwave amplifier, we can dissect or compartmentalize the problem into three separate areas, obtain SFG for each part and then create a final SFG by combining these three SFGs into one (see Chapter 1, "Solutions to Problems" section):

a. Source SFG
The SFG of a signal generator with an internal impedance (Z_S) is obtained from Figure 4.18 as follows:

$$V_g = E_S + Z_S I_g \Rightarrow V_g^+ + V_g^- = E_S + Z_S \left(\frac{V_g^+}{Z_0} - \frac{V_g^-}{Z_0} \right)$$

$$\Rightarrow \frac{V_g^-}{\sqrt{Z_0}} = \frac{V_g^+}{\sqrt{Z_0}} \left(\frac{Z_S - Z_0}{Z_S + Z_0} \right) + E_S \frac{\sqrt{Z_0}}{Z_0 + Z_S} \qquad (4.61)$$

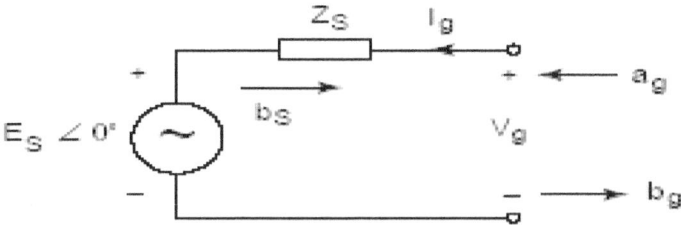

Figure 4.18 Schematic of a signal generator.

Thus Equation (4.61) can be written as:

$$b_g = \Gamma_S a_g + b_S \qquad (4.62)$$

Where

$$b_g = \frac{V_g^-}{\sqrt{Z_0}}$$

$$a_g = \frac{V_g^+}{\sqrt{Z_0}}$$

$$\Gamma_S = \frac{Z_S - Z_0}{Z_S + Z_0} \qquad (4.63)$$

$$b_S = E_S \frac{\sqrt{Z_0}}{Z_0 + Z_S} \qquad (4.64)$$

The source SFG is shown in Figure 4.19.

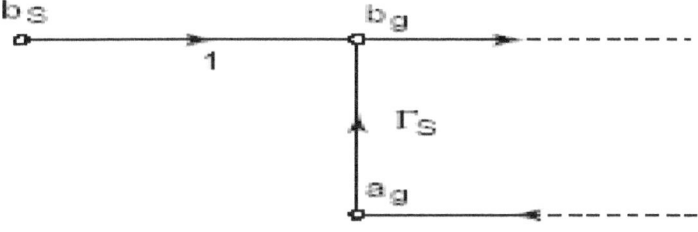

Figure 4.19 SFG of a signal generator.

b. Load SFG

The SFG of a load impedance (Z_L) is obtained from Figure 4.20 as follows:

$$V_L = Z_L I_L \Rightarrow V_L^+ + V_L^- = Z_L \left(\frac{V_L^+}{Z_0} - \frac{V_L^-}{Z_0} \right)$$

$$\Rightarrow \frac{V_L^-}{\sqrt{Z_0}} = \frac{V_L^+}{\sqrt{Z_0}} \left(\frac{Z_L - Z_0}{Z_L + Z_0} \right) \tag{4.65}$$

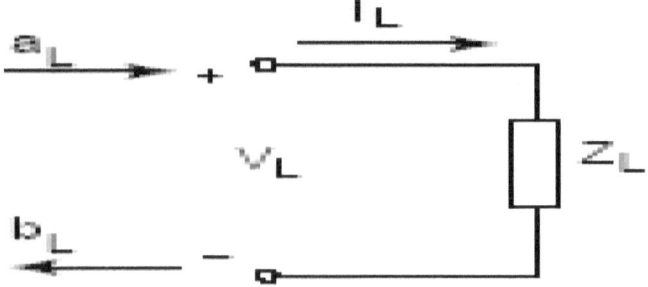

Figure 4.20 Schematic of a load.

Thus Equation (4.65) can be written as:
$b_L = \Gamma_L a_L$
Where

$$b_L = \frac{V_L^-}{\sqrt{Z_0}}$$

$$a_L = \frac{V_L^+}{\sqrt{Z_0}}$$

$$\Gamma_L = \frac{Z_L - Z_0}{Z_L + Z_0} \tag{4.66}$$

The load SFG is shown in Figure 4.21.

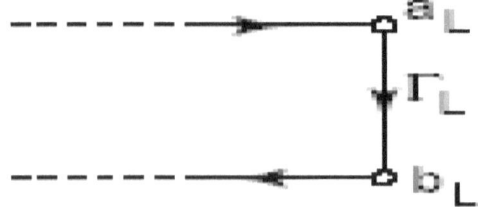

Figure 4.21 SFG of a load.

c. Linear Two-Port Network SFG
The SFG of a linear two-port network has already been obtained in example 4.9 and is shown in Figure 4.16.

d. Final SFG
Combine the SFG for part (a), (b), and (c) to obtain the final SFG as shown in Figure 4.22.

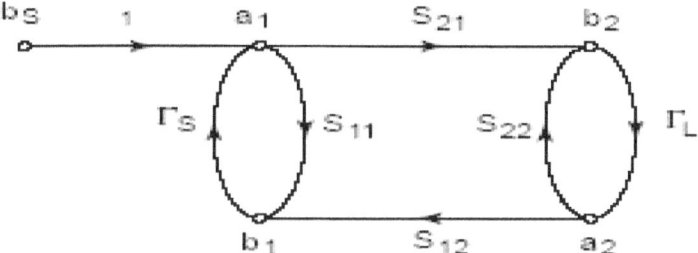

Figure 4.22 Final SFG of a two-port network.

4.9.2 Signal Flow Graph Reduction
Once a microwave network has been represented in terms of a signal flow graph, the wave amplitude ratio of any two variables can be obtained by using the following two techniques:

1. Mason's Rule (from Control System Theory)
This method has been well documented in any "control System" text and will not be repeated here.

2. Signal Flow Graph Reduction Technique
This method is worth presentation and will be further discussed in this work. In its simplest form, it consists of a reduction of a signal flow graph to a single branch using four basic decomposition rules. These rules are briefly summarized herein but each can easily be obtained by simple observation and basic application of the basics of signal flow graphs set forth earlier.

RULE #1-SERIES RULE: Two branches in series whose common node has only one incoming and outgoing wave may be combined into a single branch with a coefficient (or multiplication factor) equal to the product of the two coefficients (see Figure 4.23). That is
$$S_c = S_a S_b \qquad (4.67)$$

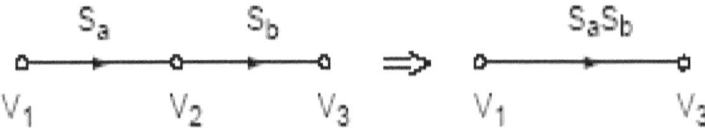

Figure 4.23 Series rule.

RULE #2- PARALLEL RULE: Two branches in parallel both going from one node to another may be combined into a single branch whose coefficient is the sum of the two coefficients (see Figure 4.24). That is

$$S_c = S_a + S_b \tag{4.68}$$

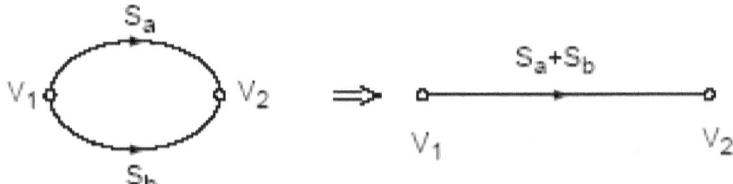

Figure 4.24 Parallel rule.

RULE #3- SELF LOOP RULE: A branch beginning and ending on the same node (called a self loop) with a coefficient S_ℓ (see Figure 4.25) can be eliminated by multiplying the coefficients of the branches (feeding that node) by $\left(\dfrac{1}{1-S_\ell} \right)$.

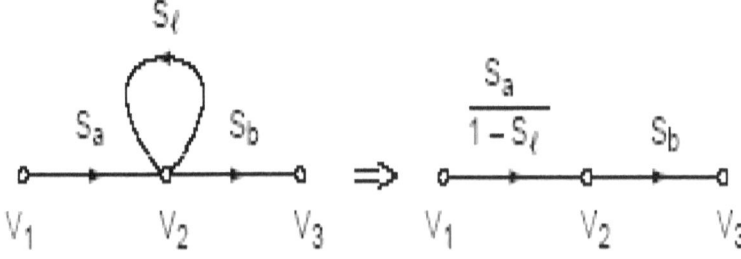

Figure 4.25 Self-loop rule.

RULE #4- SPLITTING RULE: Any node can be split into two separate nodes where each of the two nodes are connected only once to the incoming and the outgoing nodes as shown in Figure 4.26.

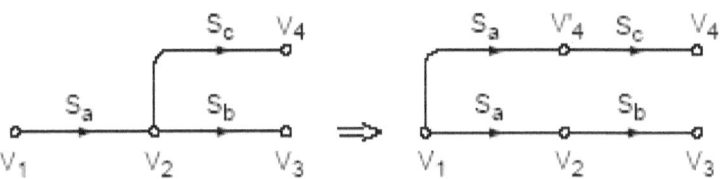

Figure 4.26 Splitting rule.

EXAMPLE 4.11
Consider a two-port network characterized by a scattering matrix [S] driven by a source with an internal impedance of (Z_S) and terminated in a load impedance (Z_L) as shown earlier in Figure 4.27. Calculate:
a. The input reflection coefficient (Γ_{in}), and
b. The output reflection coefficient (Γ_{out});
(a) and (b) are to be calculated in terms [S], the source reflection coefficient (Γ_S), and the load reflection coefficient (Γ_L).

Solution:
The overall signal flow graph was shown earlier in Figure 4.22, from which we can derive the following for each of the two cases:

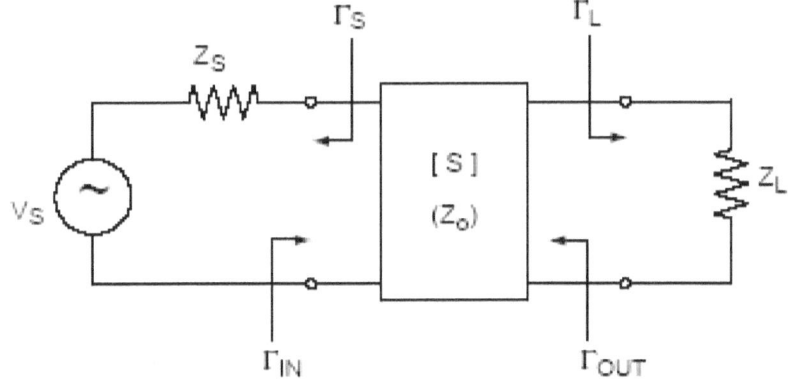

Figure 4.27 A terminated two-port network.

a. The input reflection coefficient (Γ_{in})

The signal flow graph (SFG) for this network is shown in Figure 4.28.

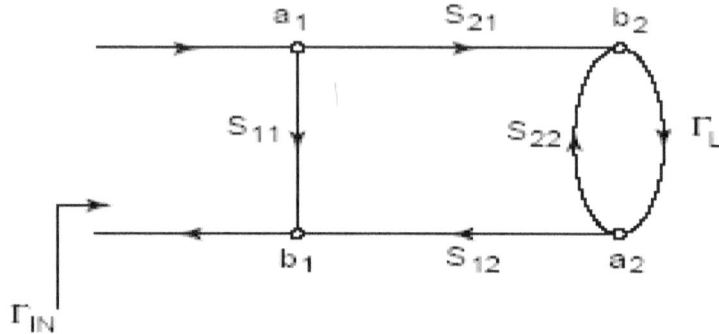

Figure 4.28 SFG for the input reflection coefficient.

Using the 4 decomposition rules stated above, we can reduce the signal flow graph step by step as shown in Figure 4.29. Using Figure 4.29(d), we can write Γ_{in} as:

$$\Gamma_{in} = S_{11} + \frac{S_{12}S_{21}\Gamma_L}{1 - S_{22}\Gamma_L} \tag{4.69}$$

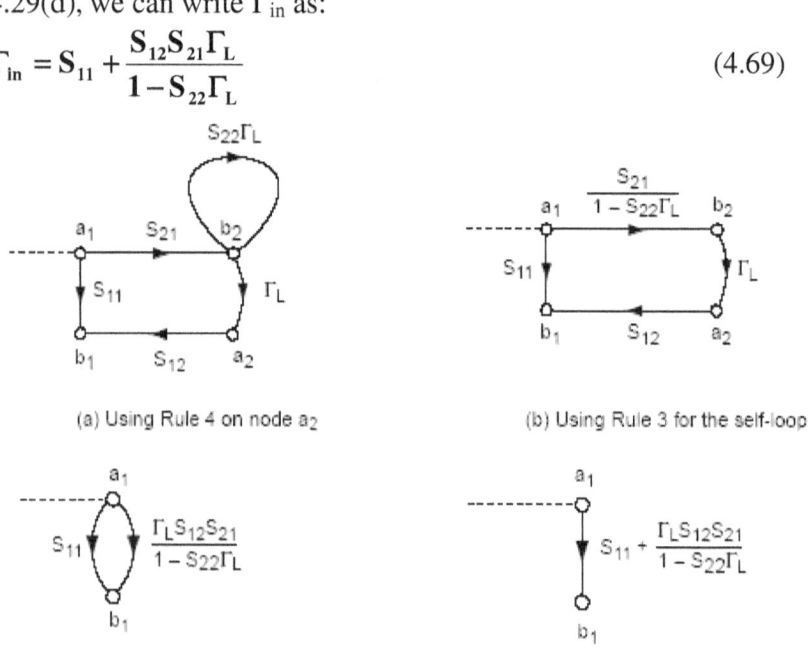

Figure 4.29 Using rules 1-4 on the flow graph.

b. The output reflection coefficient (Γ_{out})

The signal flow graph for this part is shown in Figure 4.30. Very similar to part (a) for Γ_{in}, we can reduce the signal flow graph step by step and write a similar equation for Γ_{out} as:

$$\Gamma_{out} = S_{22} + \frac{S_{12}S_{21}\Gamma_S}{1 - S_{11}\Gamma_S} \qquad (4.70)$$

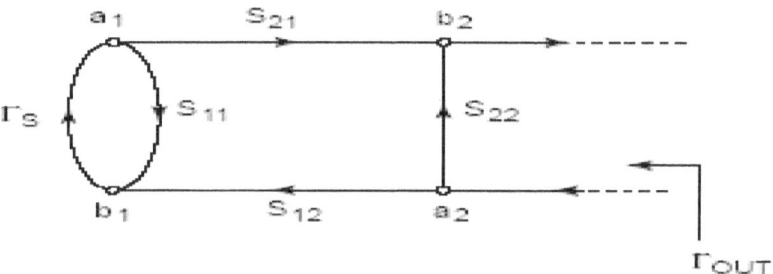

Figure 4.30 SFG for the output reflection coefficient.

4.9.3 Applications of Signal Flow Graphs (SFGs)

Consider a two-port network and its corresponding signal flow graph as shown earlier in Figures 4.17 and 4.22, respectively. We know that the square of the incident normalized wave represents the incident power while the reflected power is represented by the reflected normalized wave. Using SFG, we can find the relationship between the various wave variables and therefore the several types of power as follows:

a. P_{IN}

P_{IN} is the input power to the network given by:

$$P_{IN} = \frac{1}{2}|a_1|^2 - \frac{1}{2}|b_1|^2 = \frac{1}{2}|a_1|^2\left(1 - |\Gamma_{IN}|^2\right) \qquad (4.71)$$

b. P_{AVS}

P_{AVS} is the power available from the source, defined as the input power (P_{IN}) delivered by the source to a conjugately matched input impedance:

$$P_{AVS} = P_{IN}\,|_{\Gamma_{IN}=\Gamma_S^*} = \frac{1}{2}|b_g|^2 - \frac{1}{2}|a_g|^2 \qquad (4.72)$$

Where

$$a_g = \Gamma_s b_g$$

$$b_g = b_s + \Gamma_s a_g = \frac{b_s}{1 - |\Gamma_s|^2} \tag{4.73}$$

Thus we have:

$$P_{AVS} = \frac{|b_s|^2}{2(1 - |\Gamma_s|^2)} \tag{4.74}$$

c. P_L

P_L is the power delivered to the load given by:

$$P_L = \frac{1}{2}|b_2|^2 - \frac{1}{2}|a_2|^2 = \frac{1}{2}|b_2|^2 \left(1 - |\Gamma_L|^2\right) \tag{4.75}$$

d. P_{AVN}

P_{AVN} is the power available from the network, which is defined as the power delivered to the load when the load is conjugately matched to the network is given by:

$$P_{AVN} = P_L \, |_{\Gamma_L = \Gamma_{OUT}*} = \left[\frac{1}{2}|b_2|^2 - \frac{1}{2}|a_2|^2 \right]_{\Gamma_L = \Gamma_{OUT}*}$$

$$P_{AVN} = \frac{1}{2}|b_2|^2 \left(1 - |\Gamma_{OUT}|^2\right) \tag{4.76}$$

4.9.4 Power Gain Expressions

Using the results obtained in the previous section for various powers, we can now obtain several power gain expressions as follows:

a. TRANSDUCER POWER GAIN (G_T)

$$G_T \equiv \frac{P_L}{P_{AVS}} = \frac{|b_2|^2}{|b_s|^2} \left(1 - |\Gamma_s|^2\right)\left(1 - |\Gamma_L|^2\right) \tag{4.77}$$

Using the SFG technique, the ratio b_2/b_s is obtained to be:

$$\frac{b_2}{b_s} = \frac{S_{21}}{(1 - S_{11}\Gamma_s)(1 - S_{22}\Gamma_L) - S_{12}S_{21}\Gamma_s\Gamma_L} \tag{4.78}$$

Thus G_T can be written as:

$$G_T = \frac{1-|\Gamma_S|^2}{|1-\Gamma_{IN}\Gamma_S|^2}|S_{21}|^2 \frac{1-|\Gamma_L|^2}{|1-S_{22}\Gamma_L|^2} \qquad (4.79)$$

Or, we can write an alternate form for G_T as:

$$G_T = \frac{1-|\Gamma_S|^2}{|1-S_{11}\Gamma_S|^2}|S_{21}|^2 \frac{1-|\Gamma_L|^2}{|1-\Gamma_{OUT}\Gamma_L|^2} \qquad (4.80)$$

b. POWER GAIN (G_P)

$$G_P \equiv \frac{P_L}{P_{IN}} = \frac{|b_2/b_S|^2}{|a_1/b_S|^2}\left(\frac{1-|\Gamma_L|^2}{1-|\Gamma_{IN}|^2}\right) \qquad (4.81)$$

Using the SFG technique, the ratio a_1/b_S is obtained to be:

$$\frac{a_1}{b_S} = \frac{1-S_{22}\Gamma_L}{1-(S_{11}\Gamma_S + S_{22}\Gamma_L + S_{12}S_{21}\Gamma_S\Gamma_L)(1-S_{11}\Gamma_S)+S_{11}S_{22}\Gamma_S\Gamma_L} \qquad (4.82)$$

Thus G_P can be written as:

$$G_P = \frac{1}{1-|\Gamma_{IN}|^2}|S_{21}|^2 \frac{1-|\Gamma_L|^2}{|1-S_{22}\Gamma_L|^2} \qquad (4.83)$$

c. AVAILABLE POWER GAIN (G_A)

$\Gamma_L = \Gamma_{out}^{*}$

$$G_A \equiv \frac{P_{AVN}}{P_{AVS}} = \frac{|b_2|^2}{|b_S|^2}\left(1-|\Gamma_S|^2\right)\left(1-|\Gamma_{OUT}|^2\right) \qquad (4.84)$$

Using (4.78), the ratio b_2/b_S when $\Gamma_L = \Gamma_{out}^{*}$ we obtain:

$$\frac{b_2}{b_S} = \frac{S_{21}}{(1-S_{11}\Gamma_S)(1-|\Gamma_{OUT}|^2)} \qquad (4.85)$$

Thus G_A can be written as:

$$G_A = \frac{1-|\Gamma_S|^2}{|1-S_{11}\Gamma_S|^2} |S_{21}|^2 \frac{1}{1-|\Gamma_{OUT}|^2} \tag{4.86}$$

Note: *So far we have discussed the various power gains which are mostly encountered in Microwave amplifier design. However, in most audio and RF amplifier designs we use voltage gain. The voltage gain of an amplifier is defined to be the ratio of the total output voltage to the total input voltage as follows:*

$$A_V = \frac{V_{OUT}}{V_{IN}} = \frac{a_2 + b_2}{a_1 + b_1} \tag{4.87}$$

Dividing Equation (4.87) by b_S, we obtain:

$$A_V = \frac{a_2/b_S + b_2/b_S}{a_1/b_S + b_1/b_S} \tag{4.88}$$

Using Equations (4.78) and (4.82) for b_2/b_S and a_1/b_S and similar derivations for b_1/b_S and a_2/b_S, we obtain A_V to be given by:

$$A_V = \frac{S_{21}(1+\Gamma_L)}{(1-S_{22}\Gamma_L) + S_{11}(1-S_{22}\Gamma_L) + S_{12}S_{21}\Gamma_L} \tag{4.89}$$

4.10 CONVERSION AMONG PARAMETERS

Voltages and currents at each port provide us with four variables of interest: V_1, V_2, I_1, and I_2 at lower frequencies, which leads to four parameters of interest, i.e., Z-, Y-, h-, ABCD-parameters.

At higher frequencies we also have four variables of interest: V_1^+, V_2^+, V_1^-, V_2^- which leads to two parameters of interest, i.e., S- and T-parameters.

NOTE: *It should be noted that v and i are net voltages given by:*

$v = V^+ + V^-$,

$i = I^+ + I^-$,

A two-port network, with two voltage- and two current-variables, is shown in Figure 4.31.

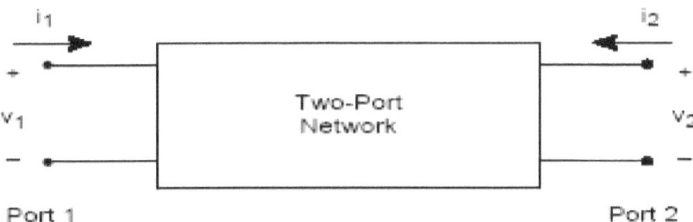

Figure 4.31 Net voltages and net currents in a two-port network.

These six parameters are defined in a matrix form as follows:

A. Z-prameters

$$[V]=[Z][I] \tag{4.90a}$$

$$[V]=\begin{bmatrix} v_1 \\ v_2 \end{bmatrix}$$

$$[I]=\begin{bmatrix} i_1 \\ i_2 \end{bmatrix}$$

$$[Z]=\begin{bmatrix} Z_{11} & Z_{12} \\ Z_{21} & Z_{22} \end{bmatrix} \tag{4.90b}$$

B. Y-PARAMETERS

$$[I]=[Y][V] \tag{4.91a}$$

where [Y] is defined as follows:

$$[Y]=\begin{bmatrix} Y_{11} & Y_{12} \\ Y_{21} & Y_{22} \end{bmatrix} \tag{4.91b}$$

C. H-PARAMETERS

$$\begin{bmatrix} v_1 \\ i_2 \end{bmatrix}=\begin{bmatrix} h_{11} & h_{12} \\ h_{21} & h_{22} \end{bmatrix} \cdot \begin{bmatrix} i_1 \\ v_2 \end{bmatrix} \tag{4.92}$$

D. ABCD-PARAMETERS

$$\begin{bmatrix} v_1 \\ i_1 \end{bmatrix}=\begin{bmatrix} A & B \\ C & D \end{bmatrix} \cdot \begin{bmatrix} v_2 \\ -i_2 \end{bmatrix} \tag{4.93}$$

E. S- PARAMETERS

$$\begin{bmatrix} V_1^- \\ V_2^- \end{bmatrix} = \begin{bmatrix} S_{11} & S_{12} \\ S_{21} & S_{22} \end{bmatrix} \cdot \begin{bmatrix} V_1^+ \\ V_2^+ \end{bmatrix}$$

(4.94)

Or,

$$[V^-]=[S][V^+]$$

Where,

$$[V^-] = \begin{bmatrix} V_1^- \\ V_2^- \end{bmatrix},$$

$$[V^+] = \begin{bmatrix} V_1^+ \\ V_2^+ \end{bmatrix}$$

and,

$$[S] = \begin{bmatrix} S_{11} & S_{12} \\ S_{21} & S_{22} \end{bmatrix}$$

(4.95)

F. T-PARAMETERS

$$\begin{bmatrix} V_1^+ \\ V_1^- \end{bmatrix} = \begin{bmatrix} T_{11} & T_{12} \\ T_{21} & T_{22} \end{bmatrix} \begin{bmatrix} V_2^- \\ V_2^+ \end{bmatrix}$$

(4.96)

These parameters can be converted one into another as shown in the table on the next page.

4.11 SUMMARY

Having defined the S-parameters and derived power gain expressions in this chapter, we will present important concepts about the design of matching networks as well as the stability of two-ports in the next chapters which will lay the foundation for active circuit design.

	S	z	y	h	ABCD
S	$S_{11}\quad S_{12}$ $S_{21}\quad S_{22}$	$S_{11}=\dfrac{(z'_{11}-1)(z'_{22}+1)-z'_{12}z'_{21}}{\Delta_1}$ $S_{12}=\dfrac{2z'_{12}}{\Delta_1}$ $S_{21}=\dfrac{2z'_{21}}{\Delta_1}$ $S_{22}=\dfrac{(z'_{11}+1)(z'_{22}-1)-z'_{12}z'_{21}}{\Delta_1}$	$S_{11}=\dfrac{(1-y'_{11})(1+y'_{22})+y'_{12}y'_{21}}{\Delta_2}$ $S_{12}=\dfrac{-2y'_{12}}{\Delta_2}$ $S_{21}=\dfrac{-2y'_{21}}{\Delta_2}$ $S_{22}=\dfrac{(1+y'_{11})(1-y'_{22})+y'_{12}y'_{21}}{\Delta_2}$	$S_{11}=\dfrac{(h'_{11}-1)(h'_{22}+1)-h'_{12}h'_{21}}{\Delta_3}$ $S_{12}=\dfrac{2h'_{12}}{\Delta_3}$ $S_{21}=\dfrac{-2h'_{21}}{\Delta_3}$ $S_{22}=\dfrac{(1+h'_{11})(1-h'_{22})+h'_{12}h'_{21}}{\Delta_3}$	$S_{11}=\dfrac{A'+B'-C'-D'}{\Delta_4}$ $S_{12}=\dfrac{2(A'D'-B'C')}{\Delta_4}$ $S_{21}=\dfrac{2}{\Delta_4}$ $S_{22}=\dfrac{-A'+B'-C'+D'}{\Delta_4}$
z	$z'_{11}=\dfrac{(1+S_{11})(1-S_{22})+S_{12}S_{21}}{\Delta_5}$ $z'_{12}=\dfrac{2S_{12}}{\Delta_5}$ $z'_{21}=\dfrac{2S_{21}}{\Delta_5}$ $z'_{22}=\dfrac{(1-S_{11})(1+S_{22})+S_{12}S_{21}}{\Delta_5}$	$z_{11}\quad z_{12}$ $z_{21}\quad z_{22}$	$\dfrac{y_{22}}{\lvert y\rvert}\quad \dfrac{-y_{12}}{\lvert y\rvert}$ $\dfrac{-y_{21}}{\lvert y\rvert}\quad \dfrac{y_{11}}{\lvert y\rvert}$	$\dfrac{\lvert h\rvert}{h_{22}}\quad \dfrac{h_{12}}{h_{22}}$ $\dfrac{-h_{21}}{h_{22}}\quad \dfrac{1}{h_{22}}$	$\dfrac{A}{C}\quad \dfrac{\Delta_8}{C}$ $\dfrac{1}{C}\quad \dfrac{D}{C}$
y	$y'_{11}=\dfrac{(1-S_{11})(1+S_{22})+S_{12}S_{21}}{\Delta_6}$ $y'_{12}=\dfrac{-2S_{12}}{\Delta_6}$ $y'_{21}=\dfrac{-2S_{21}}{\Delta_6}$ $y'_{22}=\dfrac{(1+S_{11})(1-S_{22})+S_{12}S_{21}}{\Delta_6}$	$\dfrac{z_{22}}{\lvert z\rvert}\quad \dfrac{-z_{12}}{\lvert z\rvert}$ $\dfrac{-z_{21}}{\lvert z\rvert}\quad \dfrac{z_{11}}{\lvert z\rvert}$	$y_{11}\quad y_{12}$ $y_{21}\quad y_{22}$	$\dfrac{1}{h_{11}}\quad \dfrac{-h_{12}}{h_{11}}$ $\dfrac{h_{21}}{h_{11}}\quad \dfrac{\lvert h\rvert}{h_{11}}$	$\dfrac{D}{B}\quad \dfrac{-\Delta_8}{B}$ $\dfrac{-1}{B}\quad \dfrac{A}{B}$
h	$h'_{11}=\dfrac{(1+S_{11})(1+S_{22})-S_{12}S_{21}}{\Delta_7}$ $h'_{12}=\dfrac{2S_{12}}{\Delta_7}$ $h'_{21}=\dfrac{-2S_{21}}{\Delta_7}$ $h'_{22}=\dfrac{(1-S_{11})(1-S_{22})-S_{12}S_{21}}{\Delta_7}$	$\dfrac{\lvert z\rvert}{z_{22}}\quad \dfrac{z_{12}}{z_{22}}$ $\dfrac{-z_{21}}{z_{22}}\quad \dfrac{1}{z_{22}}$	$\dfrac{1}{y_{11}}\quad \dfrac{-y_{12}}{y_{11}}$ $\dfrac{y_{21}}{y_{11}}\quad \dfrac{\lvert y\rvert}{y_{11}}$	$h_{11}\quad h_{12}$ $h_{21}\quad h_{22}$	$\dfrac{B}{D}\quad \dfrac{-\Delta_8}{D}$ $\dfrac{-1}{D}\quad \dfrac{C}{D}$
ABCD	$A'=\dfrac{(1+S_{11})(1-S_{22})+S_{12}S_{21}}{2S_{21}}$ $B'=\dfrac{(1+S_{11})(1+S_{22})-S_{12}S_{21}}{2S_{21}}$ $C'=\dfrac{(1-S_{11})(1-S_{22})-S_{12}S_{21}}{2S_{21}}$ $D'=\dfrac{(1-S_{11})(1+S_{22})+S_{12}S_{21}}{2S_{21}}$	$\dfrac{z_{11}}{z_{21}}\quad \dfrac{\lvert z\rvert}{z_{21}}$ $\dfrac{1}{z_{21}}\quad \dfrac{z_{22}}{z_{21}}$	$\dfrac{-y_{22}}{y_{21}}\quad \dfrac{-1}{y_{21}}$ $\dfrac{-\lvert y\rvert}{y_{21}}\quad \dfrac{-y_{11}}{y_{21}}$	$\dfrac{-\lvert h\rvert}{h_{21}}\quad \dfrac{-h_{11}}{h_{21}}$ $\dfrac{-h_{22}}{h_{21}}\quad \dfrac{-1}{h_{21}}$	$A\quad B$ $C\quad D$

$\Delta_1=(z'_{11}+1)(z'_{22}+1)-z'_{12}z'_{21}$
$\Delta_2=(1+y'_{11})(1+y'_{22})-y'_{12}y'_{21}$
$\Delta_3=(h'_{11}+1)(h'_{22}+1)-h'_{12}h'_{21}$
$\Delta_4=A'+B'+C'+D'$
$\Delta_5=(1-S_{11})(1-S_{22})-S_{12}S_{21}$
$\Delta_6=(1+S_{11})(1+S_{22})-S_{12}S_{21}$
$\Delta_7=(1-S_{11})(1+S_{22})+S_{12}S_{21}$
$\Delta_8=AD-BC$

$z'_{11}=z_{11}/Z_o,\ z'_{12}=z_{12}/Z_o,\ z'_{21}=z_{21}/Z_o,\ z'_{22}=z_{22}/Z_o$
$y'_{11}=y_{11}Z_o,\ y'_{12}=y_{12}Z_o,\ y'_{21}=y_{21}Z_o,\ y'_{22}=y_{22}Z_o$
$h'_{11}=h_{11}/Z_o,\ h'_{12}=h_{12},\ h'_{21}=h_{21},\ h'_{22}=h_{22}Z_o$
$A'=A,\ B'=B/Z_o,\ C'=CZ_o,\ D'=D$
$\lvert z\rvert=z_{11}z_{22}-z_{12}z_{21}$
$\lvert y\rvert=y_{11}y_{22}-y_{12}y_{21}$
$\lvert h\rvert=h_{11}h_{22}-h_{12}h_{21}$

Chapter 4- Symbol List

A symbol will not be repeated again, once it has been identified and defined in an earlier chapter, with its definition remaining unchanged.

A, B, C, D - ABCD parameters

$h_{11}, h_{12}, h_{21}, h_{22}$ – Hybrid or h-parameters

i_1 - Current into port 1 of a network

i_2 - Current into port 2 of a network

i_k – Current at branch k

SFG – Signal Flow Graph

$S_{11}, S_{12}, S_{21}, S_{22}$ – Scattering parameters (or S-parameters)

$T_{11}, T_{12}, T_{21}, T_{22}$ – Transmission parameters (or T-parameters)

v_1 - Voltage at port 1 of a network

v_2 - Voltage at port 2 of a network

v_k - voltage at branch k

$Y_{11}, Y_{12}, Y_{21}, Y_{22}$ – Admittance parameters (or Y-parameters)

$Z_{11}, Z_{12}, Z_{21}, Z_{22}$ – Impedance parameters (or Z-parameters)

CHAPTER-4 PROBLEMS

4.1)   Determine the S-parameters of the circuit shown in Figure P4.1

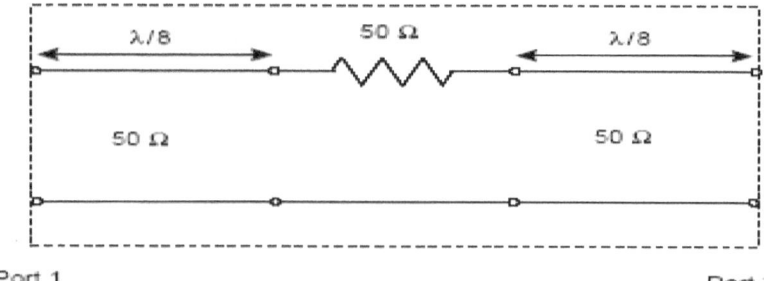

Port 1   Port 2

Figure P4.1

4.2)   Find the scattering matrix and the transmission matrix of a loss-less transmission line of length "ℓ" in a 50 Ω system when:

a. Characteristic impedance of the line is (Z_o=50 Ω).

b. Characteristic impedance of the line is (Z_o=100 Ω).

4.3)   Find the generalized scattering matrix of a two-port consisting of a junction of two lossless transmission lines as shown in Figure P4.3.

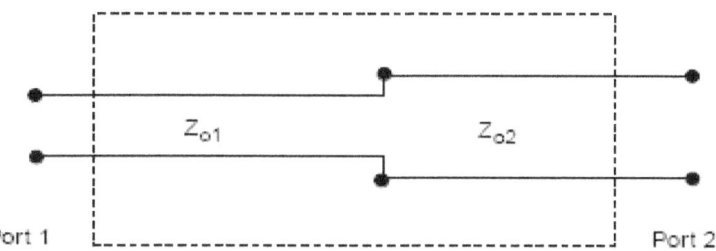

Figure P4.3

4.4)   Find the transmission matrix of the circuit shown in Figure P4.4

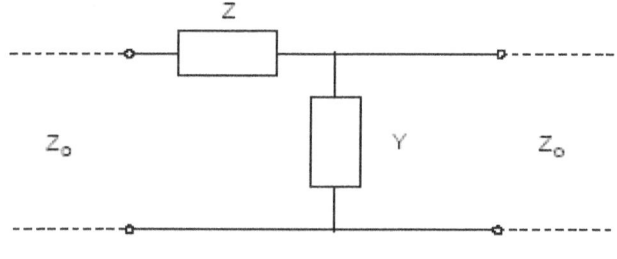

Figure P4.4

4.5) Derive the [ABCD] matrix for the step-up transformer circuit shown in Figure P4.5.

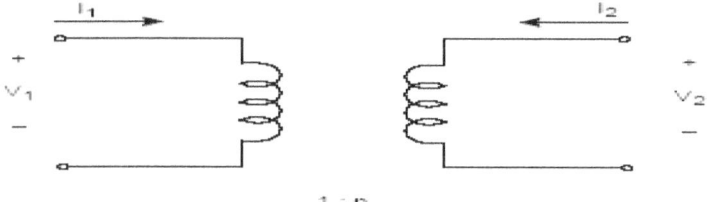

Figure P4.5

4.6)   A two-Port network has the scattering matrix as shown below. From this data:

a) Determine whether the network is reciprocal or lossless,

b) If the output terminals are shorted together, what will the input reflection coefficient be?

$$[\mathbf{S}] = \begin{bmatrix} 0.2 & j0.9 \\ j0.9 & 0.4 \end{bmatrix}$$

REFERENCES

[4.1] Chang K. *Microwave and Optical Component.* Vols I, II. New York: John Wiley & Sons, 1989.

[4.2] Cheung, W. S. and F. H. Levien. *Microwave Made Simple.* Dedham: Artech House, 1985.

[4.3] Desor, C. A. and E. S. Kuh. *Basic Circuit Theory.* Tokyo: McGraw-Hill, 1969.

[4.4] Gonzalez, G. *Microwave Transistor Amplifiers, Analysis, and Design,* 2nd ed. Upper Saddle River: Prentice Hall, 1997.

[4.5] Gardiol, F. E. *Introduction to Microwaves.* Dedham: Artech House, 1984.

[4.6] Ishii, T. K. *Microwave Engineering.* 2nd ed., Orlando: Harcourt Brace Jovanovich Publishers, 1989.

[4.7] Laverghetta, T. *Practical Microwaves.* Indianapolis: Howard Sams, 1984.

[4.8] Pozar, D. M. *Microwave Engineering,* 2nd ed. New York: John Wiley & Sons, 1998.

[4.9] Saad, T. *Microwave Engineer's Handbook.* Vols I, II. Dedham: Artech House, 1988.

[4.10] Scott, A. W. *Understanding Microwaves.* New York: John Wiley & Sons, 1993.

[4.11] Vendelin, G. D. *Design of Amplifiers and Oscillators by the S-Parameter Method.* New York: John Wiley & Sons, 1981.

[4.12] Vendelin, G. D., A. M. Pavio, and Ulrich L. Rhode. *Microwave Circuit Design.* New York: John Wiley, 1990.

CHAPTER 5

The Smith Chart

5.1 INTRODUCTION
One of the most valuable and yet pervasive graphical tools in all of microwave engineering is the "Smith chart", originally developed in 1939 by P. Smith at the Bell Telephone Laboratories. This chart is the reflection coefficient-to-impedance/admittance converter or vice versa and can greatly simplify the analysis of complex design problems involving transmission lines or lumped elements.

Furthermore, the smith chart provides valuable information about the circuit's performance when line lengths change or new elements are added to the circuit, particularly where obtaining the same amount of information through mathematical models and calculations would be very tedious and time consuming.

Over the years, it has proven itself to be a most useful tool and is thus employed frequently in all stages of circuit analysis or design whether done through manual methods or computer-aided-design (CAD) Software techniques.

5.2 A VALUABLE GRAPHICAL AID
Considering the equation for the reflection coefficient (as given earlier in Chapter 8) we have:

$$\Gamma = \frac{Z - Z_O}{Z + Z_O} = \frac{Z_N - 1}{Z_N + 1} \qquad (5.1)$$

where $Z_N = \dfrac{Z}{Z_O} = r + jx$ is the normalized impedance and Z_O is the characteristic impedance of the transmission line or a reference impedance value.

Based on (5.1), the Smith chart can be derived mathematically as discussed in the next section. This chart is a plot of Γ for different normalized resistance and reactance values, where the circuit is assumed to be passive i.e. $Re(Z) \geq 0$.

It can be shown that the loci of constant resistance values are circles centered on the horizontal (or real) axis while the loci of constant reactance values are circles centered on the vertical (or imaginary) axis offset by one unit.

5.3 DERIVATION OF SMITH CHART

The Smith chart is a plot of

$$\Gamma = \frac{Z_N - 1}{Z_N + 1} \qquad (5.2)$$

in the Γ-plane as a function of r and x. Using (5.2) and separating Γ in terms of its real part (U) and imaginary part (V) we obtain:

$Z_N = r + jx$

$$\Gamma = \frac{r + jx - 1}{r + jx + 1} = U + jV \qquad (5.3)$$

$$U = \frac{r^2 - 1 + x^2}{(r + 1)^2 + x^2} \qquad (5.4)$$

$$V = \frac{2x}{(r + 1)^2 + x^2} \qquad (5.5)$$

At this juncture we note that by using (5.4) and (5.5), we can obtain two families of circles which when superimposed on each other will make up the entire Smith chart. The procedures to obtain these two families of circles are described next.

a. Constant-r circles

The first family of circles is obtained by eliminating "x" from (5.4) and (5.5) which gives:

$$\left(U - \frac{r}{r+1}\right)^2 + V^2 = \left(\frac{1}{r+1}\right)^2 \qquad (5.6)$$

Equation (5.6) represents a family of circles with a center located at

$$(U_0, V_0) = \left(\frac{r}{r+1}, 0\right), \qquad (5.7a)$$

with a radius of

$$R = \left(\frac{1}{r+1}\right). \qquad (5.7b)$$

From Equations (5.7) we can observe that all constant-r circles are centered on the real axis with a shrinking size as "r" is increased. In this regard, we note that r=0 circle is the most-outer circle of the Smith chart while r=∞ circle is reduced to a point at (0,1). Figure 5.1 depicts this concept further.

b. Constant-x circle

The second family of circles is obtained by eliminating "r" from (5.4) and (5.5) which gives:

$$(U-1)^2 + \left(V - \frac{1}{x}\right)^2 = \left(\frac{1}{x}\right)^2 \qquad (5.8)$$

Equation (5.8) represents a family of circles with a center located at

$$(U_0', V_0') = \left(1, \frac{1}{x}\right) \qquad (5.9a)$$

with a radius of

$$R' = \left(\frac{1}{x}\right) \qquad (5.9b)$$

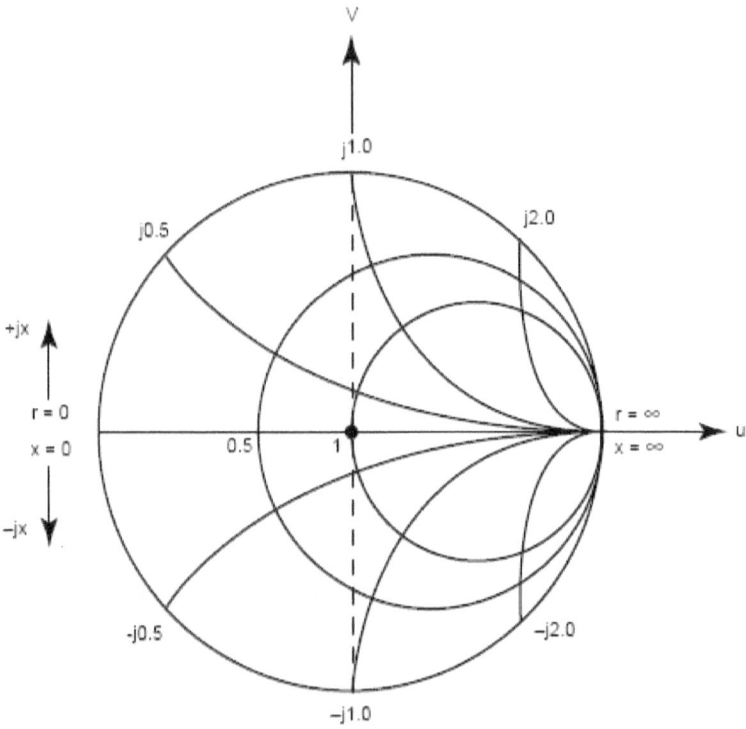

Figure 5.1 Construction of a standard Smith Chart.

From Equations (5.9) we can observe that all constant-x circles are centered on a shifted line parallel to the imaginary axis (by +1 unit to the right), with a shrinking size as "x" increases. In this regard, we note that x=0 circle is the real axis of the Smith chart while x= ±∞ circles are reduced to a point at (1,0). This case is shown in Figure 5.2.

As described earlier, plotting the two families of circles as represented by (5.6) and (5.8) for all values of (r,x) creates a circular chart commonly known as the Smith chart.

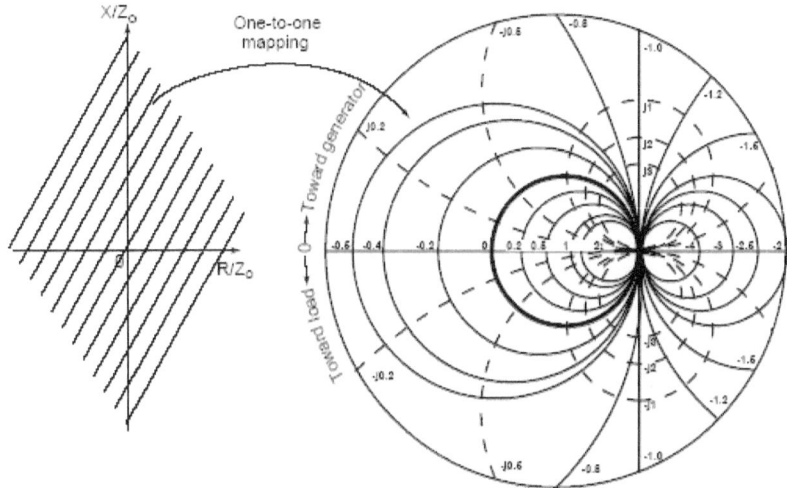

Figure 5.2 Mapping of impedance plane into a compressed Smith Chart.

The standard Smith chart is a one-to-one correspondence between points in the normalized impedance (Z_N) plane [where r=Re(Z_N)≥0] and points in the reflection coefficient (Γ) plane. The upper half of the chart represents normalized impedance values with positive reactances while the lower half corresponds to negative reactances (x ≤ 0)

NOTE: *The Smith chart could have also been developed based on normalized admittance (Y_N) as follows:*

$$Y_N = \frac{Y}{Y_0} = g + jb \qquad (5.10a)$$

Where $Y_0 = 1/Z_0$ is the normalized characteristic admittance or a reference admittance value.

Thus we can write Equation (5.2) as:

$$\Gamma = \frac{Z_N - 1}{Z_N + 1} = \frac{\dfrac{1}{Y_N} - 1}{\dfrac{1}{Y_N} + 1} = -\left(\frac{Y_N - 1}{Y_N + 1}\right) \qquad (5.10b)$$

Now using the transformation:

$$\Gamma' = \left(\frac{Y_N - 1}{Y_N + 1} \right) \tag{5.10c}$$

we obtain the same results as for impedance except that the transformation will be from Y_N-plane into Γ'-plane where

$$\Gamma' = -\Gamma = \Gamma e^{j180°} \tag{5.11}$$

Equation (5.11) indicates that Γ' and Γ are only $180°$ apart but have the same magnitude, which means that when dealing with admittances and impedances on the same chart we need to keep in mind the $180°$ phase adjustment every time we convert Z_N to Y_N or vice versa. Therefore a Smith chart can be used as an impedance chart (Z-Smith chart) or equally as an admittance chart (Y-Smith chart).

SUMMARY: *In summary, using a Smith chart requires awareness and an understanding of the following transformations:*

$Z_N \leftrightarrow \Gamma$

$Y_N \leftrightarrow \Gamma'$

$\Gamma' \leftrightarrow \Gamma e^{j180°}$

The magic of the Smith chart lies in the fact that through the use of the above transformation, a semi-infinite and an unbounded region (i.e. $0 \le r \le \infty$, $-\infty \le x \le +\infty$) is transformed into a finite and workable region (i.e., $0 \le \Gamma \le 1$) which creates easily understood graphical solutions to many complex microwave problems.

5.4 DESCRIPTION OF THE SMITH CHART

As discussed in the previous sections, instead of plotting contours of constant reflection coefficient, contours of constant normalized resistance and reactance are plotted in the Γ-plane. A selected collection of these contours (which are circles), plotted in the Γ-plane, comprise the entire smith chart (commonly known as the "compressed smith chart") which includes impedances with both positive and negative real parts (see Figure 5.1). The "compressed smith chart" is obtained when the entire impedance plane is mapped on a one-to-one basis onto the reflection coefficient plane as shown in Figure 5.2.

The "compressed smith chart", even though very general and applies to both active and passive circuits, is yet impractical and is seldom used in design. Instead, a more useful part of this chart (called a standard smith chart) is used in practice for all passive networks where Re(Z)≥0, which corresponds to mapping only the right-hand side half of the impedance plane into a circle in the reflection coefficient plane with radius |Γ|≤1 as shown in Figure 5.3.

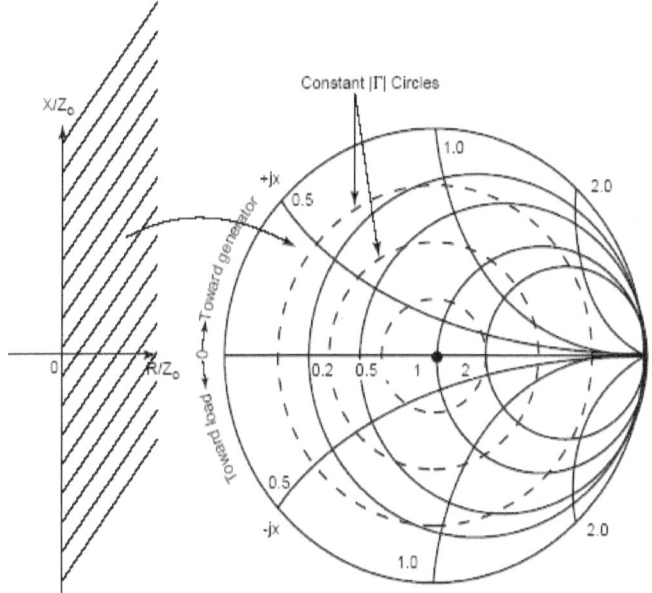

Figure 5.3 Mapping of impedance half-plane into a standard Smith Chart.

Standard Smith chart represents a graphical display of impedance-to-reflection coefficient transformation, in which all values of impedance with Re(Z)≥0 (representing a semi-infinite region of the resistance-reactance rectangular plane) is mapped one-to-one into a circle with the radius of one unit in the reflection coefficient plane. A full blown-out version of the standard smith chart is shown in Figure 5.4 where each circle is marked with its corresponding resistance or reactance value.

IMPEDANCE OR ADMITTANCE COORDINATES

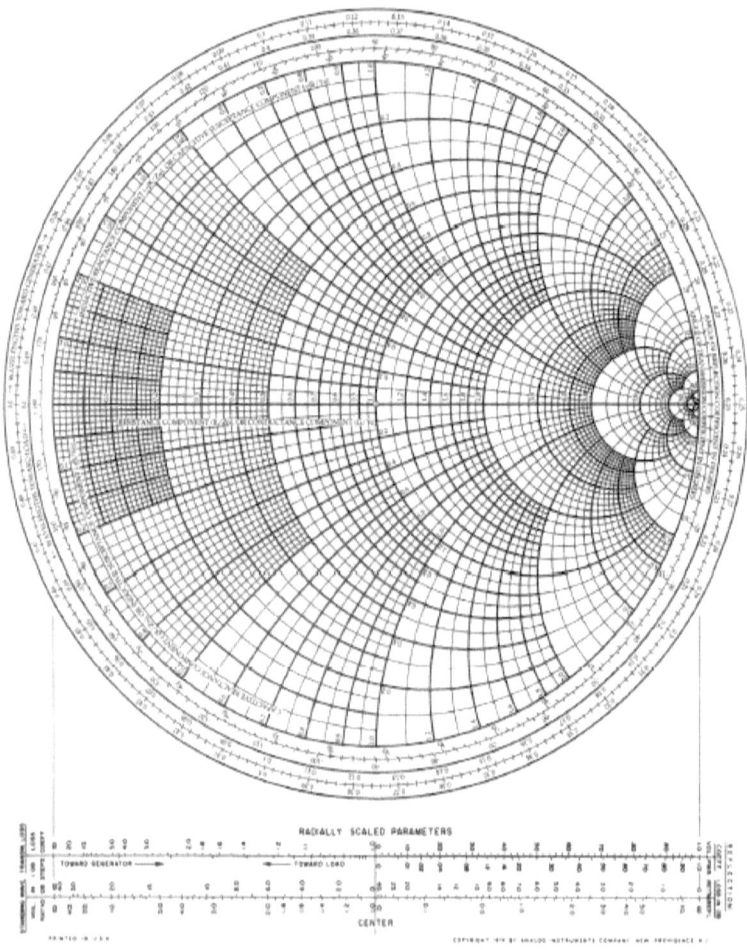

Figure 5.4 The standard Smith Chart.

Thus the Smith chart is comprised of many circles either fully or partially enclosed within the outermost circle ($|\Gamma|=1$) of the standard Smith chart.

The set of circles centered on the horizontal (or real) axis are circles of constant normalized resistance with values ranging from zero (extreme far left) to infinity (extreme far right) on the chart with each circle having a variable reactance.

On the other hand, the set of circles centered on the vertical axis which is offset by one unit from the center, represents circles of constant normalized reactances with values ranging from $-\infty$ to $+\infty$ with each circle having a variable positive resistance. These are shown as partial-circles starting from the right-hand side of the chart and going above the real axis (representing normalized positive reactances) and below (representing normalized negative reactances); the center real axis (horizontal line) represents the zero reactance circle with an infinite radius.

The markings for the positive and negative normalized reactances can be seen on the Smith chart close to the outermost circle.

The key to understanding the Smith chart is realizing that it is a polar plot of the reflection coefficient:

$$\Gamma = |\Gamma| e^{j\theta}, \quad 0 \le \theta \le 180° \tag{5.12}$$

with the reference of zero degrees at the right side of the horizontal semi-axis.

All passive networks ($|\Gamma| \le 1$) have impedance values with $Re(Z) \ge 0$ which when normalized by the characteristic impedance of the transmission line (to which they are connected) can be represented uniquely on the Smith chart.

The real usefulness of the Smith chart lies in its ability to provide a one-to-one correspondence between reflection coefficient and input normalized impedance (or admittance) values.

Furthermore, moving a distance "ℓ" toward the load along a lossless transmission line corresponds to a change in the reflection coefficient by a factor of $e^{-2j\beta\ell}$ which corresponds to a counter-clockwise rotation of $2\beta\ell$ on the Smith chart as shown in Figure 5.5.

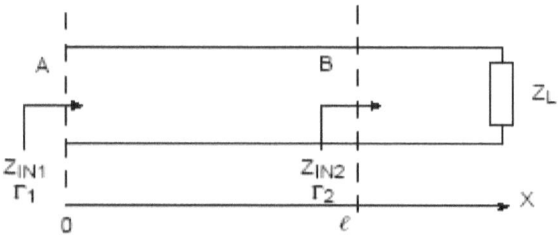

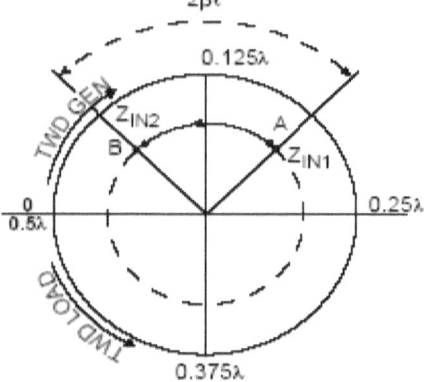

Figure 5.5 Traveling on a transmission line and its Smith Chart.

5.5 SMITH CHART'S CIRCULAR SCALES

Consider a standard smith chart as shown in Figure 5.4. Any specific normalized impedance (z=r+jx) value can be uniquely located on this chart for r≥0. The r-values would be on the resistance circles and x-values on the partial circles for reactance.   The positive reactance would be on the upper half of the chart whereas negative reactance values are plotted on the lower half of the chart.

Traveling on a transmission line (TL) toward the generator corresponds to moving clockwise on the Smith chart, whereas traveling toward the load corresponds to moving in a counter-clockwise direction as indicated by arrows on the left-hand side outer-edge of the Smith chart.

The phase relationship and electrical length along a transmission line are shown on the outer edge of the Smith chart in terms of two secondary scales: One is graduated in fractional wavelength (l/λ) and the other in degrees.

5.5.1 Wavelength Scale
Since the impedance value on a transmission line repeats itself every half wavelength, a complete revolution on the wavelength scale is equivalent to a half wavelength on the Smith chart.

5.5.2 Degree Scale
The degree scale goes through 180 degrees positive and 180 degrees negative with 0 degrees being on the right-hand semi-axis. This scale shows that in a complete revolution of the chart, the reflection coefficient's phase changes 360 degrees corresponding to a half wavelength on the wavelength scale.

These two scales (i.e. the wavelength & the degree scales) are important because they show that when a transmission line is terminated in a load impedance not equal to the line's characteristic impedance, the resulting impedance value on the line varies cyclically every half wavelength (this is of course due to the periodic nature of the standing wave pattern on the line). This fact is built into the Smith chart through these two scales and thus facilitates impedance calculations at various points along a line after the chart has been entered for a specific impedance value.

5.6 SMITH CHART'S RADIAL SCALES
A number of radially marked scales, at the bottom of the Smith chart, are placed in such a manner that they can be radially set off and their values read off from the center of the Smith chart by using a pair of dividers or compass. These scales are described as follows:

5.6.1 Reflection
Starting with the scale on the right-hand side, this scale is designed to show the ratio of the reflected wave to the incident wave and is further sub-divided into four scales in the following manner:

1. REFL. COEF. is the reflection coefficient and has the following two sub-scales:

a. **VOL:** is the voltage reflection coefficient magnitude and is defined as:

$$|\Gamma|=|V^-/V^+| \qquad (5.13)$$

This scale starts from 0 at the center and ends at 1 at the outer rim of the chart.

b. **PWR:** is the power reflection coefficient and is defined to be:

$$|\Gamma|^2=|V^-/V^+|^2 \qquad (5.14)$$

Similar to 1(a) above, this scale starts at 0 at the center and ends at 1 at the outer edge of the chart.

2. LOSS IN DB is the loss due to reflection and is expressed in dB with the following two sub-scales:

a. **RETN:** is the return loss and is defined as the ratio of the incident power to the reflected power at any point on the transmission line, expressed in dB and is equal to:

$$R_{loss}(dB)= 10 \log_{10}(P_i/P_O)=10 \log_{10} (1/|\Gamma|^2)$$
$$\Rightarrow R_{loss}(dB)= -20 \log10|\Gamma| \qquad (5.15)$$

This scale starts from 0 (corresponding to $|\Gamma|= 1$) at the outer edge of the chart and approaches infinity at the center of the chart (where $|\Gamma|=0$) which indicates that the more perfect the load, the less reflection from the load and higher the return loss.

b. **REFL:** (Reflected loss or Mismatch loss) is the loss caused by reflection and is equal to the ratio of incident power to the difference between incident and reflected power expressed in decibels as follows:

$$M_{loss}(dB)=10 \log_{10}[P_i/(P_i-P_O)]= -10 \log_{10}(1-|\Gamma|^2) \qquad (5.16)$$

This scale starts from zero at the center and approaches to infinity as $|\Gamma|$ approaches unity at the outer edge of the chart.

5.6.2 TRANSM Loss

TRANSM Loss: is the transmission loss and is used primarily for lossy transmission lines and has two scales:

a. **LOSS COEF:** is the transmission loss coefficient and is used as a correction factor for the additional line losses created in a lossy transmission line due to high VSWR. A high VSWR on a line creates peaks of high current densities alternated with high voltage density peaks. Since resistive losses are proportional to the current value squared and dielectric losses are proportional to voltage value squared, the locale where these peaks of energy lie create additional losses on the line which is not accounted for through ordinary calculations.

Thus a correction factor is needed to provide a more accurate estimation of line losses when a high VSWR exists on the lossy transmission line. The "LOSS COEF" scale provides the much-needed correction factor when the VSWR on the line is greater than unity. The correction factor provided by this scale would increase the calculated line losses which will affect the attenuation factor calculations.

For example, when the VSWR is 1 (i.e. A matched case) the correction factor from this scale is read to be one. On the other hand, when the VSWR of the line is increased to 4 (due to a load mismatch), then the correction factor is read off to be approximately 2.1, which means that the line losses have more than doubled due to this high VSWR.

b. 1 DB STEPS: is the transmission loss in 1-dB steps and is used to calculate VSWR on a lossy transmission line. Graphically, a lossy line can no longer be represented by a constant VSWR circle, instead by a spiral on the Smith chart due to the attenuation of both the incident wave's amplitude traveling "toward the load" and the reflected wave's amplitude back to the generator. This power loss is shown in Figure 5.6. From Figure 5.6, it can be seen that:

$$|\Gamma_g|^2 = P_r/P_i \quad \text{(at the source end)} \tag{5.17}$$

$$|\Gamma_L|^2 = P_r'/P_i' \quad \text{(at the load end)} \tag{5.18}$$
$$\Rightarrow |\Gamma_g| < |\Gamma_L|,$$

This would hold true as long as the line remains lossy.

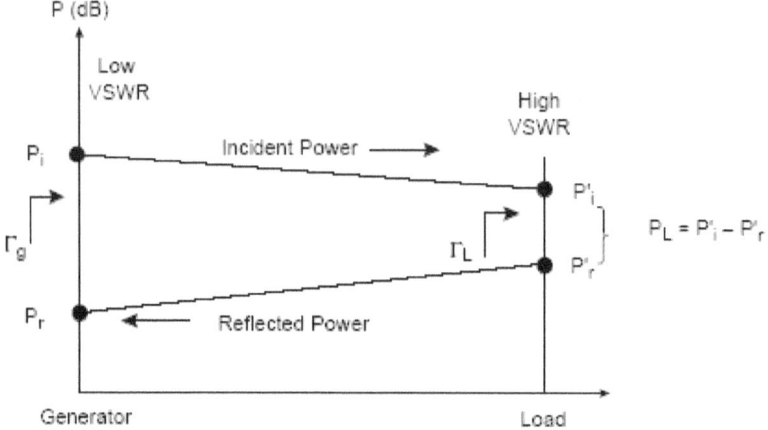

FIGURE 9.6   Power attenuation on a lossy transmission line.

Figure 5.6 Power attenuation on a lossy TL.

It is important to note that as the measurement plane moves toward the generator and away from the load, the reflection coefficient becomes smaller and thus VSWR is reduced as illustrated in the next example.

EXAMPLE 5.1

Consider an unknown load connected to a lossy 50 Ω cable (with 2 dB of insertion loss) connected to a generator. The VSWR at the generator end is measured to be 2.0. What is the VSWR at the load?

Solution:

We first plot the VSWR circles at the generator end by dropping a vertical line from the constant-VSWR circle to intersect the "1 dB Step" scale at point A (see Figure 5.7). Now we add 2 dB correction to this value "toward load" as indicated on the scale to obtain point B. The radius related to point B is that of the load VSWR and is found by drawing a vertical line from point B to intersect the left-hand semi-axis on the Smith chart. By swinging this radius around, it is seen that the new VSWR circle has a VSWR=3.2 at the load end.

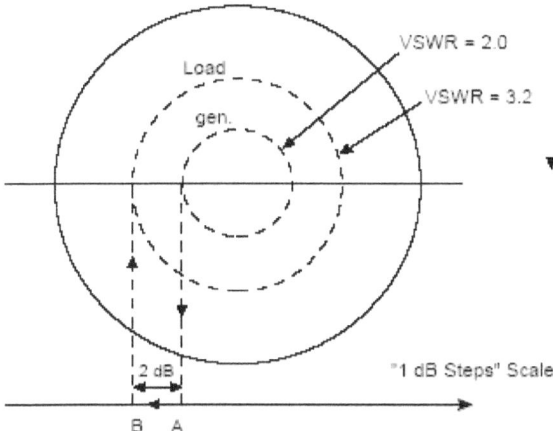

Figure 5.7 Use of "1 dB steps" scale in a Smith Chart.

As can be seen from Example 5.1, a lossy line can improve the VSWR at the generator end at the expense of power loss, which may not always be desirable.

We also note from this example that moving away from the center (i.e. higher $|\Gamma|$) on this scale is labeled as "Toward Load" while moving toward the center of the chart (i.e. lower $|\Gamma|$) is labeled as "Toward generator".

NOTE: *Incidentally, it is interesting to note that the values on "the transmission loss (in one dB steps)" are one-half of the values on the "Return loss in dB" scales. This factor of "One-half" is caused by the fact that the "return loss scale" indicates two-way power attenuation through a given piece of cable, whereas the transmission loss is defined as merely a one-way attenuation loss.*

5.6.3 Standing Wave

This scale shows the voltage standing wave ratio (VSWR) as follows:

1. **VOL. RATIO**: (voltage ratio) this scale plots the VSWR as a ratio of maximum voltage to minimum voltage as given by the following equation:

$$\text{VSWR} = \frac{V_{max}}{V_{min}} = \frac{1+|\Gamma|}{1-|\Gamma|}$$
(5.19)

The "VOL. RATIO" scale progresses from 1 at the center of the chart ($|\Gamma|= 0$) to infinity at the left-hand margin ($|\Gamma|= 1$).

2. **IN DB**: This scale expresses VSWR in dB by the relation:
   $(\text{VSWR})_{dB}=20 \log_{10} (\text{VSWR})_{ratio}$
(5.20)

EXAMPLE 5.2
What does VSWR= 2.0 on the "Voltage ratio" scale correspond in dB?
Solution:
Using the adjacent dB scale, we read a value of 6.0 dB on it.

In the next chapter we will discuss the applications of the smith chart which are of great importance to the design of RF and microwave circuits.

5.7 THE NORMALIZED ZY SMITH CHART
By superimposing two Smith charts, with one 180° rotated, we obtain a normalized impedance-admittance Smith chart (also known as a ZY Smith chart) as shown in Figure 5.8. The rotated represents the admittance, whereas the other chart represents impedance. The proof for 180° chart rotation to obtain admittance values, is presented in Chapter 10 (see application #4).

The ZY Smith chart has therefore two markings: one for impedance chart and another for the admittance chart. Symbols $-X_S$ and $+X_S$ are used on the left-hand side for the impedance chart and $-B_P$ and $+B_P$ are used for admittance chart, respectively. From these markings, we can see that positive reactances ($+X_S$) are on the upper half of the chart while negative reactances ($-X_S$) are on the lower half of the chart, respectively. This situation is reversed for the admittance chart, where positive susceptances ($+B_P$) are located on the lower half and the negative susceptances ($-B_P$) are on the upper half of the chart.

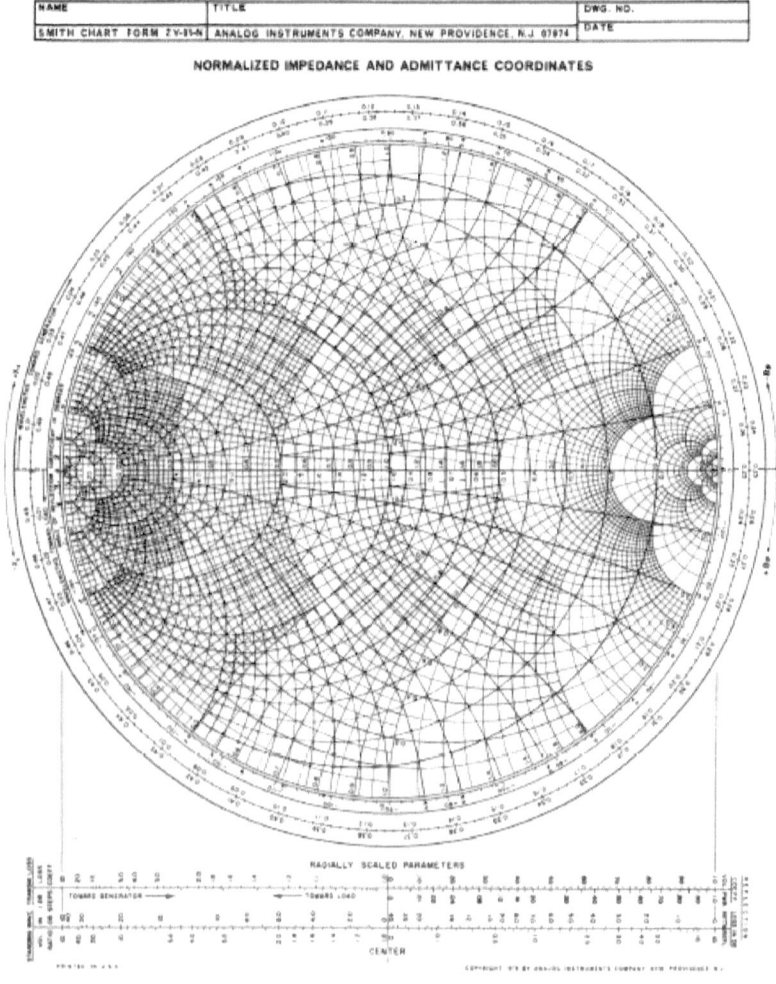

Figure 5.8 The ZY Smith Chart.

Each point on a ZY Smith chart represents the impedance and the corresponding admittance value simultaneously, whereby one can read off these values by a simple glance at the chart. This means that given an impedance (or admittance) value, its corresponding admittance (or impedance) value can readily be read off from the chart without any resort to calculations. This is an important feature and a major improvement over a standard Smith chart, since it greatly facilitates the circuit design process, particularly where complicated designs are desired. As will be seen in Chapter 11, the ZY Smith chart is an essential analytical tool and will be extensively used for RF/Microwave circuit design applications

Chapter 5- Symbol List

A symbol will not be repeated again, once it has been identified and defined in an earlier chapter, with its definition remaining unchanged.

P_i – Power incident

P_r – power reflected

PWR – Power reflection coefficient

REFL – Reflection loss or mismatch loss

REFL COEF – Reflection coefficient

RETN – Return loss

R_{loss} - Return loss

R_N - Normalized resistance

VOL – Voltage reflection coefficient magnitude

VSWR – Voltage Standing Wave Ratio

X_N - Normalize reactance

Z_0 – Characteristic impedance

Z_N – Normalized impedance

Γ_g - Reflection coefficient at the generator/source

Γ_L - Reflection coefficient at the load

CHAPTER-5 PROBLEMS

5.1)   What is a standard Smith chart? What range of resistor and reactive values is mapped into a standard Smith chart?

5.2)   What resistor values get mapped into a compressed Smith chart? Show by drawing a diagram.

5.3)   A lossless transmission line (Z_O=50 Ω) is connected to a load Z_L=100+j100 Ω. Using a Smith chart:

a) Determine the reflection coefficient at the load

b) Calculate the return loss.

c) Find the VSWR on the line.

d) Determine the reflection coefficient and the input impedance $\lambda/8$ away from the load.

5.4)   Using a smith chart find Z_L for:

a) $\Gamma=0.6\ e^{j45°}$

b) $\Gamma= -0.3$

c) $\Gamma=0.4+j0.4$

5.5)   VSWR on a lossless transmission line ($Z_O=50\ \Omega$) is measured to be 3.0. Using a Smith chart determine:

a) The magnitude of the reflection coefficient in "ratio" and in "dB".

b) The return loss in dB.

c) The mismatch loss in dB

d) If the load is resistive($R>Z_O$) and is located $3\lambda/8$ away from the source, determine the load impedance value, the input impedance of the transmission line and the reflection coefficients at the load and at the source (see Figure P5.5).

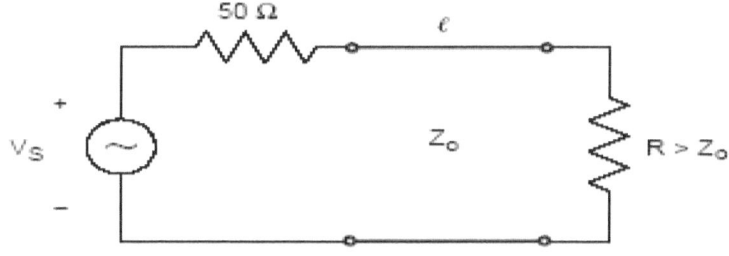

Figure P5.5

REFERENCES

[5.1] Cheng, D. K. *Field and Wave Electromagnetics,* 2nd ed., Reading: Addison Wesley, 1989.

[5.2] Cheung, W. S. and F. H. Levien. *Microwave Made Simple, Principles and Applications.* Norwood: Artech House, 1985.

[5.3] Ginzton, E. L. *Microwave Measurements,* New York:

McGraw-Hill, 1957.

[5.4] Gonzalez, G. *Microwave Transistor Amplifiers, Analysis and Design,* 2nd ed. Upper Saddle River: Prentice Hall, 1997.

[5.5] Kosow, I. W. and Hewlett-Packard Engineering Staff, *Microwave Theory and Measurement.* Englewood Cliffs: Prentice Hall, 1962.

[5.6] Reich, H. J., F. O. Phillip, H. L. Krauss, and J. G. Skalnik, *Microwave Theory and Techniques,* New York: D. Van Norstrand Company, Inc., 1953.

[5.7] Schwarz, S. E. *Electromagnetics for Engineers,* Orlando: Saunders College Publishing, 1990.

[5.8] Smith, P. H. *Transmission-Line Calculator,* Electronics, 12, pp 29-31, Jan. 1939.

[5.9] Smith, P. H. *An Improved Transmission-Line Calculator,* Electronics, 17, pp 130, Jan. 1944.

CHAPTER 6

Smith Chart Applications

6.1 INTRODUCTION

The Smith chart applications in the analysis or design of RF and microwave circuits can be subdivided into three categories:

a. Circuits containing primarily "distributed elements", particularly transmission lines (TL's).

b. Circuits containing "lumped elements".

c. Circuits containing "Distributed and lumped elements" in combination.

6.2 DISTRIBUTED CIRCUIT APPLICATIONS

The most common distributed circuit element is a transmission line (TL) and the Smith chart can be used effectively for calculation of values of its different parameters. Before we proceed into different Smith chart applications, it would serve us well, at the outset, if we define, the following notations which will be used throughout this book:

Impedance: $\mathbf{Z=R+jX}$ (Ω) (6.1)

Admittance: $\mathbf{Y=G+jB}$ $(\mathbf{S})$ (6.2)

The normalized values are given by:

$(\mathbf{Z})_N = \mathbf{Z/Z_O} = \mathbf{R/Z_O + jX/Z_O} = \mathbf{r+jx}$ (6.3)

$(Y)_N=Y/Y_O=G/Y_O +jB/Y_O=g+jb$ (6.4)

Where,

$r=R/Z_O,$ (6.5)
$x=X/Z_O,$ (6.6)
$g=G/Z_O,$ (6.7)
$b=B/Z_O$ (6.8)
$Y_O=1/Z_O$ (6.9)

And "Z_O" is the characteristic impedance of the transmission line or a reference impedance value.

6.2.1 Application #1:

INPUT IMPEDANCE (Z_{IN}) DETERMINATION USING A KNOWN LOAD (Z_L)

The input impedance (Z_{in}) at any point on a transmission line, a distance "l" away from the load (Z_L), can be calculated by the following procedure:

a. Plot the normalized load impedance $[(Z_L)_N=Z_L/Z_0)]$ on the Smith chart,

b. Draw the constant VSWR circle that goes through $(Z_L)_N$,

c. Starting from $(Z_L)_N$, move "toward generator" on the constant VSWR circle a distance "l/λ",

d. Read off the normalized input impedance value (Z_{in}/Z_o) from the chart as shown in Figure 6.1.

This process can be reversed easily when the input impedance (Z_{in}) is known and the load impedance is unknown (Z_L). In this case, starting from $(Z_{in})_N$, one moves "Toward load" a distance "l/λ" on the constant VSWR circle to arrive at $(Z_L)_N$.

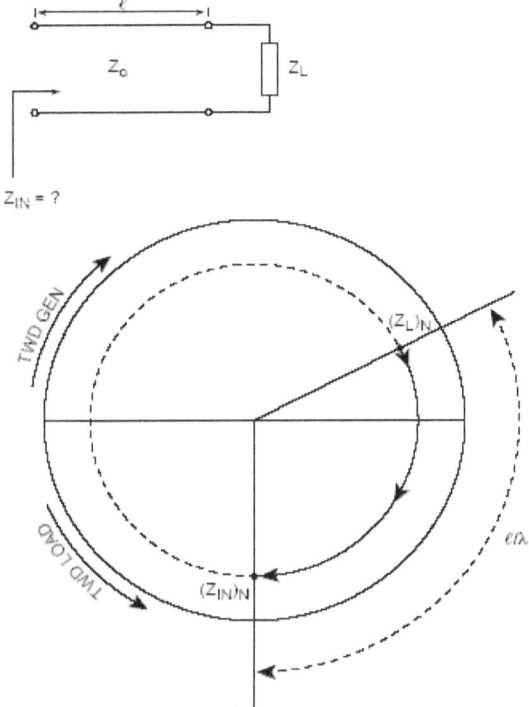

Figure 6.1 Smith Chart plot.

EXAMPLE 6.1

Find the input impedance of a transmission line ($Z_O=50\ \Omega$) that has a length of $\lambda/8$ and is connected to a load impedance of $Z_L=50+j50\Omega$?

Solution:

a. Locate $(Z_L)_N=Z_L/Z_O=1+j1$ on the smith chart.

b. Draw the constant VSWR circle as shown in Figure 6.2.

c. Now move "toward Generator" on the constant VSWR Circle a distance of $\lambda/8$ (or $90°$) to obtain:

$(Z_{in})_N=2-j1 \Rightarrow Z_{in}=Z_O(Z_{in})_N=100-j50\ \Omega$.

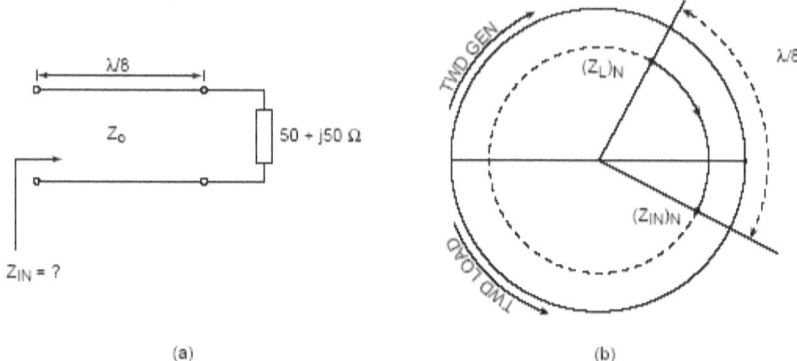

(a)   (b)

Figure 6.2 Circuit diagram and its Smith Chart.

6.2.2 Application #2

INPUT   IMPEDANCE   DETERMINATION   USING   THE   **INPUT**
REFLECTION COEFFICIENT ($|\Gamma_{IN}| \leq 1$)

When the reflection coefficient at any point on a transmission line is known, the input impedance at that point can be calculated as follows:

a. Locate $\Gamma_{IN} = |\Gamma_{IN}| e^{j\theta}$ on the Smith chart; The magnitude of $|\Gamma|$ can be read off the "Reflection coefficient voltage" radial scale at the bottom of the chart while "θ" is read off the circular scale (See Figure 6.3).

b. Normalized values of resistance and reactance (r,x) can be read off the Smith chart at point "A", giving Z_{in} as:
$$Z_{in} = Z_o(r+jx)$$

NOTE 1: *If conversely, the input impedance (Z_{in}) is known and the corresponding reflection coefficient is desired to be found, the procedure would be as follows:*

a. Plot the normalized input impedance (Z_{in})$_N$ on the Smith chart and read off the angle "θ" on the circular scale.
b. Draw the constant VSWR circle,
c. The intersection of this circle with the right-hand horizontal axis is found and dropped off onto the "reflection coef." radial scale at the bottom and the $|\Gamma|$ value is read off as shown in Figure 6.3.

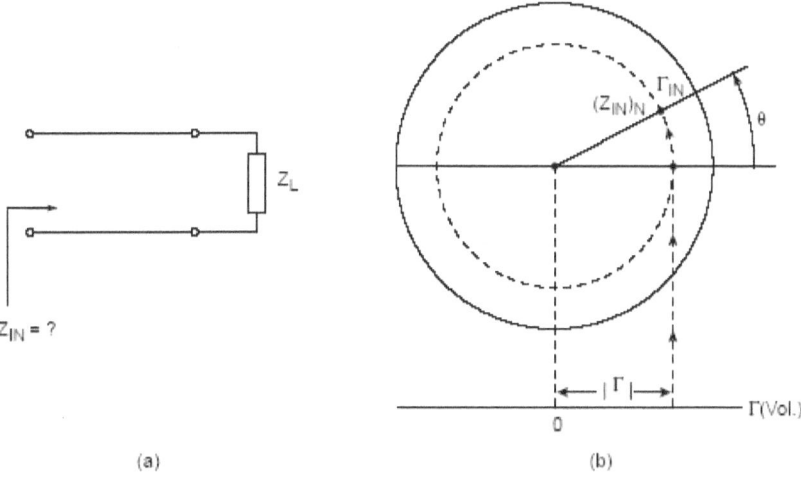

Figure 6.3 Impedance on the Smith Chart.

NOTE 2: *If the value of Z_{in} at a distance ℓ from the reflection coefficient location is sought, one needs to use the procedure described in application #1.*

6.2.3 Application #3

IMPEDANCE DETERMINATION USING REFLECTION COEFFICIENT WHEN $|\Gamma| > 1$

When the magnitude of the reflection coefficient is greater than unity, the corresponding impedance has a negative resistance value and thus maps outside the standard Smith chart. In this case, another type of chart called a compressed Smith chart (as discussed earlier) should be used. This chart includes the standard Smith chart ($|\Gamma| \leq 1$) and ($|\Gamma| > 1$) region which corresponds to the negative resistance region.

An alternate way of determining an impedance (Z) having ($|\Gamma| > 1$), is by using a standard smith chart with the help of the following procedure:

a. Obtain the complex conjugate of the reflection coefficient at point "B", ($\Gamma^* = |\Gamma| \angle -\theta$).

b. Plot $1/\Gamma^*$ on the standard Smith chart (see point "C" in Figure 6.4)

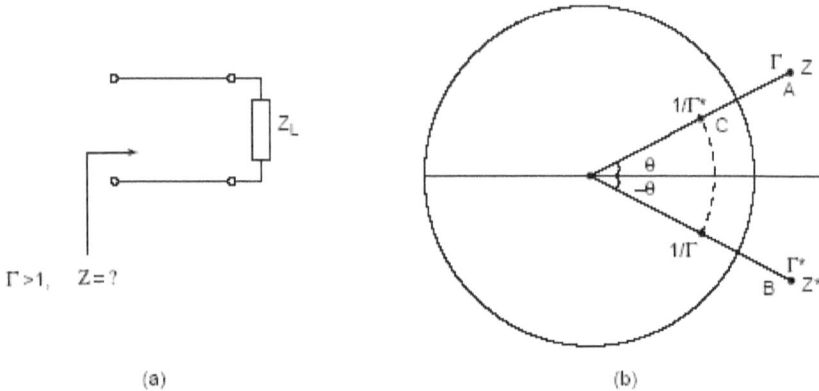

Figure 6.4 Smith Chart solution.

c. Read off the normalized impedance value (r+jx, corresponding to $1/\Gamma^*$) on the Smith chart.

d. The impedance (Z) value corresponding to Γ is obtained by negating "r" and keeping "x" intact, i.e.,
   $$Z=Z_O(-r+jx)$$

This procedure can be proven as shown below.

PROOF:
Assuming Z_O is a real number, the normalized impedance (Z/Z_O) corresponding to Γ is given by (see point A in Figure 6.4):
$$Z/Z_O= -r+jx, \quad r>0 \tag{6.10}$$
Where "r" and "x" are normalized values of resistance and reactance, respectively.

Knowing that $\Gamma=|\Gamma|\angle\theta$, we can write:
$$\Gamma \leftrightarrow -r+jx$$
$$\Gamma=(Z-Z_o)/(Z+Z_O)=(-r+jx-1)/(-r+jx+1) \tag{6.11}$$
$$\Gamma^* =|\Gamma|\angle-\theta$$
$$=(r+jx+1)/(r+jx-1) \tag{6.12}$$

Thus we have:
$$1/\Gamma^* =\frac{1}{|\Gamma|}\angle\theta \tag{6.13a}$$

$1/\Gamma^* =(r+jx-1)/(r+jx+1) =(Z'/Z_o-1)/(Z'/Z_o+1)$ (6.13b)

Where $Z'/Z_o=r+jx$ is the impedance corresponding to $1/\Gamma^*$, that is,

$1/\Gamma^* \leftrightarrow r+jx$ (6.14)

From Equation (6.13a) we can see that $1/\Gamma^*$ (shown at point C in Figure 6.4) has the same angle as Γ, namely, they are on the same vector. Therefore from Equations (6.11) and (6.14) we conclude that the impedance corresponding to Γ is simply obtained by reversing the sign of the real part of Z', i.e.,

$Z=Z'|_{(r,x) \rightarrow (-r,x)}$ (6.15)

EXAMPLE 6.2

What is the impedance (Z_D) of a device having $\Gamma_D=2.23\angle 26.5°$? Assume $Z_O=50\ \Omega$.

Solution:
a.   We find $\Gamma_D^* =2.23\angle -26.5°$
b.   Plot $1/\Gamma_D^* =0.447\angle 26.5°$ on the smith chart. From the chart we obtain:
     $Z_D'=50(2+j1)=100+j50\ \Omega$
c.   Using Z_D' from step (b), we can write Z_D as:
     $Z_D= -100+j50\ \Omega$

6.2.4 Application #4
DETERMINATION OF ADMITTANCE (Y) FROM IMPEDANCE (Z)

As discussed in Chapter 7, we know that the reflection coefficient $[\Gamma(x)]$, the normalized input impedance $[Z_N(x)]$ and the normalized input admittance $[Y_N(x)]$ at any point on the line are given by:

$Z_N(x)=[1+\Gamma(x)]/[1-\Gamma(x)]$ (6.16)

and,

$Y_N(x)= 1/Z_N(x)=[1-\Gamma(x)]/[1+\Gamma(x)]$ (6.17)

Where,

$\Gamma(x)= \Gamma_L\ e^{j2\beta x}$ (6.18)

is the reflection coefficient at any point (x) on the transmission line.

From the expression for $\Gamma(x)$, we note that for every phase change of $2\beta l = \pi$ (i.e. every $\ell = \lambda/4$), $\Gamma(x)$ changes sign which leads to the inversion of the expressions given in (6.17) and (6.18) causing $Z_N(x)$ to become $Y_N(x)$ and vice versa. This observation indicates that Y_N is located 180 degrees opposite to Z_N on the VSWR circle as shown below in Figure 6.5.

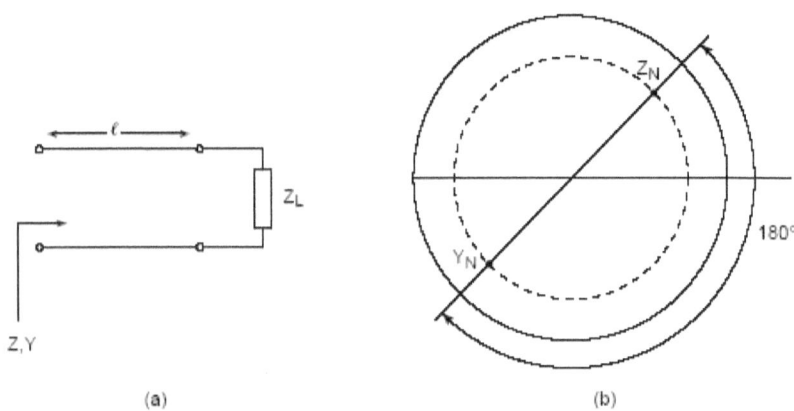

(a)   (b)

Figure 6.5 Smith Chart solution.

Example 6.3

Find the admittance value for an impedance value of Z= 50+j50 Ω, in a 50 Ω system.

Solution:

$Z_O = 50 \ \Omega \Rightarrow Y_O = 1/50 = 0.02 \ S$

$Z_N = Z/Z_O = 1+j1$

Using the smith chart, Y_N can be read off at 180° away on the constant VSWR circle:

$Y_N = 0.5-j0.5$

$Y = Y_O Y_N \Rightarrow Y = 0.01-j0.01 \ S$

NOTE 1: *Z to Y conversion can also be obtained by rotating the Z-chart by 180° and super-imposing it on the original chart, which will give a ZY-Smith chart. The Y-chart has negative susceptance on the* *upper half and positive susceptance on the lower half, exactly opposite of the Z-chart. The Z-Y chart is shown in Figure 6.6.*

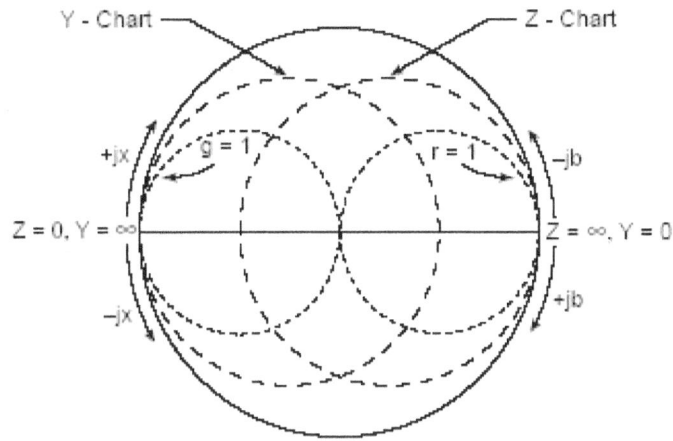

Figure 6.6 ZY Smith Chart.

NOTE 2: *The standard Smith chart may be considered to be a Y- or Z-chart depending on the first time of entrance of values in it, being either admittance (Y) or impedance (Z).*

NOTE 3: *When working with **series elements,** the concept of impedance becomes important and we need to use the **Z-chart.** On the other hand when working with **parallel (or shunt) elements,** the concept of admittance becomes paramount and therefore we switch to the **Y-chart.**

6.2.5 Application #5
DETERMINATION OF Z_{MAX} AND Z_{MIN} FROM A KNOWN LOAD (Z_L)

a. Z_{max} and Z_{min} Value and location
Given a known load $(Z_L)_N$, the VSWR circle can be drawn (see Figure 6.7). Furthermore, using the results for Γ and Z_{in} from Chapter 7, we can write:

$$\Gamma(x) = \Gamma_L\, e^{j2\beta X}, \tag{6.19}$$

Where $\Gamma_L = |\Gamma_L|\, e^{j\theta}$.
Therefore we can write:

$\Gamma(x) = |\Gamma_L| e^{j\phi(x)}$, $\phi(x) = 2\beta x + \theta$ (6.20)

and,

$[Z_{in}(x)]_N = Z_{in}(x)/Z_O = [1+\Gamma(x)]/[1-\Gamma(x)]$ (6.21)

From Equation (6.21) we note that maximum input impedance $(Z_{max})_N$ occurs when the numerator is maximum and denominator is minimum. By observation, this condition occurs when $\Gamma(x) = |\Gamma_L| e^{j\phi(x)}$ is a positive real number i.e. $\phi(x) = 0$, which gives $(Z_{max})_N$ as:

$(Z_{max})_N = [1+|\Gamma_L|]/[1-|\Gamma_L|]$ (6.22)

This value can be read off the chart at the intersection of the VSWR circle with the right-hand horizontal axis (where $\phi=0$) as shown in Figure 6.7.

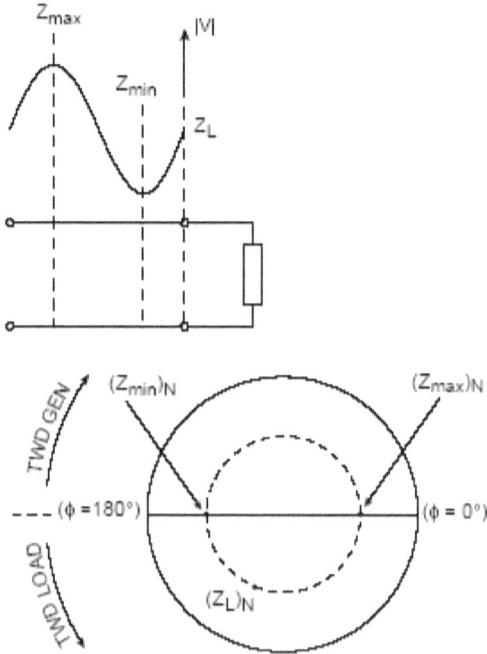

Figure 6.7 Smith Chart solution.

From Equation (6.21) we can observe that minimum input impedance $(Z_{min})_N$, being the inverse of $(Z_{max})_N$, occurs when $\Gamma(x)$ is a negative real number (i.e., $\phi=180°$):

$(Z_{min})_N = 1/(Z_{max})_N$ $= [1-|\Gamma_L|]/[1+|\Gamma_L|]$ (6.23)

This value can be read off the chart where the VSWR circle intersects the left semi-axis, (see Figure 6.7)

NOTE: *Since $(Z_{min})_N= 1/(Z_{max})_N=(Y_{max})_N$, thus the value and location of $(Z_{min})_N$ could have easily been found (using the application #4) by locating it 180° away from $(Z_{max})_N$ on the VSWR circle.*
b. Z_{max} and Z_{min} location (distance from load)

- Z_{max} distance from load
Starting from the load on the VSWR circle, we now move "Toward generator" to arrive at $\phi = 0°$ where $(Z_{max})_N$ is located. The distance "ℓ_{max}" can now be read off using the circular scale on the outer edge of the Smith chart as shown in Figure 6.8.

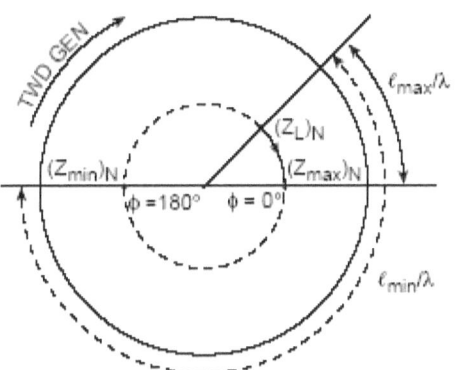

Figure 6.8 Smith Chart solution.

-Z_{min} distance from load
Starting from the load, we now travel "Toward generator" on the VSWR circle to arrive at $\phi = 180°$ where $(Z_{min})_N$ is located. The distance traveled is "ℓ_{min}/λ" which can be read off on the outer circular scale (see Figure 6.8).

From the Smith Chart, we can observe that the length difference between the locations of Z_{max} and Z_{min} is $\lambda/4$, i.e.
$$|\ell_{max}-\ell_{min}|=\lambda/4 \tag{6.24}$$
This observation is further confirmed by our earlier discussion of the transmission line and Smith chart where we noted that the input

impedance at any point on a line (e.g. Z_{max}) repeats itself every $\lambda/2$. Since Z_{min} is located one half of the distance between the two repeating maxima, thus the distance between Z_{max} and Z_{min} should be $\lambda/4$ (as shown in Figure 6.9), as indicated by (6.15).

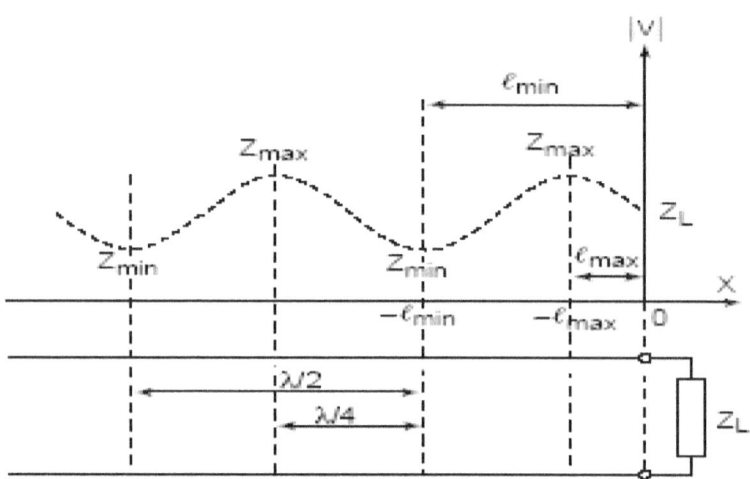

Figure 6.9 Voltage standing wave plot on a TL.

For example, if ℓ_{max} is known and is nearest to the load, then ℓ_{min} is simply given by:

$$\ell_{min} = \ell_{max} + \lambda/4 \tag{6.25}$$

Without having to resort to the chart; and vice versa if ℓ_{min} is known and is nearest to the load then :

$$\ell_{max} = \ell_{min} + \lambda/4 \tag{6.26}$$

A SPECIAL CASE: LINES WITH PURELY RESISTIVE LOADS

When a transmission line with a real characteristic impedance (Z_O) is terminated in a resistive load (i.e. has no reactive component, $Z_L = R_L$), then there are two possible cases:

Case I: $R_L > Z_O$

In this case, the maximum impedance on the line equals the load value and is located at the load repeating every $\lambda/2$, that is:

$$Z_{max} = R_L, \tag{6.27}$$

$\ell_{max}=n(\lambda/2), \quad n=0,1,2,...$ (6.28)

The minimum line impedance (Z_{min}) is located $\lambda/4$ away from the load and repeats itself every $\lambda/2$, i.e.,

$\ell_{min}=\lambda/4 + n(\lambda/2), \quad n=0,1,2,...$ (6.29)

Furthermore, we can write:

$(Z_{min})_N=1/(Z_{max})_N \Rightarrow Z_{min}/Z_O=Z_O/Z_{max}$

$Z_{min}=Z_O^2/Z_{max}$ (6.30)

Or,

$Z_{min}=Z_O^2/R_L$ (6.31)

These are shown in Figure 6.10.

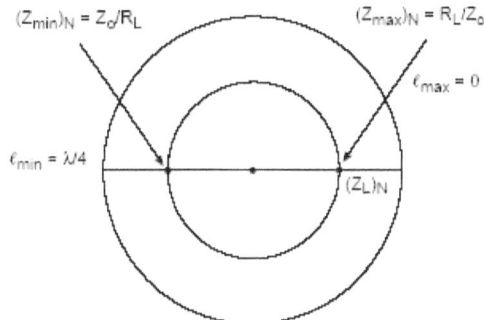

Figure 6.10 Smith Chart solution.

EXAMPLE 6.4

A microwave signal is traveling on a transmission line which has $Z_O= 50 \; \Omega$ and a load value of $Z_L = 100 \; \Omega$. Find the values of Z_{max} and Z_{min} and plot their locations on the transmission line.

Solution:

Since $R_L>Z_O$, the maximum voltage and thus maximum impedance occurs at the load, i.e.,

$Z_{max}=R_L=100 \; \Omega$

$\ell_{max}=0,$

The minimum impedance occurs $\lambda/4$ away from the load :

$\ell_{min}=\lambda/4$

$Z_{min}=50^2/100=25 \; \Omega$

This is shown in Figure 6.11.

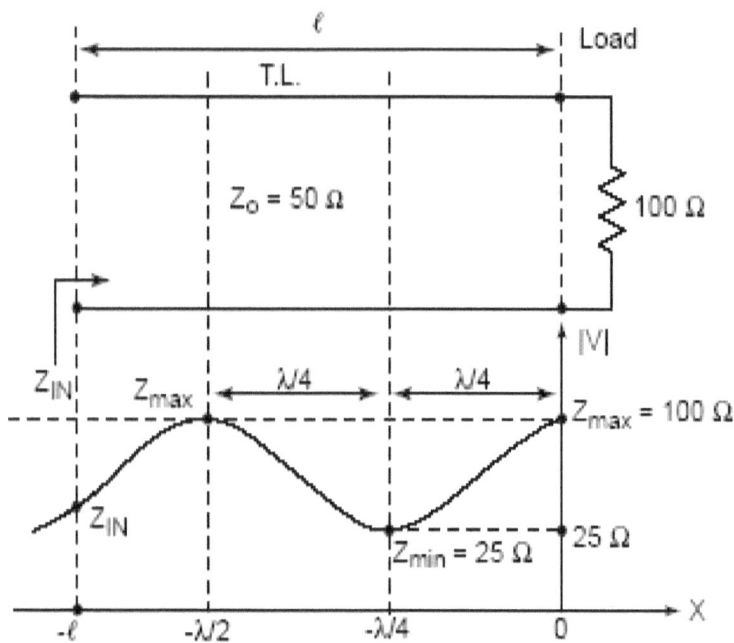

Figure 6.11 Impedance on a circuit diagram and its plot.

Case II: $R_L < Z_O$

In this case, the minimum impedance on the line is located at the load and repeats every $\lambda/2$, i.e.,

$$Z_{min} = R_L,$$ (6.32)

$$\ell_{min} = n(\lambda/2), \quad n = 0,1,2,...$$ (6.33)

The maximum line impedance is located $\lambda/4$ away from the load and also repeats every $\lambda/2$ as shown in Figure 6.12. Thus using Equation (6.22) we can write:

$$(Z_{min})_N = 1/(Z_{max})_N$$

$$Z_{max} = Z_O^2/Z_{min} = Z_O^2/R_L,$$ (6.34)

$$\ell_{max} = \lambda/4 + n(\lambda/2), \quad n = 0,1,2,...$$ (6.35)

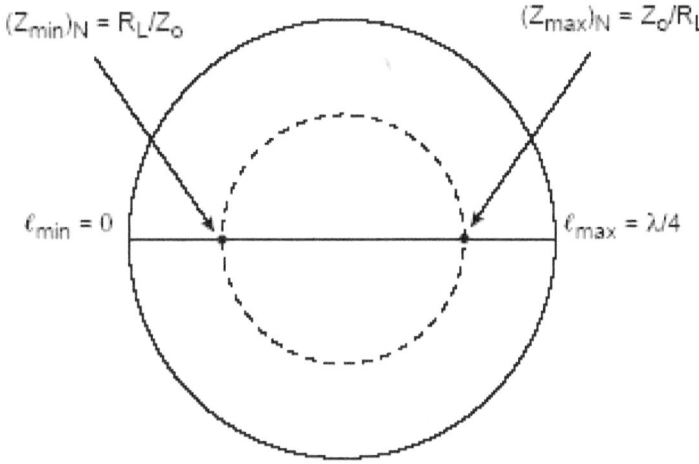

Figure 6.12 Smith Chart solution.

EXAMPLE 6.5

A microwave signal, at a frequency of f=1 GHz, is traveling on a transmission line having Z_O= 50 Ω, and terminated in a load of Z_L = 20 Ω. Find the values of Z_{max} and Z_{min} and their location on the transmission line.

Solution:

Since $R_L<Z_O$, the minimum voltage or impedance on the line occurs at the load, i.e.,

Z_{min}=20 Ω,

ℓ_{min}=0.

From Equation (6.34) and (6.35), the value and location of Z_{max} is given by:

$Z_{max}=Z_O^2/R_L=50^2/20=125\ \Omega$

$\lambda=c/f \Rightarrow \lambda(cm)=30/f(GHz)=30$ cm

The first maximum occurs at: $\ell_{max}= \lambda/4=7.5$ cm away from the load as shown in Figure 6.13.

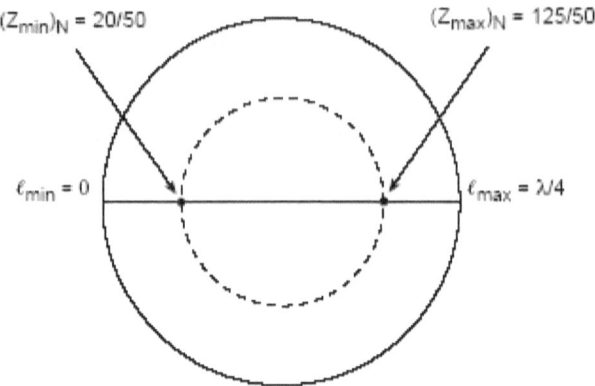

Figure 6.13 Smith Chart solution.

The standing wave pattern is plotted in Figure 6.14.

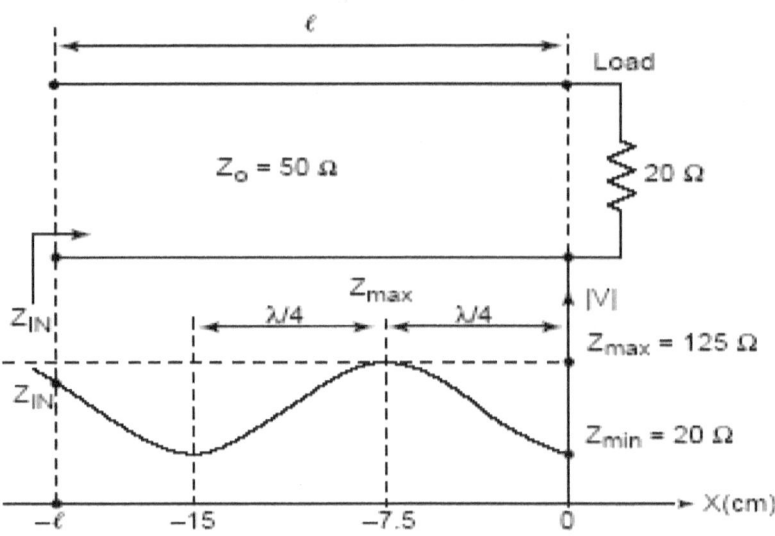

Figure 6.14 Standing-wave plot on a TL.

6.2.6 Application #6

DETERMINATION OF VSWR USING A KNOWN LOAD

There are two methods to find the VSWR on a transmission line with a given load (Z_L or Γ_L) as follows:

Method #1:

Using $|\Gamma_L|$ as the radius, draw the constant VSWR circle. From the intersection of this circle with the left-hand horizontal axis drop a vertical line onto the VSWR scale on the bottom of the chart to find the VSWR as shown in Figure 6.15.

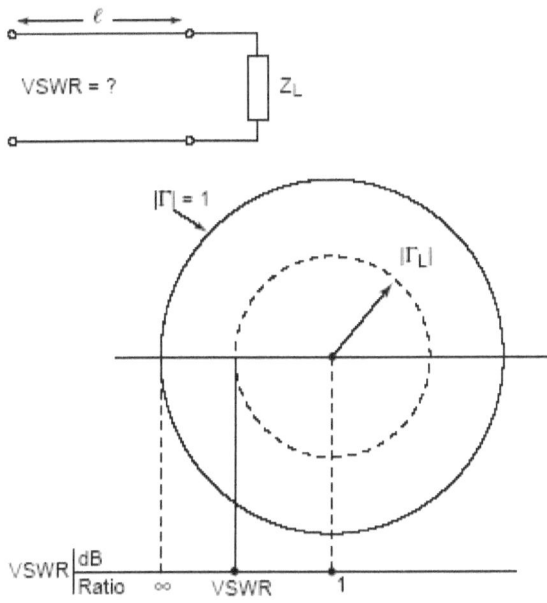

Figure 6.15 Smith Chart Solution.

Method #2:

For a lossless transmission line we know that $|\Gamma|=|\Gamma_L|$ anywhere on the line, which is the radius of the VSWR circle. Therefore VSWR can be calculated by:

$$VSWR = \frac{1+|\Gamma_L|}{1-|\Gamma_L|} \qquad (6.36)$$

$$|\Gamma_L| = |\frac{Z_L - Z_O}{Z_L + Z_O}| \qquad (6.37a)$$

From Application #5 we have:

$$(Z_{max})_N = 1/(Z_{min})_N = \frac{1+|\Gamma_L|}{1-|\Gamma_L|} = VSWR \qquad (6.37b)$$

Thus:

$$Z_{max} = Z_O(VSWR), \qquad (6.38a)$$

and,

$$Z_{min}=Z_O/VSWR \hspace{4cm} (6.38b)$$

The VSWR value can be read off from the VSWR circle intersection with the horizontal semi-axis at $\theta = 0°$ (where Z_{max} is located) as shown in Figure 6.16.

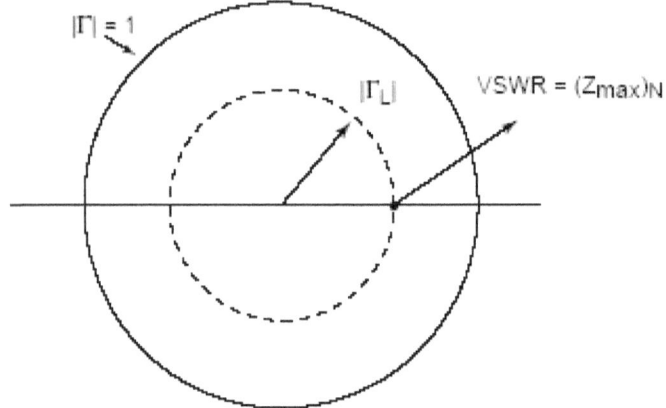

Figure 6.16 Smith Chart Solution.

6.2.7 Application #7

PLOT OF STANDING WAVE PATTERN USING A KNOWN LOAD

As discussed in Chapter 7, when the incident wave encounters a discontinuity of any kind (such as a load), which is different than the characteristic impedance of the propagating media, a portion (or all) of the wave will be reflected.

The reflected wave when combined with the incident wave will create a standing wave pattern in voltage and current.

Voltage and current waves simultaneously coexist and each has its own standing wave pattern with peaks and valleys occurring at different points along the line as shown in Figure 6.17.

The use of the Smith Chart will help us determine the exact standing wave pattern for voltage and current for a known load with a relatively good degree of accuracy.

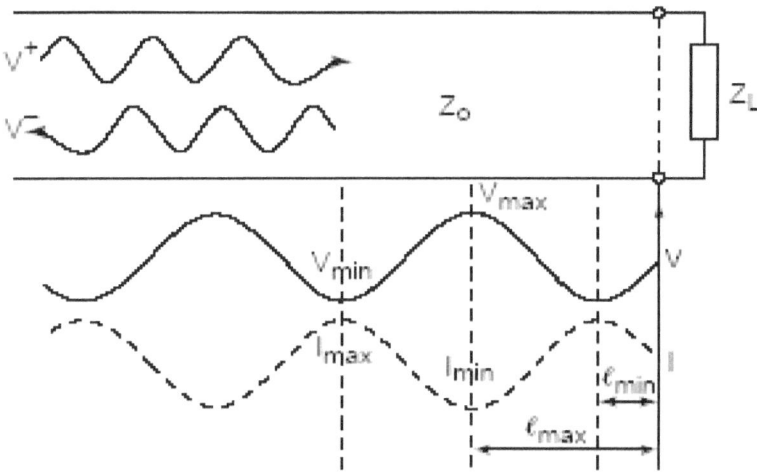

Figure 6.17 Standing waves on a TL.

To determine the standing wave pattern let us consider an incident voltage wave (V^+) causing an incident current wave (I^+) traveling toward the load Z_L on a lossless transmission line (Z_O). The reflected voltage and current waves (V^-, I^-) will interact with the incident voltage and current waves (V^+, I^+) to create the standing wave pattern.

To get an exact pattern determination in terms of its peak and valley magnitude and location, the following procedure can be used:

a. Locate $(Z_L)_N = Z_L/Z_O$ on the Smith chart, draw the constant-VSWR circle and determine the VSWR on the line.

b. Using Application #5, determine the location and value of Z_{max} (ℓ_{max} away from the load). The location of Z_{max} corresponds to the location of the peak of that voltage standing wave pattern (V_{max}) and the valley of the current standing wave pattern (I_{min}) because:
$$Z_{max} = V_{max}/I_{min} \tag{6.39}$$

Furthermore, to calculate the magnitude of the maximum voltage (V_{max}) on the line, we note that:
$$V_{max}=|V^+|+|V^-|=|V^+|(1+|V^+/V^-|)=|V^+|(1+|\Gamma_L|) \tag{6.40}$$
From Equation (6.37) we can write:

$$|\Gamma_L| = \left(\frac{VSWR - 1}{VSWR + 1}\right) \tag{6.41}$$

Substituting (6.41) in (6.40), we get:

$$V_{max} = |V^+| \left(\frac{2VSWR}{VSWR + 1}\right) \tag{6.42}$$

Similarly, I_{min} is given by:

$$I_{min} = V_{max}/Z_{max} = |I^+| - |I^-| = |V^+|/Z_O - |\Gamma_L||V^+|/Z_O$$

$$= \frac{|V^+|}{Z_O}(1-|\Gamma_L|)/Z_O = \frac{|V^+|}{Z_O}\left(\frac{2}{VSWR + 1}\right) \tag{6.43}$$

Thus the location and magnitude of V_{max} and I_{min} can easily be determined once the load value (Z_L, or Γ_L) and the incident voltage value ($|V^+|$) are known.

c. Similarly, the location and value of Z_{min} (ℓ_{min} away from the load) can be determined from the Smith chart. The location of Z_{min} corresponds to the valley of the voltage standing wave pattern (V_{min}) and the peak of the current standing wave pattern (I_{max}), because:

$$Z_{min} = V_{min}/I_{max} \tag{6.44}$$

To calculate the value of V_{min} and I_{max} in terms of the magnitude of the incident voltage $|V^+|$ and the load, we note that:

$$V_{min} = |V^+| - |V^-| = |V^+|(1-|V^+/V^-|) = |V^+|(1-|\Gamma_L|) \tag{6.45}$$

Using Equation (6.41) in (6.45), we get:

$$V_{min} = |V^+|\left(\frac{2}{VSWR + 1}\right) \tag{6.46}$$

Similarly, I_{max} is given by:

$$I_{max} = V_{min}/Z_{min} = |I^+| + |I^-| = |V^+|/Z_O + |\Gamma_L||V^+|/Z_O$$

$$= |V^+|(1+|\Gamma_L|)/Z_O$$

$$= \frac{|V^+|}{Z_O}\left(\frac{2VSWR}{VSWR + 1}\right) \tag{6.47}$$

NOTE: *Comparing (6.42) and (6.47), we can see that:*

$I_{max}=V_{max}/Z_O$

And similarly, from Equations (6.43) and (6.45) we can write:

$I_{min}=V_{min}/Z_O$

This would give an alternate way to find I_{max} and I_{min}.

Knowing the value and location of V_{max}, V_{min}, I_{max}, and I_{min}, the patterns for voltage and current can now be plotted easily. Figure 6.18 shows the standing wave pattern for voltage and current on a transmission line.

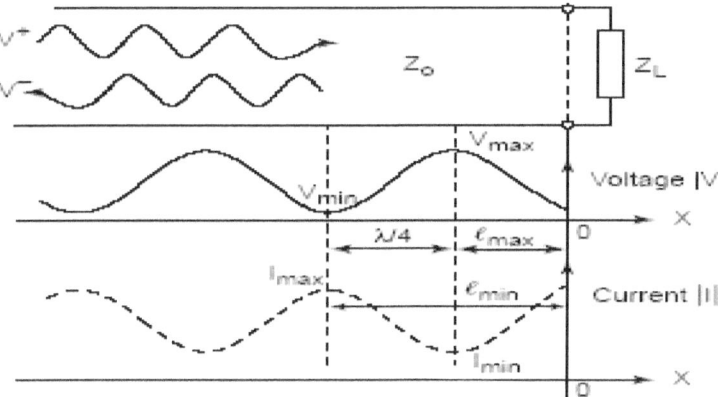

Figure 6.18 Standing waves on a TL.

The following example may further help to illustrate the concept of standing waves.

EXAMPLE 6.6

Determine the standing wave pattern on a transmission line ($Z_O=50$ Ω) terminated in $Z_L=100+j100$ Ω with an incident voltage of $V^+=1\angle0°$ as shown in Figure 6.19.

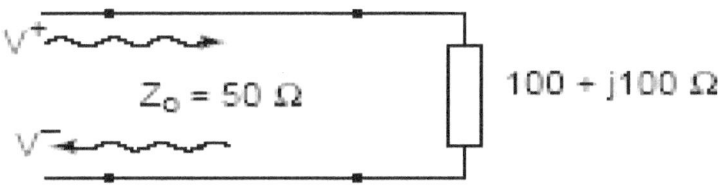

Figure 6.19 Circuit diagram for example 6.6.

Solution:

a. Locate $(Z_L)_N = Z_L/Z_O = 2+j2$ on the smith chart and draw the constant VSWR circle as shown in Figure 6.20.

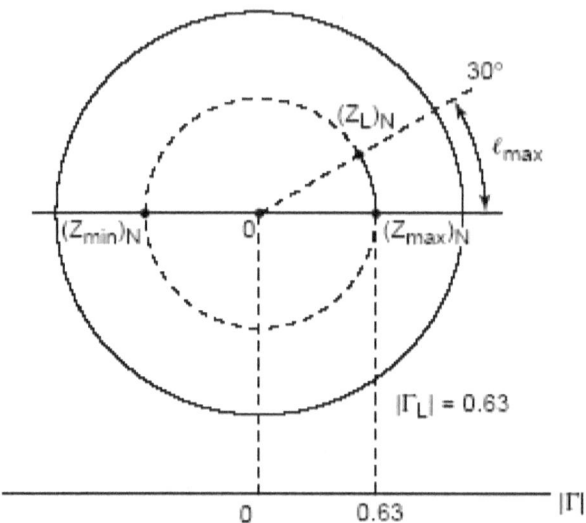

Figure 6.20 Smith Chart solution.

From this Figure we can read off [VSWR $=4.4$ at $(Z_{max})_N$] and calculate the following:

VSWR$=(Z_{max})_N=4.4$

$Z_{max}=4.4 \times 50 = 220\ \Omega$

$\ell_{max}=0.292-0.250=0.042\lambda$

Thus from (6.39) to (6.43), V_{max} and I_{min} are given by:

$V_{max}=[2\times 4.4/(1+4.4)]=1.63$ V

$I_{min}=V_{max}/Z_{max}=1.63/220=7.4$ mA

b) Similarly, from the VSWR circle we can write:

$(Z_{min})_N=1/(Z_{max})_N=1/4.4$

$Z_{min}=50(1/4.4)=12\ \Omega$

$\ell_{min}=0.042\lambda+\lambda/4=0.292\lambda$

Thus from (6.44) to (6.47), V_{min} and I_{max} are given by:

$V_{min}=[2/(1+4.4)]=0.37$ V

$I_{max}=V_{min}/Z_{min}=0.37/12=30.9$ mA

The final standing wave pattern is plotted in Figure 6.21.

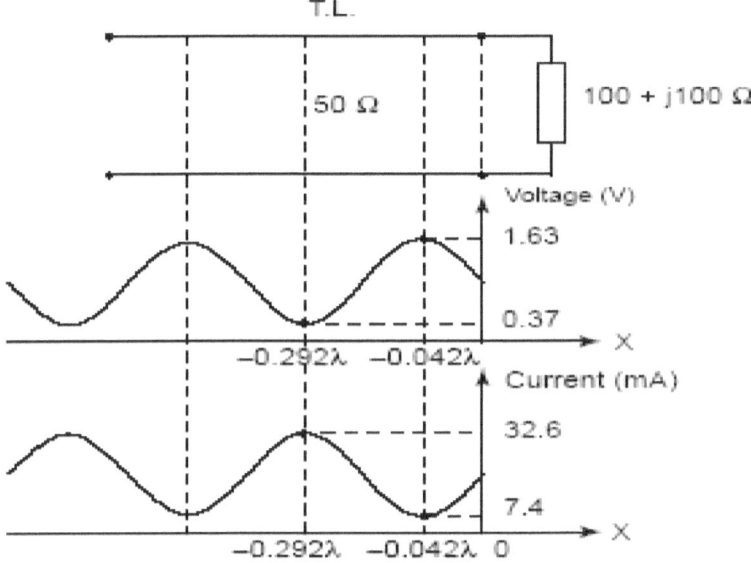

Figure 6.21 Standing waves on a TL.

NOTE: *An alternate method would be to calculate V_{max} and V_{min} using the VSWR value and then find I_{max} and I_{min} as follows:*
VSWR=4.4
V_{max}=2x4.4/(1+4.4)=1.63 v
V_{min}=2/(1+4.4)=0.37 v
I_{max}= V_{max}/Z_O=1.63/50=30.9 mA
I_{min}= V_{min}/Z_O=0.37/50=7.4 mA
These are the same values that were obtained earlier.

6.2.8 Application #8
INPUT IMPEDANCE DETERMINATION USING SINGLE STUBS
DEFINITION-STUB: *A stub is defined to be a short section of a transmission line (usually terminated in either an open or a short) often connected in parallel and sometimes in series with a feed transmission line in order to transform the load to a desired value.*

In general, the stub can have any general termination (Z_L'), however in practice as explained above, Z_L' is either a short or an open circuit as shown in Figures 6.22, 6.23 and 6.24.

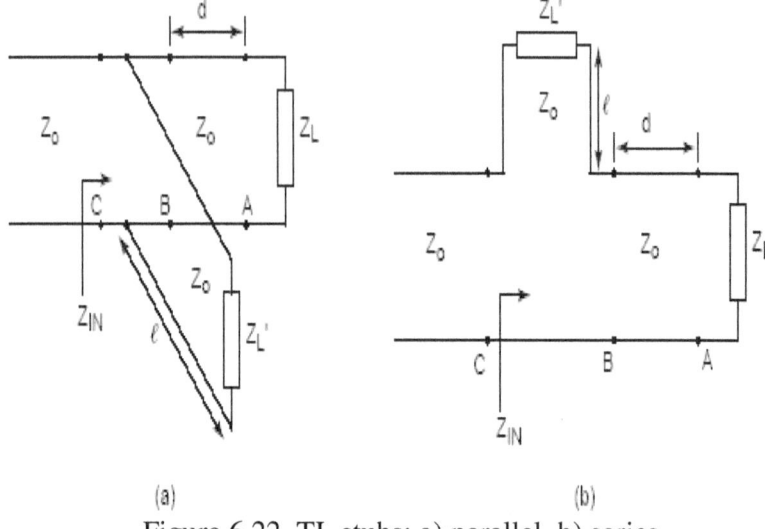

(a)   (b)

Figure 6.22 TL stubs: a) parallel, b) series.

There are two cases that will be considered separately as follows:
a. Parallel stubs, and
b. Series stubs

a. Parallel (or Shunt) Stubs

Consider the stub located a distance "d" away from a load (Z_L) as shown in Figure 6.22a. We would like to determine the input impedance of the combination.

Before we proceed to find the input impedance, we need to determine the stub's susceptance. Since the stub is connected in parallel, we use the smith chart as a Y-chart. The stub has a length (l) which can be used to determine its input admittance (or susceptance).

If the stub is terminated in a short, we use the Y-chart and start from $Y = \infty$ (see point "A" in Figure 6.23c) and travel "l" toward the generator to arrive at point "B". We read off the stub's susceptance from the chart. In a similar fashion, an open stub's susceptance can be found except we should start at $Y=0$ on the opposite side (see Figure 6.23 and 6.24).

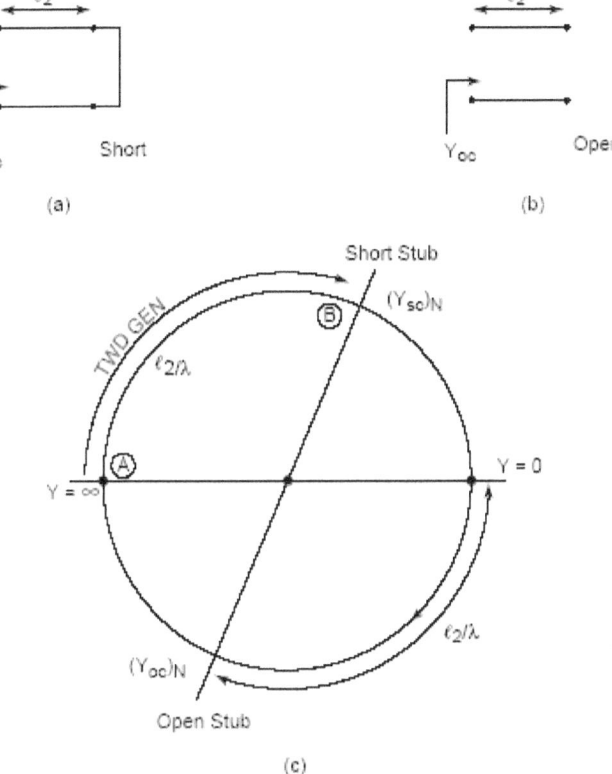

Figure 6.23 Smith Chart solution.

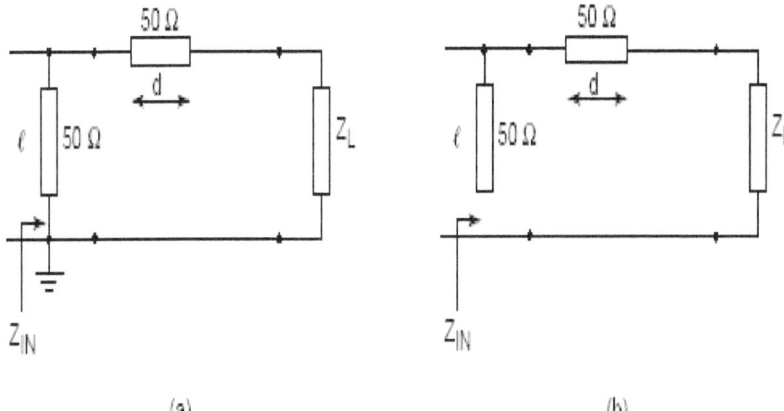

Figure 6.24 Shorthand schematics for shunt stubs.

To find the input impedance, the following steps are carried out:
1. Locate Z_L on the Smith chart (use a ZY-chart)at point "A" in Figure 6.25.

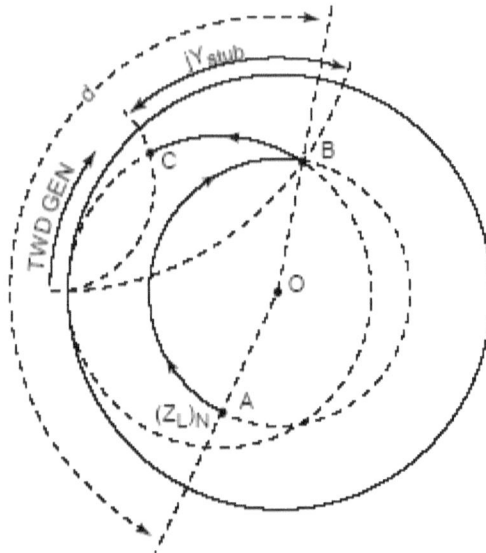

Figure 6.25 Smith Chart solution.

2. Draw the constant VSWR circle.

3. Travel a distance (d) toward the generator on the VSWR circle to arrive at point "B",

4. Now since we are adding the parallel stub, we must switch to the Y-chart and travel on a constant conductance circle an amount equal to the susceptance of the stub to arrive at point "C", as shown in Figure 6.25.

5. To find the input impedance, we switch back to the Z-chart and read off the normalized values (r,x) at point "C" corresponding to $(Z_{in})_N$. The total input impedance is given by:
$Z_{in}=Z_O(Z_{in})_N$

b. Series Stubs
Consider a series stub located a distance (d) away from the load (Z_L) as shown in Figure 6.26. Similar to the parallel stub case, we need

to know the series stub's reactance (jX) based on its electrical length ($\beta\ell$).

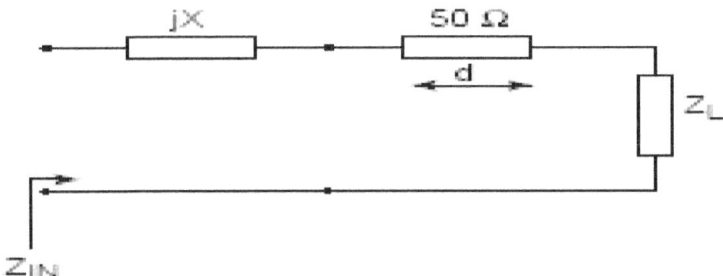

Figure 6.26 A series stub in circuit.

Since the stub is in series, we use the smith chart as a Z-chart. If the stub is terminated in a short, start from Z = 0 (point "A" in Figure 6.27c) and travel a distance of l/λ "toward generator" to arrive at point "B". Read off the normalized stub's reactance ($jx=jX/Z_O$) from the chart as shown in Figure 6.27b.

Similarly, an open stub's reactance can be determined by following the above procedure except by starting from Z=∞ on the chart as shown in Figure 6.27c.

To find the input impedance, the following steps are carried out:
1. Locate $(Z_L)_N$ on the Smith chart at point "A" as shown in Figure 6.28 (use a Z-chart).

2. Draw the constant VSWR circle,

3. From $(Z_L)_N$, travel a distance (d) toward the generator on the VSWR circle to arrive at point "B",

4. Now, since we are adding the series stub, we travel on a constant resistance circle an amount equal to the reactance of the stub, jx, to arrive at point "C".

5. The input impedance is read off at point "C" in Figure 6.28

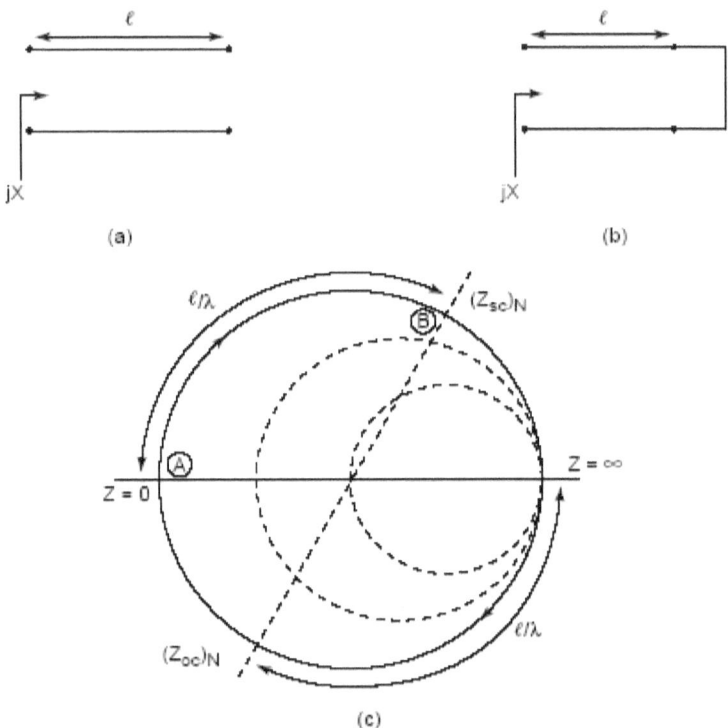

Figure 6.27 Smith Chart solutions for the reactance of series stub.

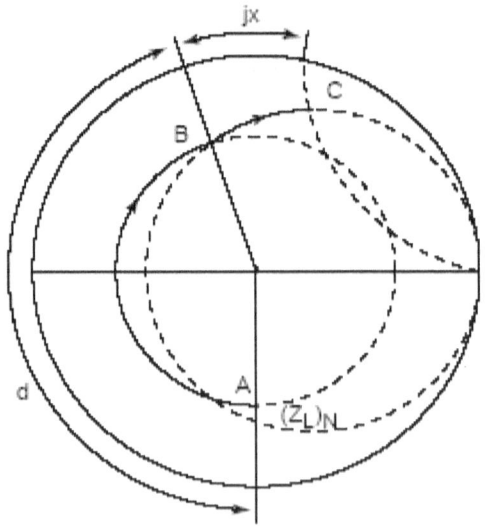

Figure 6.28 Smith Chart solution for a series stub.

EXAMPLE 6.7

Consider a transmission line ($Z_0=50$ Ω) terminated in a load $Z_L=15+j10$ Ω as shown in Figure 6.29. Calculate the input impedance of the line where the shunt open stub is located a distance of d=0.044λ From the load and has a length of ℓ=0.147λ.

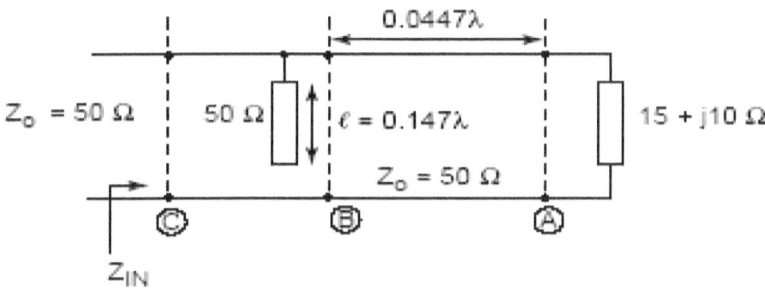

Figure 6.29 Circuit for example 6.7.

Solution:

a.   The susceptance of the open stub is first calculated by moving on a smith chart from Y=0 and moving a distance of 0.147λ toward generator to arrive at $(Y_{OC})_N=j1.33$ as shown in Figure 6.30.

FIGURE 10.30   Shunt open stub.

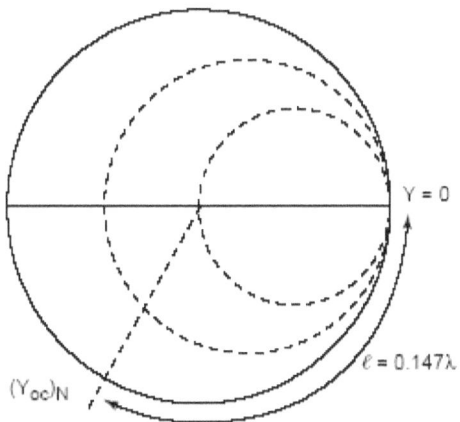

Figure 6.30 Smith Chart solution.

Next, the input impedance is found by :

b. Locate$(Z_L)_N=(15+j10)/50=0.3+j0.2$ on the smith chart (see point "A" in Figure 6.31):
c. Draw the constant VSWR circle.
d. From Z_L, travel a distance of 0.044λ to arrive at point "B". The admittance is read off to be:
    $(Y_B)_N=1-j1.33$ (point "B" in Figure 6.31)

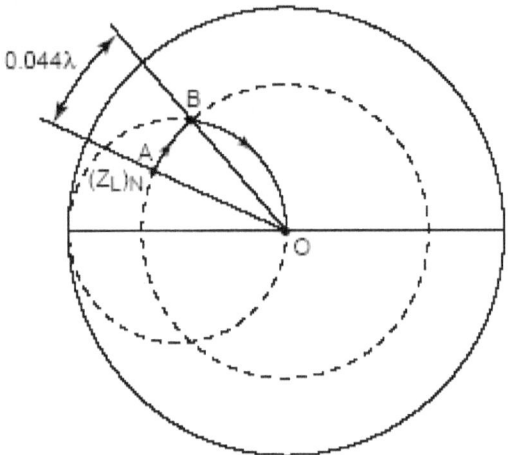

Figure 6.31 Smith Chart solution.

e. Adding an open shunt stub of length $\ell=0.147\lambda$ with $(Y_{OC})_N=j1.33$ gives:
    $(Y_{in})_N=(Y_B)_N+(Y_{OC})_N=(1-j1.33)+j1.33=1$
    $(Z_{in})_N=1/(Y_{in})_N=1 \Rightarrow Z_{in}=Z_O=50 \,\Omega$
    Adding the shunt stub on the smith chart results in arriving at point "O", which is obtained by moving on r=1 constant resistance circle by -j1.33.

NOTE: *Use of Application #8 in the design of circuits to bring about reflection-less loads are widely explored in the next chapter, where matching circuits are treated in depth.*

EXAMPLE 6.8

Consider a transmission line $(Z_O=50\,\Omega)$ with $Z_L=100\,\Omega$ as shown in Figure 6.32. Calculate the input impedance of the line where the shorted series stub is located a distance of $d=\lambda/4$ from the load and has a length $l=\lambda/8$.

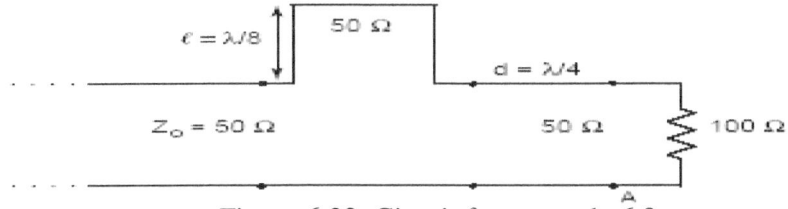

Figure 6.32 Circuit for example 6.8.

Solution:

a. The reactance of the series shorted stub is first calculated by moving on a smith chart from Z=0 a distance of 0.125λ toward generator to arrive at $(Z_{SC})_N=j1$ as shown in Figure 6.33. Next, to find the input impedance we perform the following steps.

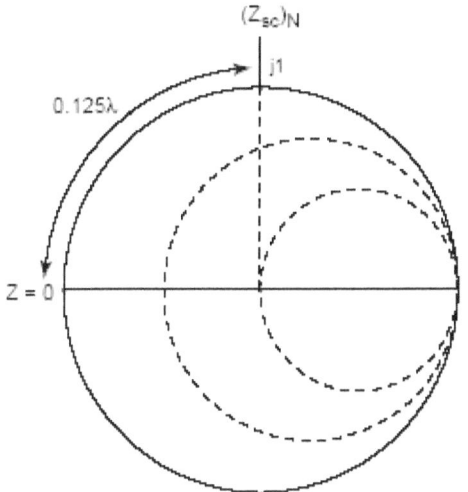

Figure 6.33 Smith Chart solution.

b. Locate$(Z_L)_N$=100/50=2 on the smith chart (see point "A" in Figure 6.34).

c. Draw the constant VSWR circle. From Z_L, travel a distance of 0.25λ to arrive at point "B". The impedance is read off to be: $(Z_B)_N$=0.5 (at point "B")

NOTE: *Since the load is resistive and has a value more than Z_O, the $(Z_L)_N$ value and location corresponds to $(Z_{max})_N$ (at point "A") and Z_B corresponds to $(Z_{min})_N$ (for more details, see application #5).*

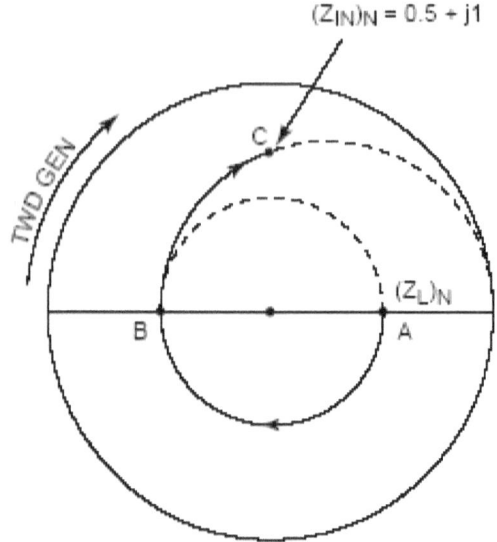

$(Z_{IN})_N = 0.5 + j1$

Figure 6.34 Smith Chart solution.

d. From point "B", move toward generator on a constant resistance circle to 0.5+j1 (point "C" in Figure 6.34) which corresponds to adding a series stub of length $\ell=0.125\lambda$ or $(Z_{SC})_N=j1$, giving:
$(Z_{in})_N=(Z_B)_N+(Z_{SC})_N=0.5+j1$
$Z_{in}=Z_O(Z_{in})_N =25+j50\ \Omega$

6.3 LUMPED ELEMENT CIRCUIT APPLICATIONS

Lumped elements are usually employed in the design of microwave circuits. These elements are mostly lossless reactive elements (such as inductors or capacitors) and are added either in series or in parallel in the circuit.

6.3.1 Application #9

INPUT IMPEDANCE FOR A SERIES LUMPED ELEMENT

Consider the circuit shown in Figure 6.35 where a load (Z_L) is in series with a series element (Z_S). The lumped element can be reactive (lossless), resistive (lossy) or a combination of both. In this application we consider a very general lumped element consisting of both resistive and reactive components.

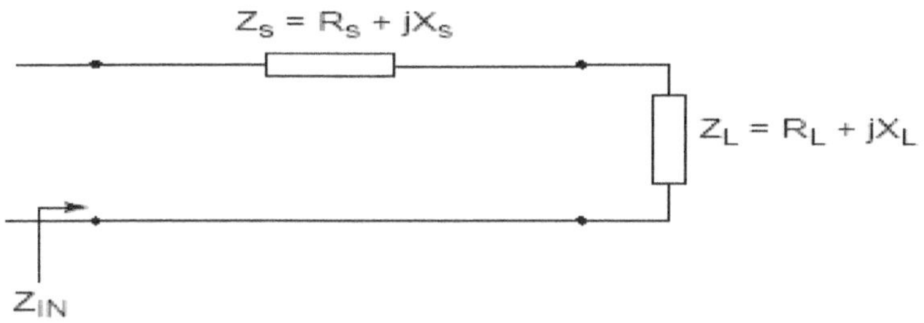

Figure 6.35 Circuit for series lumped element.

Since the lumped element is in series with the load, we need to consider only the Z-chart markings of the ZY-Smith chart (or only a Z-chart), in order to determine Z_{in}. We know mathematically that:

$Z_{in}=Z_L+Z_S$

Thus:

$(Z_{in})_N=(r_L+r_S)+j(x_L+x_S)$ $\hspace{3cm}$ (6.48)

The purpose of this application is to show how to achieve this result graphically where the exact steps are delineated below:

a. Locate $(Z_L)_N$ on the Smith chart(see point "A" in Figure 6.36)

b. Moving on the constant resistance circle that passes through Z_L, add a reactance of jx_S to arrive at point "B".

c. Now moving on a constant reactance circle that passes through point "B", add a resistance of r_S to arrive at point "C".

d. The input impedance value is read off at point "C", using the Z-chart markings.

ALTERNATE PROCEDURE:
Point "C" could have equally been reached by the following steps (see Figure 6.36):

   a. Move on a constant reactance circle (that passes through "A" and add the resistance of r_S to arrive at point " B' " (see Figure 6.36)

   b. Now moving on a constant resistance circle (that passes through point " B' "), add the reactance of jx_S to arrive at point "C".

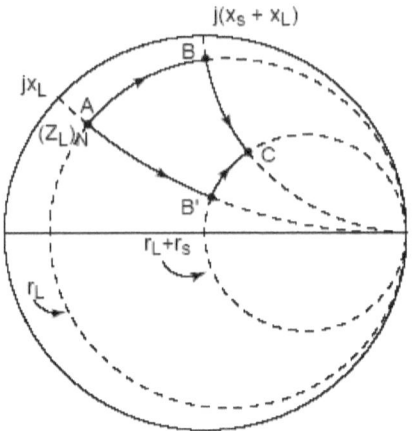

Figure 6.36 Smith Chart solution.

6.3.2 Application #10

INPUT ADMITTANCE FOR A SHUNT LUMPED ELEMENT

Consider the circuit shown in Figure 6.37, where a load (Y_L) is in parallel with a shunt element (Y_P). In general, The lumped element is considered to have both resistive and reactive components (similar to Application #9).

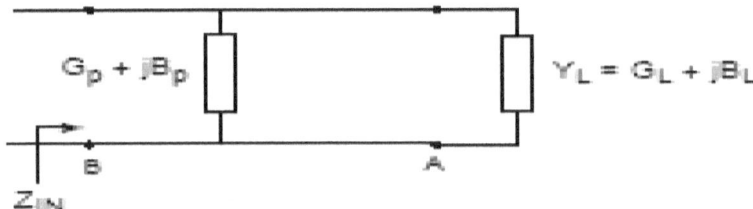

Figure 6.37 Circuit for shunt lumped element.

Since the lumped element is in parallel with the load, only the Y-chart markings of the ZY-Smith chart need to be considered. The total admittance is given mathematically by:

$Y_{in}=Y_L+Y_P$

$(Y_{in})_N=(g_L+g_P)+j(b_L+b_P)$

Similar to the application #9, we now present the procedure to determine $(Y_{in})_N$ graphically:

a. Locate $(Y_L)_N$ on the Y-chart at point "A" in Figure 6.38.

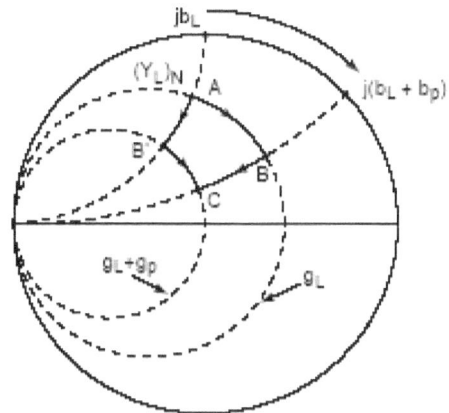

Figure 6.38 Smith Chart solution.

b. Move on the constant conductance circle [that passes through $(Y_L)_N$] and add a susceptance of "jb_P" to arrive at point "B".

c. Move on the constant susceptance circle (passing through

"B") by adding a conductance of "g_P" to arrive at point "C".

d. The input admittance is read off at point "C" using the Y-chart markings.

ALTERNATE PROCEDURE:
Similar to Application #9, the input admittance equally could have been determined by:
a. Moving on a constant susceptance circle and adding g_P to arrive at point " B' " as shown in Figure 6.38.
b. Now add jb_P on a constant conductance circle to arrive at point "C".

6.3.3 Application #11
INPUT IMPEDANCE OF SINGLE SHUNT/SERIES REACTIVE ELEMENTS
This is a special case of applications #9 and #10 where the series or the shunt elements are lossless (i.e. purely reactive). In this case, there are 4 possible combinations (see Fig. 6.39) as follows:

1. Series L
2. Series C

3. Shunt L
4. Shunt C

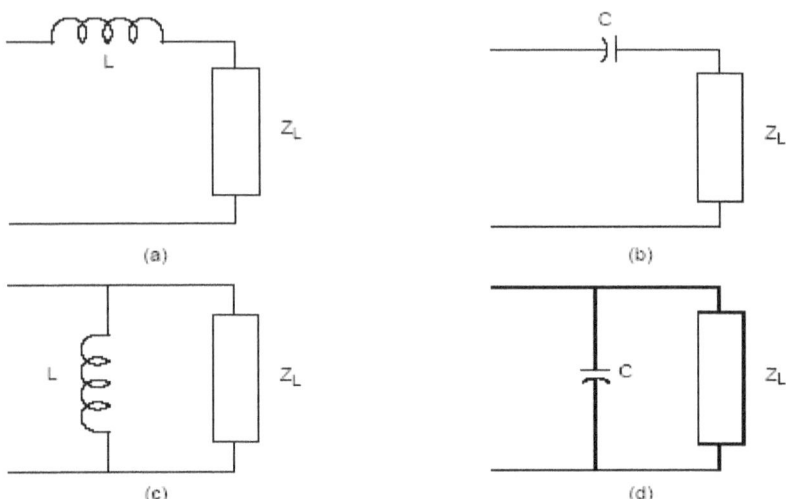

Figure 6.39 Four possible combinations.

This case has already been discussed in the "RF Electronics" chapter under the heading of "L-network matching" which is now revisited and treated with the help of the smith chart.

To find the input impedance, we first calculate the normalized series reactance $(jx=jX/Z_O)$ or normalized shunt susceptance value $(jb=jB/Y_O)$ of the lumped element before entering the Smith chart. Next, we locate $(Z_L)_N$ on the chart as point "A" (see fig. 6.40). Now starting from point "A", the following steps are applied:

1. To add a series L: on a constant resistance circle, move up by $jx_S=j\omega L/Z_O$.
2. To add a Series C: on a constant resistance circle, move down by $jx_S=-j/\omega C Z_O$.
3. To add a shunt L: on a constant conductance circle, move up by $jb_P=-j/\omega L Y_O$.
4. To add a shunt C: on a constant conductance circle, move down by $jb_P=j\omega C/Y_O$.

These are all shown in Fig. 6.40.

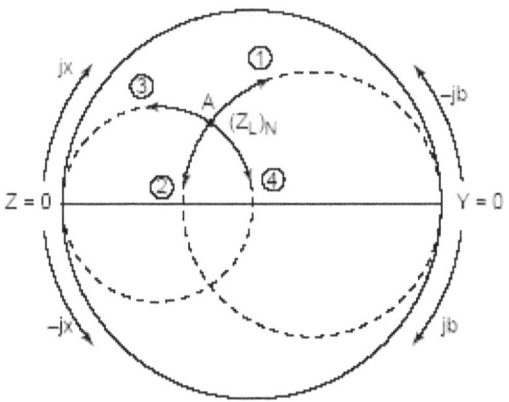

Figure 6.40 Smith Chart solution.

RULE OF THUMB

Upon close observation of these four cases, it appears that for the majority of load values, adding series (or shunt) inductor would move point "A" upward on the constant resistance (or conductance) circle while adding a series (or shunt) capacitance would move point A downward.

However, it should be noted that the above is a good rule of thumb to follow when dealing with purely reactive elements, but should never be generalized outside the scope of this discussion. This rule of thumb is limited but workable and will never actually replace the reasoning and the understanding that goes into making it.

Example 6.9

Calculate the total input admittance of a combination of a load $Z_L=50+j50$ Ω with a shunt inductor of $L=8$ nH at $f_o=1$ GHz as shown in Fig. 6.41. Assume a 50 Ω system.

Solution:

$Z_O=50$ Ω $\Rightarrow Y_O=0.02$ S

a. We first find the susceptance of the shunt inductor:

$jB_P=-j/(\omega_O L)=-j0.02$ S $\Rightarrow jb_P=jB_P/Y_O=-j1$

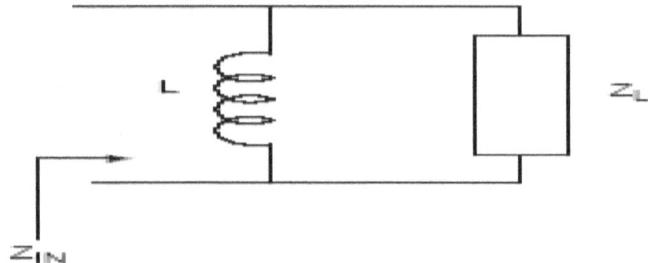

Figure 6.41 Circuit for example 10.9.

b. Locate $(Z_L)_N=Z_L/Z_O=1+j1$ on the smith chart at point "A" in Fig. 6.42.

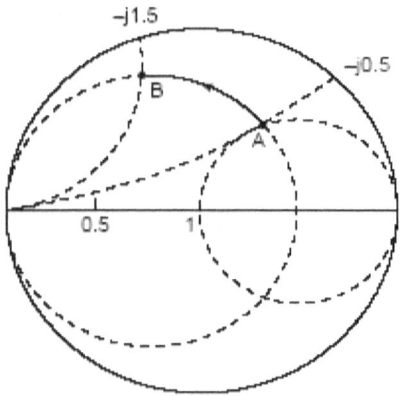

Figure 6.42 Smith Chart solution.

c. Since this is an inductor, we need to move upwards from point "A" on a constant conductance circle by -j1 to arrive at point "B".
d. The normalized input admittance is read off at point "B" as:
$(Y_{in})_N=0.5-j1.5$
$Y_{in}=Y_O(Y_{in})_N=0.01-j0.03$ S
Or,
$Z_{in}=1/Y_{in}=10+j30$ Ω

6.3.4 Application #12
COMBINATION OF SERIES AND SHUNT ELEMENTS
In this application, we will consider the case where there are several series and shunt elements in combination with the load (as shown in Figure 6.43).

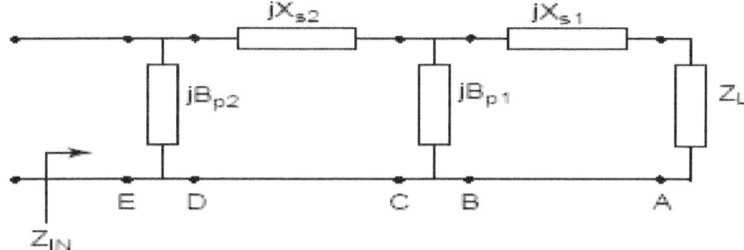

Figure 6.43 Combination of series and shunt elements.

Application #11 can be used repeatedly to arrive at the total input impedance as described in the following steps:

a. Since the first element adjacent to the load is connected in series, we start with $(Z_L)_N$ and locate it on the Z-chart (see point "A" in Figure 6.44),

b. On the constant resistance circle passing through $(Z_L)_N$, a reactance of $jx_{S1}= jX_{S1}/Z_O$ is added to arrive at point "B".

c. Now switching to the Y-chart, we move on the constant conductance circle and add a susceptance of $jb_P=jB_{P1}/Y_O$ to arrive at point "C".

d. Since the next element is in series, we switch back to the Z-chart and move on a constant resistance circle by adding a reactance of $jx_{S2}=jX_{S2}/Z_O$ to arrive at point "D".

e. The final element is in parallel, so we switch to the Y chart and add a susceptance of $jb_{P2}=jB_{p2}/Y_O$ to arrive at "E".

f. The total impedance is now read off on the Z-chart at point "E" as shown in Figure 6.44.

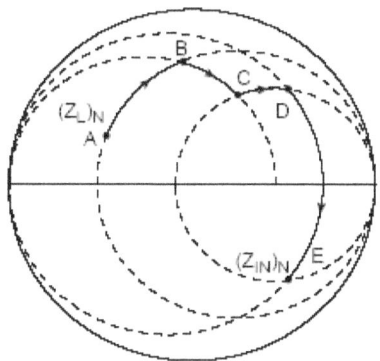

Figure 6.44 Smith Chart solution.

Example 6.10

Find the input impedance at f=100 MHz for the circuit shown in Figure 6.45.

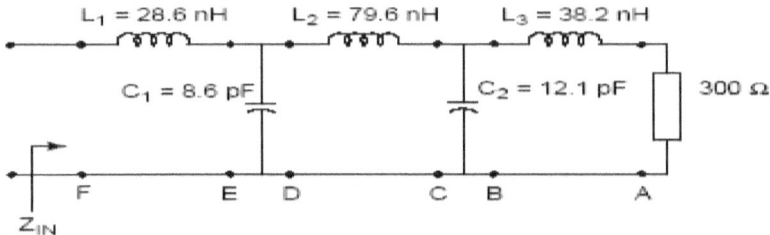

Figure 6.45 Smith Chart solution.

Solution:

First, we choose the normalizing factor arbitrarily to be:

Z_O=50 Ω,

And,

Y_O=0.02 S.

Then we normalize all impedance and admittance values:

$jx_{S1}=(jX_1)_N=j\omega L_1/Z_O=j0.36$

$jb_{P1}=(jB_1)_N=j\omega C_1/Y_O=j0.27$

$jx_{S2}=(jX_2)_N=j\omega L_2/Z_O=j1.0$

$jb_{P2}=(jB_2)_N=j\omega C_2/Y_O=j0.38$

$jx_{S3}=(jX_3)_N=j\omega L_3/Z_O=j0.48$

$(Z_L)_N=300/50=6$

a)   Locate $(Z_L)_N$ on the smith chart (point "A" in Figure 6.46).

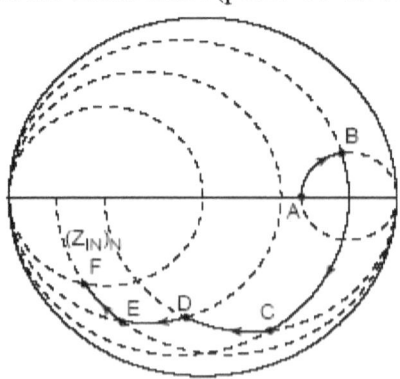

Figure 6.46 Smith Chart solution.

b) Since the first element (L_3) adjacent to the load is a series inductor, we move upward from point "A" on a constant resistance circle by a reactance of j0.36 to arrive at point "B".

c) Now switch to constant conductance circle, add the next shunt element by moving downward by j0.27 to arrive at point "C".

d) For the next series inductor, switch to the constant resistance circle and move upward by j1.0 to arrive at point "D".

e) Next, for the shunt capacitor, switch to a constant conductance circle and move downward by j0.38 to arrive at point "E".

f) Finally, for the series inductor, switch to the constant conductance circle and move upward by j0.48 to arrive at point "F".

g) Now we read off the value of the normalized input impedance at point "F" as:

$(Z_{in})_N = Z_{in}/Z_O = 0.4 - j1.0 \Rightarrow Z_{in} = 20 - j50$ Ω

6.3.5 Application #13
FREQUENCY RESPONSE OF "RLC" CIRCUITS.
The Smith chart can be used effectively to plot out the course of input impedance variation over a frequency range ($f_2 - f_1$) as frequency is increased from f_1 to f_2. Vice versa, from the frequency response plot of a complex circuit in the Smith chart, one can develop an equivalent circuit model.

Of all possible combinations of three elements (R, L and C), there are 4 non-trivial combinations that are of interest and therefore are discussed here:

CASE 1. Series RC + Shunt L Combination
This combination is shown in Figure 6.47.

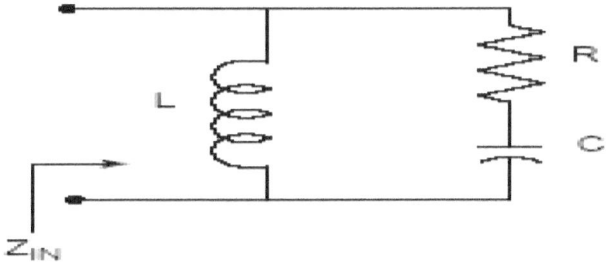

Figure 6.47 Series RC+Shunt L combination.

In the "series RC + shunt L" combination, we see that "series RC", being a capacitive load, is located on the lower half of the Smith chart at $f=f_1$ (point "A" on the Smith chart in Figure 6.48).

As frequency is increased (from f_1), the capacitive reactance magnitude decreases and point "A" moves on a constant resistance circle toward the real axis to arrive at point "B" where $f=f_2$.

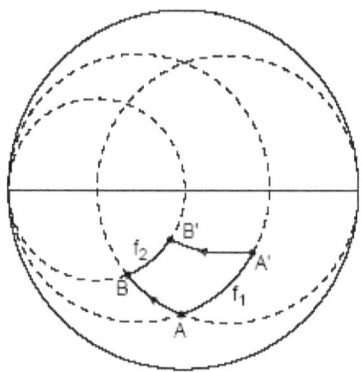

Figure 6.48 Smith Chart solution.

Now adding a shunt inductor (with a susceptance of $1/\omega L$) would move points A and B more toward the real axis (to points A' and B') as frequency is increased from f_1 to f_2.

These points move on the constant conductance circles that pass through points A and B to arrive at points A' and B'. Connecting points A' and B' would map the course of the frequency response for this RLC combination in the f_2-f_1 frequency range (see Figure 6.48).

CASE 2. Series RL + Shunt C combination

In the "series RL + shunt C" combination, the "series RL", being an inductive load, is located on the upper half of the ZY-chart as shown at point "A" in Figure 6.50). This combination is shown in Figure 6.49.

As frequency is increased from f_1 to f_2, the inductive reactance magnitude (ωL) increases and point "A" at $f=f_1$ moves on a constant resistance circle away from the real axis to point "B" at $f=f_2$. Adding a shunt capacitor (with a susceptance value of ωC) would move points A (for $f=f_1$) and B (for $f=f_2$) on constant conductance circles to points A' and B' as shown in Figure 6.50.

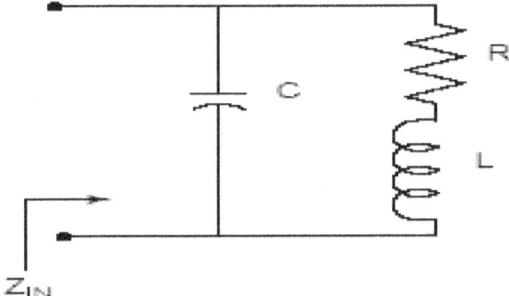

Figure 6.49 Series RL+Shunt C combination.

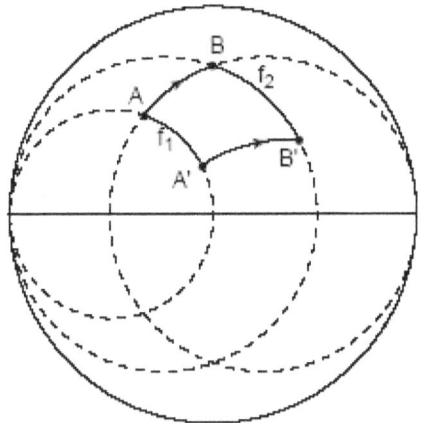

Figure 6.50 Smith Chart solution.

Similar to case 1, connecting points A' and B' would yield the input impedance (or admittance) variation in the frequency range (f_2-f_1).

CASE 3. Shunt RC + Series L Combination
For the "parallel RC+ series L" combination, we need to consider the Y-chart (see Figure 6.51).

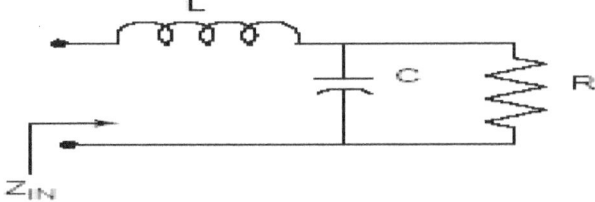

Figure 6.51 Series RL+Shunt C combination.

We are interested in its performance in the frequency range (f_2-f_1).

Thus we first plot the admittance of the parallel RC combination at $f=f_1$ at point "A" as shown in Figure 6.52 (located on the lower half of the Y-chart, due to being a capacitive load).

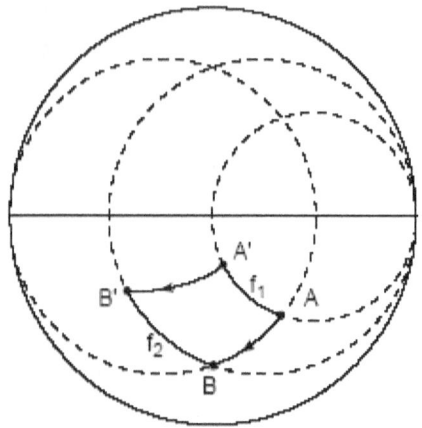

Figure 6.52 Smith Chart solution.

As frequency increases from f_1 to f_2, the capacitive susceptance magnitude (ωC) increases and point "A" (at $f=f_1$) moves on a constant conductance circle to point "B" (at $f=f_2$). Adding a series L (having a reactance "ωL") will move points A and B on constant resistance circles to points A' and B'. Connecting points A' and B' provides the plot of input impedance frequency response on the Smith chart as shown in Figure 6.51.

CASE 4.   Shunt RL + Series C Combination
For the "parallel RL + series C" combination, we need to consider the Y-chart (see Figure 6.53).

We are interested in its performance in the frequency range (f_2-f_1). Thus we first plot the admittance of the parallel RL combination at $f=f_1$ at point "A" as shown in Figure 6.54 (located on the upper half of the Y-chart, due to being an inductive load).

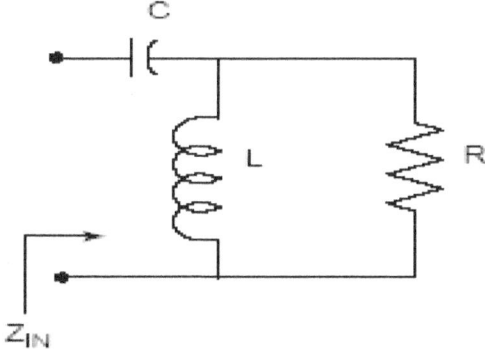

Figure 6.53 Shunt RL+series C combination.

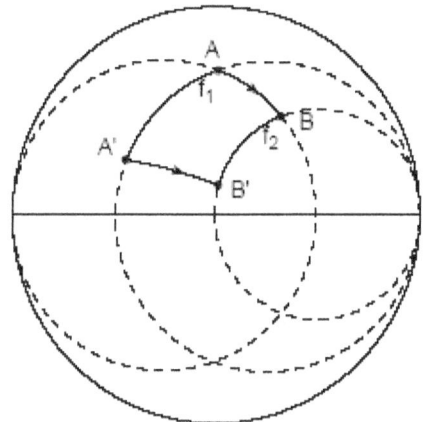

Figure 6.54 Smith Chart solution.

As frequency increases from f_1 toward f_2, point "A" would move on a constant conductance circle toward the real axis to point "B". This is because the magnitude of the inductive susceptance ($1/\omega L$) decreases with frequency.

Now adding a series C (with a reactance of $-1/\omega C$) will move points A and B on constant resistance circles to points A' and B'. Connecting A' to B', provides the plot of the input impedance frequency response of this RCL combination on the Smith chart (see Figure 6.54).

OBSERVATION: *By looking at the Smith chart plots (cases 1 through 4) we note that the input impedance moves clockwise in the Smith* *chart as the frequency is increased. This observation is actually based upon a much deeper concept which is the "Foster's Reactance Theorem". This theorem is briefly discussed in the next section.*

6.4   FOSTER'S REACTANCE THEOREM

Foster's reactance theorem can be stated as:

For a passive lossless one-Port network, the reactance and susceptance are strictly and monotonically increasing functions of the frequency.

This theorem, in essence, states that:

The slope of the reactance function $x(\omega)$ or the susceptance function $B(\omega)$, is always positive, i.e.,

$\partial X(\omega)/\partial \omega > 0$ for $0 < \omega < \infty$

$\partial B(\omega)/\partial \omega > 0$ for $0 < \omega < \infty$

This positive-slope condition means that the impedance or admittance frequency response of a passive lossless one-port network moves in a clockwise direction in the Smith chart as frequency is increased.

This is a general concept and applies to both lumped or distributed elements as illustrated in the next two examples.

Example 6.11 (Lumped circuit)
Plot the frequency response (of the input admittance) for a parallel LC lumped circuit, as shown in Figure 6.55.

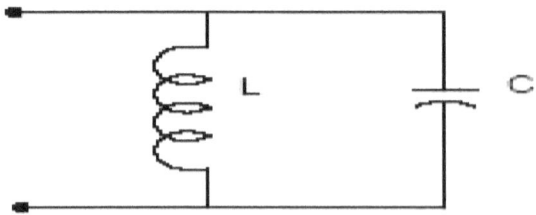

Figure 6.55 Circuit for example 10.11.

Solution:
$Y_{in} = jB = j(\omega C - 1/\omega L)$
$\partial B(\omega)/\partial \omega = C + 1/\omega^2 L > 0$

The input admittance (or susceptance) as a function of ω is sketched in Figure 6.56.

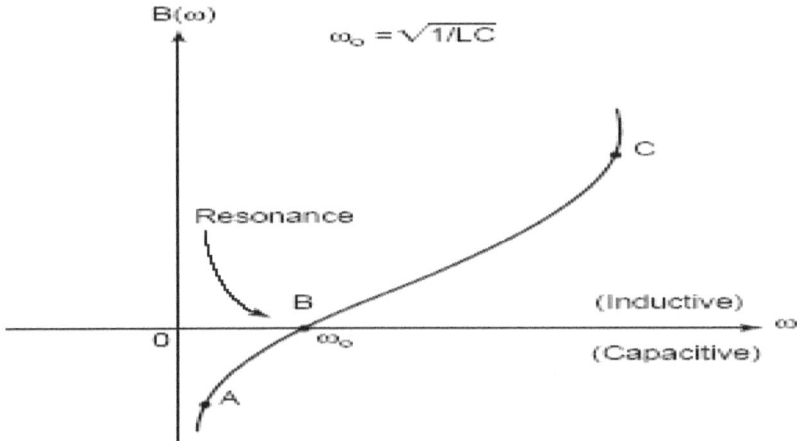

Figure 6.56 Frequency response.

Setting $B(\omega)$ equal to zero, yields the resonant frequency (ω_0) as:

$$\omega_0 = 1/\sqrt{LC}$$

Furthermore, as the frequency increases the impedance goes from point "A", through point "B"(resonance) and then to point "C". This can be plotted on a smith chart on the outermost circle (r=0) circle as shown in Figure 6.57.

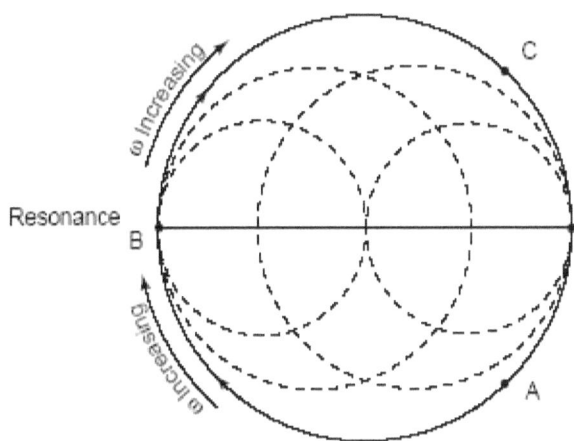

Figure 6.57 Smith Chart solution.

Example 6.12 (Distributed Circuit)
Plot the frequency response of the input impedance of a shorted transmission line as shown in Figure 6.58.

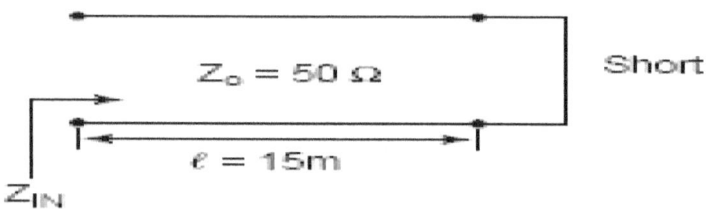

Figure 6.58 Circuit for example 10.12.

Solution:
$Z_{in} = jX_{in} = jZ_O \tan\beta\ell \Rightarrow X_{in} = 50 \tan\beta\ell$
Where,
$\beta = 2\pi/\lambda$,
$\lambda = c/f = 300/f(MHz)$ (m)
Therefore we can write:
$$\beta\ell = (2\pi/\lambda)\ell = 2\pi\ell(f/c) = 2\pi f\ell/300 = 2\pi \times 15/300 = (\pi/10)f \quad (E6.1)$$

$$X_{in} = 50 \tan\beta\ell = 50 \tan (\pi/10)f \quad (E6.2)$$

In order to get resonance at a frequency $f=f_O$, "X_{in}" must approach infinity (∞), therefore we can write:
$$\tan\beta\ell = \infty \Rightarrow \beta\ell = (2n+1)\pi/2, \quad n=0,1,2,3 \quad (E6.3)$$
Substituting for "$\beta\ell$" from (E6.1) in (E6.3), we have:
$(\pi/10)f_O = (2n+1)\pi/2 \Rightarrow f_O = 5(2n+1)$ MHz, n=0,1,2,3

Thus there are infinite number of frequencies at which the circuit resonates. The resonant frequencies occur at: f=5,15,25,....(MHz). These resonant frequencies are shown in Figure 6.59 in the frequency domain and on the smith chart in Figure 6.60.

NOTE: *This example shows that there are an infinite number of resonances that can occur for a "distributed circuit" which is in contrast with "lumped circuits" that have a finite number of resonances.*

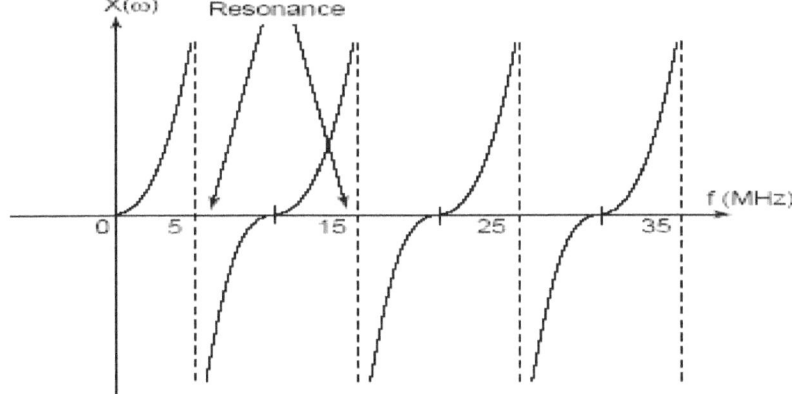

Figure 6.59 Reactance plot.

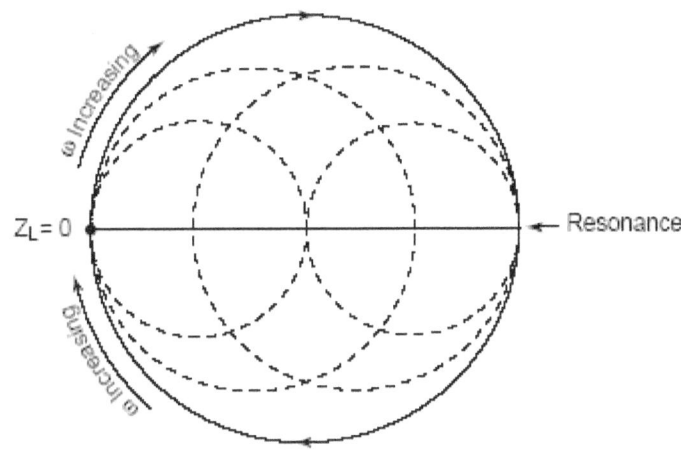

Figure 6.60 Frequency response on the Smith Chart.

6.4.1 Application #14
INPUT IMPEDANCE (OR ADMITTANCE) FOR A COMBINATION OF DISTRIBUTED AND LOSSLESS LUMPED ELEMENTS

This final Application deals with circuits having distributed elements (such as transmission lines) and lossless lumped elements (such as capacitors and inductors).

To obtain the input impedance (or admittance) we use the following two rules:

a. *When dealing with distributed elements, for ease and convenience we start from the load end. Then we travel on a constant VSWR circle a length (ℓ) towards the generator.*

b. *When dealing with lossless lumped elements, we also start from the load end but move on a constant resistance (or conductance circle) depending on whether the lumped element is in series (or shunt) with the rest of the circuit.*

The overall procedure is the same as delineated in the previous applications. The example below will illustrate this concept further.

Example 6.13
In the circuit shown below (Figure 6.61), determine the input impedance at f = 10 GHz.

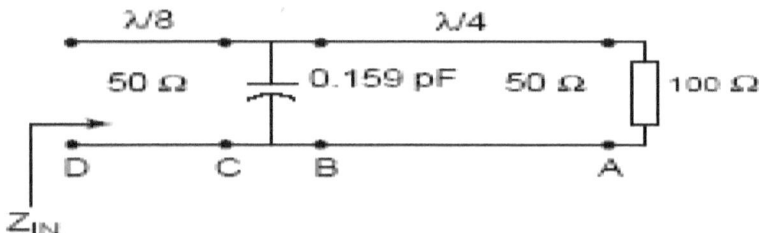

Figure 6.61 Circuit for example 6.13.

Solution:
To find Z_{in} we perform the following steps:
a) Locate $(Z_L)_N = 100/50 = 2$ on the smith chart (see Figure 6.62).
b) Since the first element adjacent to the load is a series transmission line we draw the constant VSWR transmission line.
c) Starting from $(Z_L)_N$, at point "A", we move on this circle a length of $\lambda/4$ "toward generator" to arrive at point "B".
d) Now since the next element is a shunt capacitor, we switch to the Y-chart and move on the constant conductance circle to arrive at point "C". The shunt capacitor has a susceptance of:
$jB = j2\pi \times 10^{10} \times 0.159 \times 10^{-12} = j0.01\ \Omega$

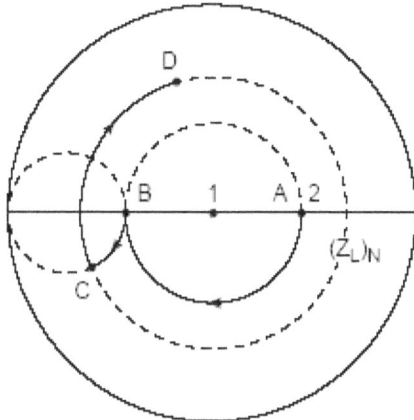

Figure 6.62 Smith Chart solution.

e) The next element is a series transmission line, so we switch back to the Z-chart and draw the constant VSWR circle that passes through "C".

f) Now from point "C" we move a distance of $\lambda/8$ "toward generator" to arrive at point "D" as shown in Figure 6.62.

g) The value of the input impedance is read off at point "D" as:

$(Z_{in})_N=0.4+j0.55 \Rightarrow Z_{in}=20+j27.5 \ \Omega$

Chapter 6- Symbol List

A symbol will not be repeated again, once it has been identified and defined in an earlier chapter, with its definition remaining unchanged.

b - Normalized susceptance, $b=B/Z_0$

b_L - Normalized susceptance at the load

b_P - Normalized susceptance at the parallel element

g - Normalized conductance, $g=G/Z_0$

g_L - Load Normalized conductance

g_P - Shunt Normalized conductance

I_{max} - Maximum current on a transmission line.

I_{min} - Minimum current on a transmission line.

ℓ_{max} – location of Z_{max} on the transmission line

ℓ_{min} – location of Z_{min} on the transmission line

r - Normalized resistance, $r=R/Z_0$

x- Normalized reactance, $x=X/Z_0$

V_{max} - Maximum voltage on a transmission line.

V_{min} - Minimum voltage on a transmission line

Y_0 - Characteristic admittance

Z_0 - Characteristic impedance

Z_D - Device impedance

Z_{in} - Input impedance

$(Z_{in})_N$ - Normalized input impedance

Z_{max} - Maximum impedance, corresponding to the location of the peak of the voltage and the valley of the
    current in a standing wave pattern on a transmission line.

Z_{min} - Minimum impedance, corresponding to the location of the valley of voltage and the peak
    of the current in a standing wave pattern on a transmission line.

Γ_D -Device reflection coefficient

CHAPTER-6 PROBLEMS

6.1) The normalized impedance of an unknown device is measured (from 1 to 2 GHz) to have a frequency response as plotted in Figure P6.1. Determine an equivalent circuit for the unknown device with all element values correctly calculated (assume $Z_0=50\ \Omega$, f=1 GHz).

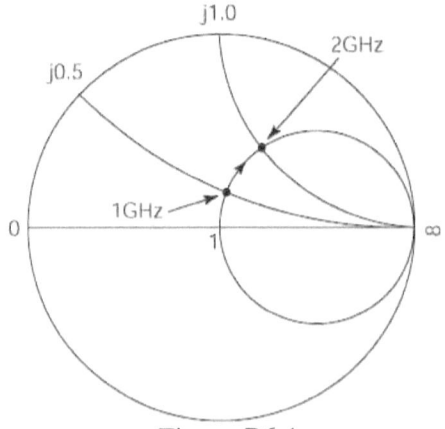

Figure P6.1

6.2) Using a Smith chart, determine Z_{in} for the circuit shown in Figure P6.2. Assume $f_o=1$ GHz.

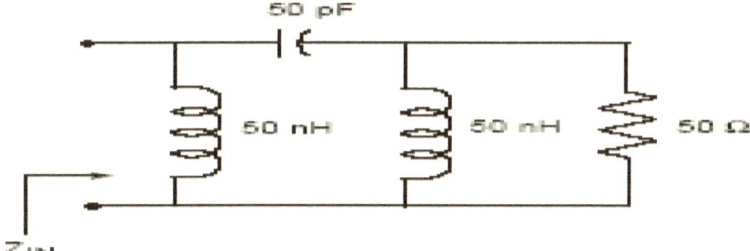

Figure P6.2

6.3) A lossless coaxial line (Z_O=50 Ω) is terminated in a

100 Ω load.   If the incident voltage wave has an rms magnitude of 10 V, determine:

a. The reflection coefficient and the VSWR on the line

b. The magnitude and location of V_{MAX}, V_{MIN}, Z_{MAX} and Z_{MIN} on the line.

c. The magnitude and location of I_{MAX} and I_{MIN} on the line.

d. Determine the power absorbed by the load.

e. Plot the voltage standing wave pattern on the line for both voltage and current.

6.4) A lossless transmission line (Z_O=50 Ω) is terminated in a load

(Z_L=100+j100 Ω).   A single shorted stub (ℓ=λ/8, 50Ω) is inserted λ/4 away from the load as shown in Figure P6.4. Using a Smith chart, determine the line's input impedance (Z_{IN}).

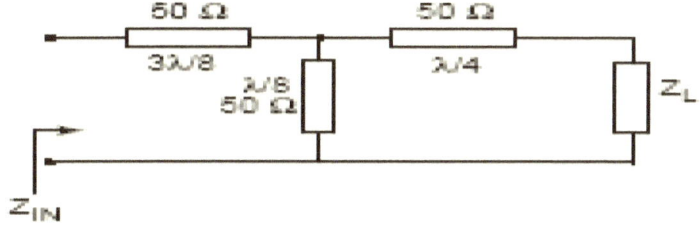

Figure P6.4

6.5) A lossless transmission line (Z_O=75 Ω) is terminated in an unknown load. Determine the load if the VSWR on the line is found to be 2 and the adjacent voltage maxima are at x=-15 cm and -35 cm where the load is located at x=0.

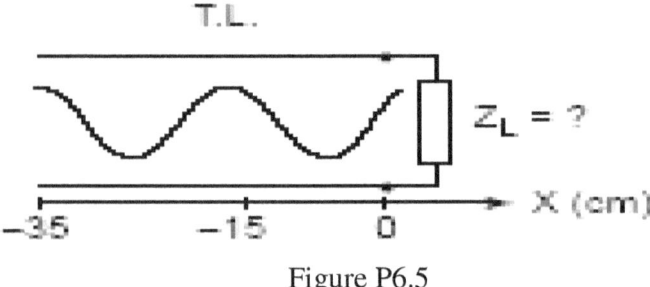

Figure P6.5

6.6) Determine the input impedance of a transmission line (Z_O=50 Ω) at a distance of 2 cm from the load impedance (Z_L=25+j25 Ω) if the wavelength on the transmission line is found to be 16 cm. What is the VSWR on the line?

6.7) Determine the input impedance of the lumped-element network shown in Figure P6.7 (all values are in Ω).

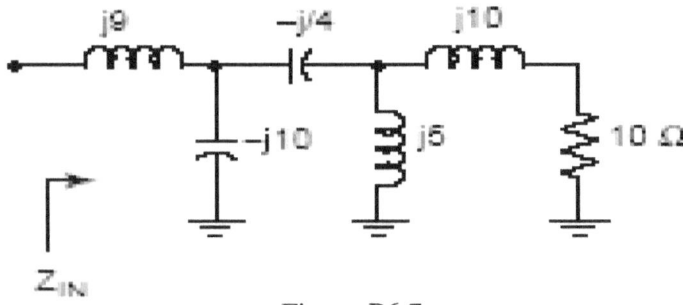

Figure P6.7

6.8) Find the input impedance of the transmission line circuit in a 50 Ω system (as shown in Figure P6.8) for Z_L=25+j25 Ω, d_1=3λ/8, d_2=λ/4 and l=λ/8. What is the VSWR at the input terminals?

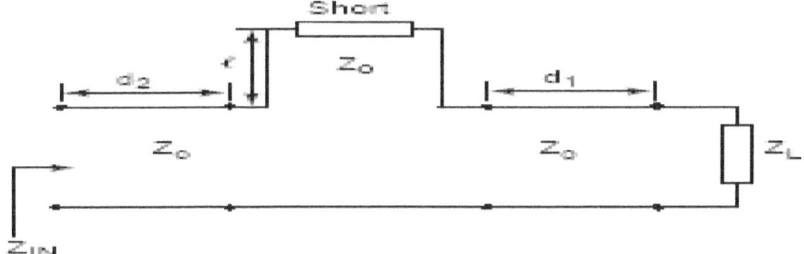

Figure P6.8

6.9) Find the input impedance of a double-stub shunt tuner as shown in Figure P6.9. Assume that the stubs are short circuited and $l_1=0.23\lambda$, $l_2=0.1\lambda$, $d=\lambda/8$ and $Z_O=50\ \Omega$. What is the reflection coefficient at the input terminals when f=1 GHz?

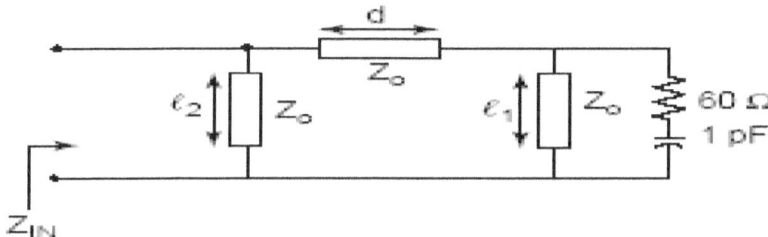

Figure P6.9

REFERENCES
[6.1] Anderson, E. M. *Electric Transmission Line Fundamentals.* Reston: Prentice Hall, 1985.

[6.2] Cheng, D. K. *Field and Wave Electromagnetics,* 2nd ed., Reading: Addison Wesley, 1989.

[6.3] Cheung, W. S. and F. H. Levien. *Microwave Made Simple, Principles and Applications.* Norwood: Artech House, 1985.

[6.4] Ginzton, E. L. *Microwave Measurements,* New York: McGraw-Hill, 1957.

[6.5] Gonzalez, G. *Microwave Transistor Amplifiers, Analysis and Design,* 2nd ed. Upper Saddle River: P. Hall, 1997.

[6.6] Kosow, I. W. and Hewlett-Packard Engineering Staff. *Microwave Theory and Measurement.* Englewood Cliffs: Prentice Hall, 1962.

[6.7] Liao, S. Y. *Microwave Circuit Analysis and Amplifier*

Design. Upper Saddle River: Prentice Hall, 1987.

[6.8] Pozar, D. M. *Microwave Engineering,* 2nd ed. New York: John Wiley & Sons, 1998.

[6.9] Radmanesh, M. M. *The Gateway to Understanding: Electrons to Waves and Beyond,* AuthorHouse, 2005.

[6.10] Radmanesh, M. M. *Cracking the Code of Our Physical Universe,* AuthorHouse, 2006.

[6.11] Reich, H. J., F. O. Phillip, H. L. Krauss, and J. G. Skalnik. *Microwave Theory and Techniques,* New York: D. Van Norstrand Company, Inc., 1953.

[6.12] Schwarz, S. E. *Electromagnetics for Engineers.* New York: Saunders College Publishing, 1990.

[6.13] Smith, P. H. *Transmission-Line Calculator,* Electronics, 12, pp. 29-31, Jan. 1939.

[6.14] Smith, P. H. *An Improved Transmission-Line Calculator,* Electronics, 17, pp. 130, Jan. 1944.

PART II

CIRCUIT DESIGN ESSENTIALS

CHAPTER 7

Design of Matching Networks

7.1 INTRODUCTION

Having studied the Smith chart in full detail and seen the ease and simplicity that it brings to the analysis of distributed or lumped element circuits, we now turn to the design of matching networks.

Applications #1 through #14 in Chapter 10 have in reality set the stage for most of the possible ways a Smith chart could be used as an essential tool in RF/microwave circuit analysis and more importantly in network design.

Many of these applications will be cited as references throughout the rest of this chapter in order to simplify and further speed up the process of the design of a matching network, which is an essential part of any modern active circuit.

7.2 DEFINITION OF IMPEDANCE MATCHING

At the outset of this section, we will define an important nomenclature:

DEFINITION- MATCHING: *is defined to be connecting two circuits (source and load) together via a coupling device or network in such* *a way that the maximum transfer of energy occurs between the two circuits.*

This is one of the most important design concepts in amplifier and oscillator design as shown in Figure 7.1.

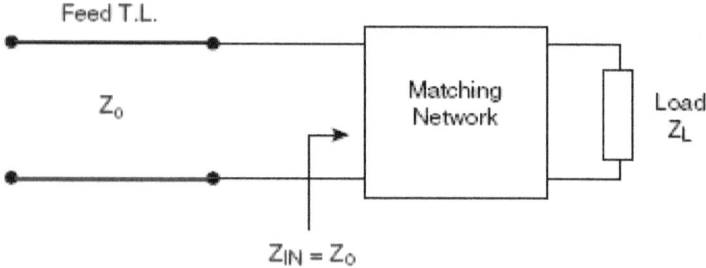

Figure 7.1 The concept of matching.

The concept of Impedance matching (also referred to as "tuning") is the third step in the overall design process (see Chapter 5, Figure 5.9) and is a very important concept at RF/microwave frequencies since it allows:

a. Maximum power transfer to occur from source to load, and
b. Signal-to-noise ratio to be improved because matching
    causes an increase in the signal level.

(a) and (b) are the primary reasons to employ tuning in practically all RF/microwave active circuit design. To get a conceptual understanding of why a matching network is needed in a circuit in general, we can visualize an active circuit in which a load impedance is different from the transmission line characteristic impedance causing power reflections back to the source.   To alleviate this problem and bring about zero power reflection from the load (i.e. maximum power transfer) a matching network needs to be inserted between the transmission line and the load.

Ideally, the matching network is lossless to prevent further loss of power to the load. It acts as an intermediary circuit between the two non-identical impedances in such a way that the feeding transmission line sees a perfect match (eliminating all possible reflections) while the multiple reflections existing between the load and the matching network will be unseen by the source.

7.3 SELECTION OF A MATCHING NETWORK

The selection of a lossless matching network is always possible as long as the load impedance is not purely imaginary and has in fact a non-zero real part.

There are many considerations in selecting a matching network including:

a. Simplicity-The simplest design is usually highly preferable since simpler matching networks have fewer elements, require less work to manufacture, are cheaper, are less lossy and more reliable compared to a more complicated and involved design.

b. Bandwidth--Any matching network can provide zero reflection at a single frequency, however, to achieve impedance matching over a frequency band more complex designs need to be used. Thus, there is a trade-off between design simplicity and matching bandwidth and eventually the network price as shown in Figure 7.2

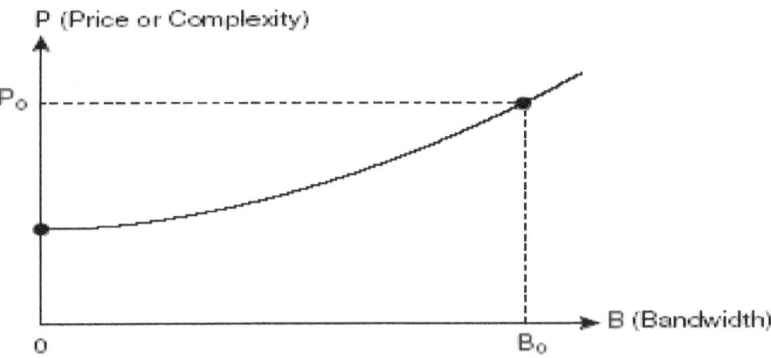

Figure 7.2 Price vs. bandwidth.

c. Feasibility of Manufacturing

To manufacture a certain design, one needs to consider first the type of transmission line technology with which the matching network will be implemented. This means that before the matching circuit is designed one needs to know before-hand whether it is a microstrip or coaxial line type technology, So that the matching circuit will be designed properly to integrate most efficiently with the rest of the network.

For example, in microstrip line technology due to its planar configuration, the use of quarter-wave transformers, stubs and chip lumped elements for matching are feasible. On the other hand, in waveguide technology, implementing tuning stubs for matching purposes is more predominant than lumped elements or $\lambda/4$ multi-section transformers.

d. Ease of Tunability
Variable loads require variable tuning. Thus the matching network design and implementation should account for this. To implement such an adjustable matching network may require a more complex design or even switching to a different type of transmission line technology in order to accommodate such a requirement.

These four considerations form the backbone of all design criteria. However, for the sake of clarity and ease, we will focus only on the first consideration, i.e., "simplicity", for the rest of this work and leave the other three considerations to more advanced texts.

NOTE: *There are cases where the matching network has two or more solutions for the same load impedance. The preference of one design over the other would greatly depend upon bringing the other three considerations into view, which will place the ensuing discussions outside the scope of this work.*

7.4 THE GOAL OF IMPEDANCE MATCHING
The most important design tool in amplifier and oscillator design is the concept of impedance matching. The goal of impedance matching in all of its different forms can be summed up into one issue, and that is:

THE GOAL OF IMPEDANCE MATCHING: *Making the input impedance of the load and the added matching network theoretically equal to the characteristic impedance of the feeding transmission line, thus allowing the maximum amount of power to transfer to the load.*

Next, section will delineate conditions under which maximum power transfer does take place.

7.4.1 Maximum Average Power Transfer

Consider the circuit (shown in Figure 7.3) which is a problem of great practical importance. In this circuit, source impedance (Z_S) is a known and fixed value, V_S is the phasor representation of the source sinusoidal voltage at the angular frequency (ω):

$$V_S = \text{Re}(|V_S| e^{j\omega t}) \tag{7.1}$$

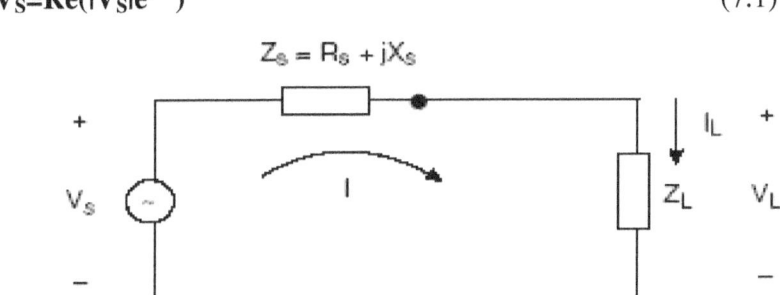

Figure 7.3 A general circuit.

The problem is to select the load impedance (Z_L) such that the maximum average power (P_{av}), at steady state, is obtained from the source and fed to the load. This problem can be easily solved with the help of the following theorem:

MAXIMUM POWER TRANSFER THEOREM
Consider the general circuit having a known source impedance and an unknown load as shown in Figure 7.3. The maximum power transfer theorem states that the maximum power that can be delivered to a load is only feasible when the load has an optimum impedance value $(Z_L)_{opt}$ equal to the complex conjugate of the source impedance value ($Z_{S,}$), i.e.,

$$(Z_L)_{opt} = Z_S^* \tag{7.2}$$

From Equation (7.2) we can see that:

$R_L=R_S$

$X_L=-X_S$

Considering the maximum power transfer as the cornerstone of matching, we can make the following conclusive observation about the goal of matching:

The goal of matching is adding a matching network to a load (Z_L) in such a way that the input impedance of the total combination will be located at the center of the Smith chart.

PROOF:

The average power delivered to the load can be written as:

$V_L=Z_L I_L$

$$I_L = \frac{V_S}{Z_S + Z_L}$$

$$P_{av} = \frac{1}{2}Re(V_L I_L^*) = \frac{1}{2}Re(Z_L|I_L|^2) = |V_S|^2 \frac{Re(Z_L)}{2|Z_S + Z_L|^2} \qquad (7.3)$$

Where V_S, $Z_S=R_S+jX_S$ and $Z_L=R_L+jX_L$ are the source voltage phasor, source impedance and load impedance, respectively.

Substitution in Equation (7.3) gives:

$$P_{av} = |V_S|^2 \frac{R_L}{2[(R_S + R_L)^2 + (X_S + X_L)^2]} \qquad (7.4a)$$

In Equation (7.4), V_S, R_S and X_S are given, R_L and X_L are to be chosen such that their value will maximize P_{av}. The reactance X_L is found by differentiating P_{av} with respect to X_L and setting it to zero, which yields:

$$\partial P_{av}/\partial X_L=0 \quad \Rightarrow X_L=-X_S \qquad (7.4b)$$

With this choice, the term $(X_L+X_S)^2$ in the denominator becomes zero, which minimizes the denominator and maximizes the expression with respect to X_L. Thus Equation (7.4) can now be written as:

$$P_{av} = |V_S|^2 \frac{R_L}{2(R_S + R_L)^2} \qquad (7.5)$$

Now to determine optimum R_L, we set the partial derivative of P_{av} with respect to R_L equal to zero, i.e.,

$$\frac{\partial P_{av}}{\partial R_L} = 0 \tag{7.6}$$

Upon differentiation and setting it equal to zero, we obtain:
$R_L = R_S$ Q.E.D. (7.7)

Using $Z_L = Z_S^*$ (referred to as a conjugately matched load), from Equation (7.4) we obtain the maximum power delivered to the load $(P_{av})_{max}$ as:

$$\left(P_{av}\right)_{max} = \frac{\left|V_S\right|^2}{8R_S} \tag{7.8}$$

Furthermore, under these conditions the power produced by the source is given by:

$$P_S = \frac{1}{2}\operatorname{Re}\left(V_S I_L^*\right) = \frac{1}{2}\operatorname{Re}\left(\frac{\left|V_S\right|^2}{Z_S + Z_L}\right) = \frac{\left|V_S\right|^2}{4R_S} \tag{7.9}$$

Thus we have:

$$P_S = \frac{\left|V_S\right|^2}{4R_S} = 2\left(P_{av}\right)_{max} \tag{7.10a}$$

From Equation (7.10), we can observe that the efficiency of a conjugately matched load is 50%, i.e.

$$\frac{\left(P_{av}\right)_{max}}{P_S} = 0.5 = 50\% \qquad \text{for } Z_L = Z_S^* \tag{7.10b}$$

NOTE 1: *For RF and microwave engineers this fact (i.e. 50% efficiency) is of much significance since the energy in the incoming electromagnetic waves would have been lost if it were not absorbed by a conjugately matched load (which is the first or "front-end stage" of a receiver).*

NOTE 2: *For power engineers and electric power companies, this situation is never allowed to occur and in fact the reverse is desired.*

This is because they are extremely interested in efficiency and want to deliver as much of the average power as possible to the load (i.e. the customer). Thus, huge power generators are never conjugately matched.

NOTE 3: *The "maximum power transfer theorem" assumes that the source impedance (Z_S) is a fixed and known quantity while the load impedance is a variable and unknown quantity (Z_L), whose value can be varied to the complex conjugate of Z_S, to achieve a maximum power transfer. Under this condition due to the complex conjugate condition, the total resistance in the circuit is given by:*

$Z_{tot}= Z_S+Z_L =2R_S=2R_L$

NOTE 4: *If the reverse is true, namely the load impedance (Z_L) is a known and fixed quantity and the source impedance (Z_S) a variable quantity, the requirement that source and load impedances be the complex conjugate of each other is no longer valid and does not apply in this case.*

Furthermore, to obtain maximum power transfer from the source to the load for this case, it can easily be observed that we need to have a minimum loss in the source. Thus, we can write the following for the source:

$R_S=0$
$X_S=-X_L$

Under this condition, since the reactances cancel out, the total resistance in the circuit is given by:
$Z_{tot}= Z_S+Z_L=R_L$

7.5 DESIGN OF MATCHING CIRCUITS USING LUMPED ELEMENTS

Considering the size of most modern RF/microwave circuits, actual discrete lumped element capacitors and inductors are used in the design process at low RF/microwave frequencies (up to around 1-2 GHz) or at higher frequencies (up to 60 GHz) if the circuit size is much smaller than the wavelength ($\ell<\lambda/10$)

Although microwave integrated circuit (MIC) technology has pushed the frequency limitation of lumped elements into the high microwave range, there are a large number of circuits whose size has become comparable with the signal wavelength at higher frequency ranges where using lumped elements would become completely impractical.

Thus one of the biggest limitations of the use of lumped elements is in circuits whose size has become comparable with the signal wavelength.

Furthermore, if the length (ℓ) of the lumped component is below ($\lambda/10$) as mentioned above, then they can be used in hybrid or "Monolithic" MICs at frequencies up to 60 GHz.   At these high frequencies, electrical elements can be realized via several methods. These methods for each component can be summarized as follows:

1. CAPACITORS:
a. A single-gap capacitor (C<0.5 pF)
b. An inter-digital gap capacitor in a microstrip line (C<0.5 pF)
c. A short or open transmission line stub (C<0.1 pF)
d. A chip capacitor
e. A metal-insulator-metal(MIM) capacitor (C<25 pF)

These are shown in Figure 7.4.

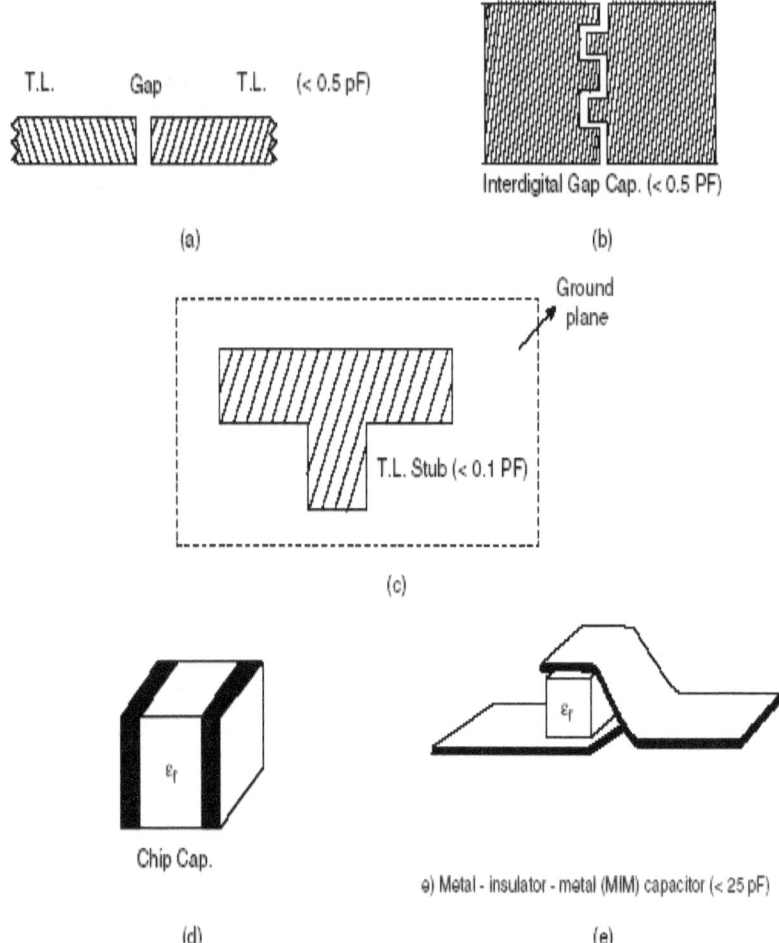

Figure 7.4 Several types of capacitors.

2. RESISTORS:
a. **Planar Resistor:** Thin film technology using NiChrome or doped semiconductor material,
b. **Chip resistor.**

These are shown in Figures 7.5 (a) and (b).

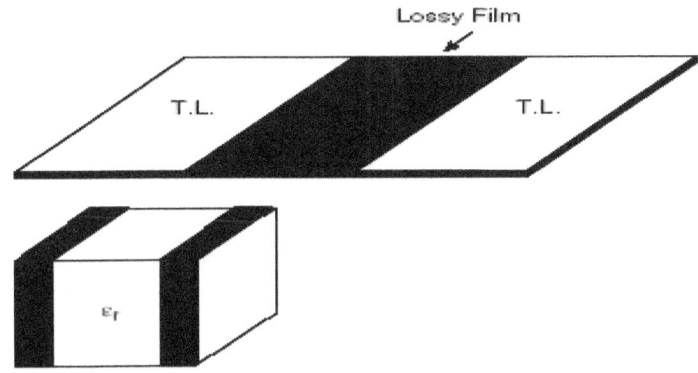

Figure 7.5 The two types of resistors.

3. Inductors

a. A loop of a transmission line,

b. A short length of a transmission line,

c. A spiral inductor using an air bridge.

These are shown in Figures 7.6 (a), (b), and (c).

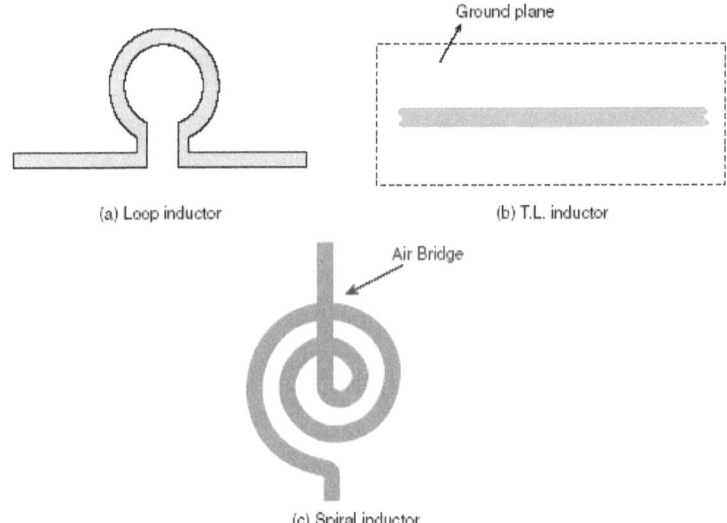

(a) Loop inductor

(b) T.L. inductor

(c) Spiral inductor

Figure 7.6 Three types of inductors.

7.5.1 Matching Network Design Using L-Sections

As already discussed in Chapter 2, the simplest type of matching network is the L(ell)-section consisting of two reactive elements that match a load to a transmission line. The actual configuration is of the form of an inverted L. The two possibilities are shown in Figure 7.7.

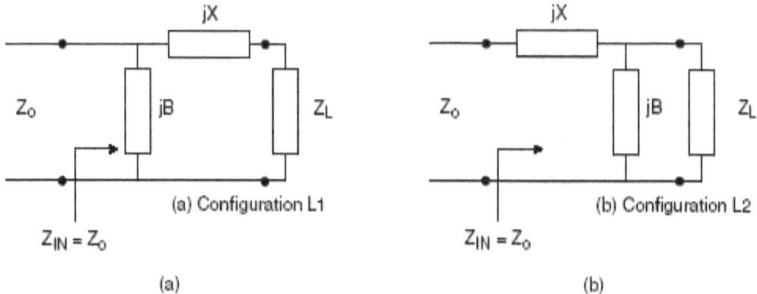

(a)   (b)

Figure 7.7 Two possibilities of an L-section matching network.

Considering the fact that either of the two reactive elements can be an inductor or a capacitor, circuit configurations L1 and L2 provide a total of eight different possibilities for a given load. The location of the load on the "smith chart" determines the useful configuration as discussed in the next section.

7.5.2 Design Based on the Load Location

Depending on the location of the normalized load impedance on the Smith chart, one or both of the two configurations L1 and L2 may become practical.

The location of the load becomes crucial in the choice of a circuit configuration for the purpose of matching. The load location can have three distinct possibilities:

CASE I. THE LOAD IS LOCATED INSIDE THE (1+jx) CIRCLE (RESISTANCE UNITY CIRCLE)
In this case, we can see from Figure 7.8 that the first element has to be a shunt element thus configuration L2 is the only practical one.

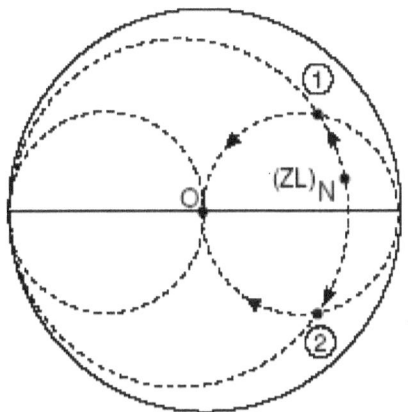

Figure 7.8 Smith Chart Solution.

Using configuration L2, two possible solutions exist:

- **Solution (1): shunt L & series C**
- **Solution (2): shunt C & series L**

These are shown in Figure 7.9.

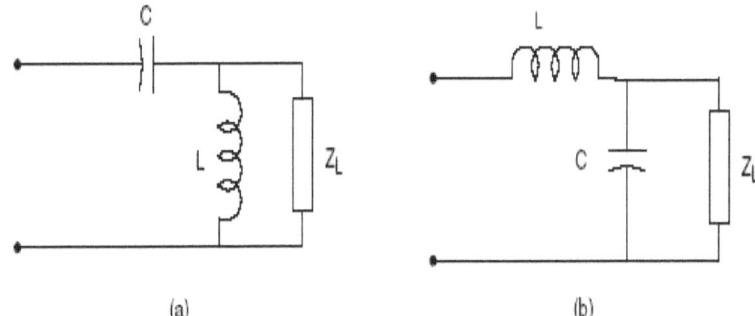

Figure 7.9 Two possible solutions for case I.

Case II. The load is located inside the (1+ jb) circle (conductance unity circle)
In this case, from Figure 7.10 we can see that the first element has to be a series element thus only configuration L1 becomes useful. Similar to case 1, there are two solutions possible as shown in Figure 7.10:

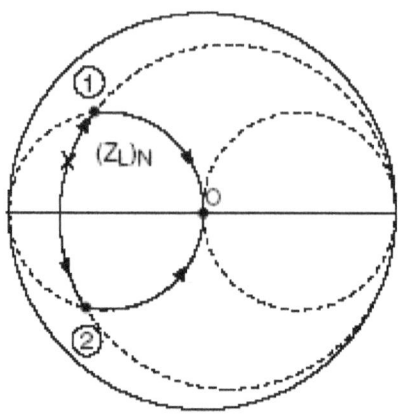

Figure 7.10 Smith Chart Solution.

- **Solution (1): series L & shunt C**
- **Solution (2): series C & shunt L**

The circuits for these two solutions are shown in Figure 7.11.

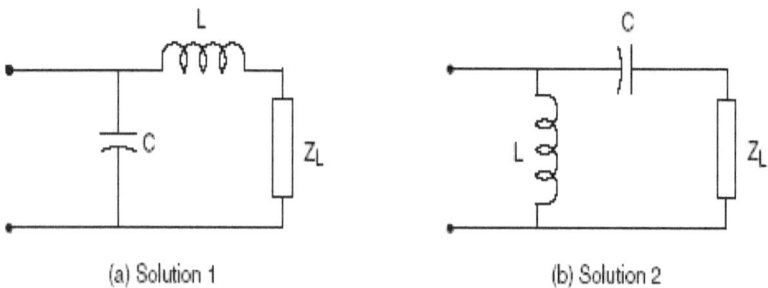

(a) Solution 1   (b) Solution 2

Figure 7.11 Two possible solutions for case II.

CASE III. THE LOAD IS LOCATED OUTSIDE THE (1+ jx) AND (1+ jb) CIRCLE

In this case, there are four solutions possible as shown in Figure 7.12. These four possibilities are described below:

Solutions (1) and (2): both require a series element first which makes configuration L1 the only practical one. Thus the solutions are:

- **Solution (1): series C & shunt C**

- **Solution (2): series C & shunt L**

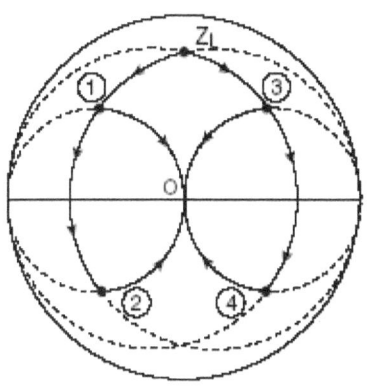

Figure 7.12 Smith Chart Solution.

The circuits for the two solutions are shown in Figure 7.13.

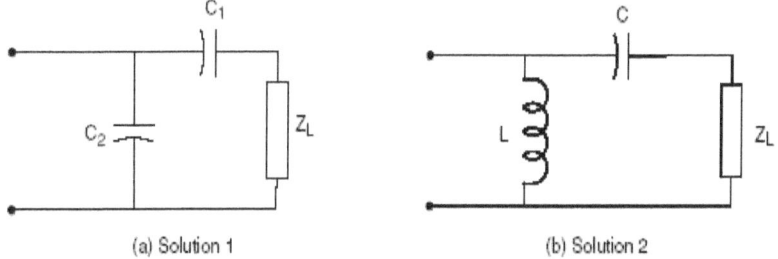

(a) Solution 1   (b) Solution 2

Figure 7.13 Solutions 1 and 2 for case III.

Solutions (3) and (4) below, Both require a shunt element inserted first, which makes configuration L2 useful. These two solutions are:

- **Solution (3): shunt C & series C**
- **Solution (4): shunt C & series L**

These are shown in Figure 7.14.

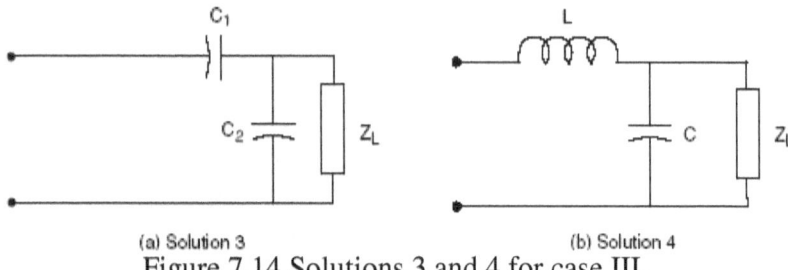

(a) Solution 3   (b) Solution 4

Figure 7.14 Solutions 3 and 4 for case III.

7.5.3 Design Flexibility

Considering all three cases (i.e. I, II and III), it appears that case III has the highest flexibility since when the load is located outside the unity circle, we have the highest amount of design flexibility to suit the designer's needs.

If the load falls inside any of the unity circles, one may be able to add a reactive element to the load in such a way as to bring the combined load to the outside of the unity circle and then take advantage of the matching possibilities that are available at the outside of these two unity circles.

EXAMPLE 7.1

Consider a load $(Z_L)_N$ located inside the $(1+jb)$ circle. Discuss the matching possibilities for this load.

Solution:

This load obviously has two matching possibilities (1 and 2) as discussed earlier in case II.

Let us add a series "L" to take the load Z_L to Z_L' outside the $(1+jB)$ circle as shown in Figure 7.15.

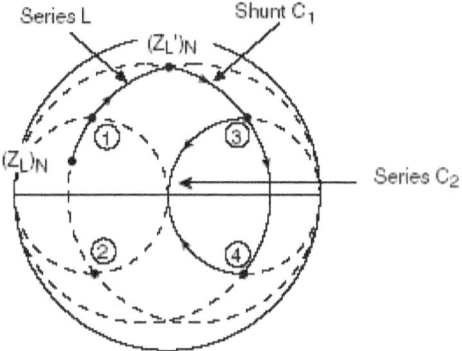

Figure 7.15 Smith Chart Solutions.

Now since $(Z_L')_N$ is outside, we have two additional possibilities (3 and 4) which may be more suitable for our design needs. Selecting solution "3", we can see that the final matching circuit will have the following three elements as shown in Figure 7.16.

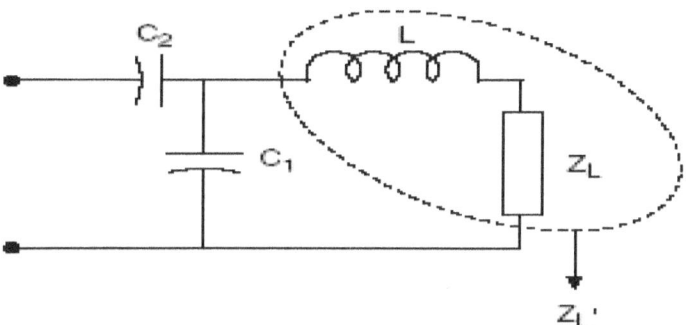

Figure 7.16 Matching solutions.

As discussed in Section 7.3, there seem to be four main considerations that govern the matching circuit design process i.e.: simplicity, bandwidth, the feasibility of manufacturing, and ease of tunability. These four criteria heavily influence one's decision in the choice of the matching circuit's design.

Therefore, even though from the simplicity point of view solution (3) shown in the previous example seems to be more complex than a simple L-design, there are instances where the other three considerations would become paramount and thus make this design a valuable one.

7.5.4 Design Rules for Matching Networks (Lumped Elements)

Based on the discussion presented in the previous two sections, there are certain rules that if followed would simplify and even speed up the matching circuit design process. These rules can be summarized as follows:

Rule #1. Use a ZY Smith chart at all times.

Rule #2. Always start off from the load end and travel "toward Generator" (in order to prevent uncertainty and confusion about the starting point).

Rule #3. Always move on a constant-R or constant-G circle in such a way as to arrive eventually at the center of the Smith chart.

Rule #4. Each motion along a constant-R or constant-G circle gives the value of a reactive element.

Rule #5. Moving on a constant-R circle yields series reactive elements, whereas moving on a constant-G circle yields shunt reactive elements.

Rule#6. The direction of travel (or motion) on a constant-R or constant-G circle determines the type of element to be used, i.e. a capacitor or an inductor.

Rules 5 and 6 lead to the following additional two rules.

Rule #7. When the motion is upward, in most cases it corresponds to a series or a shunt inductor.

Rule #8. When the motion is downward, in most cases it corresponds to a series or a shunt capacitor.

NOTE: *These rules are merely a guideline to be followed in the matching circuit design process. They will never replace the theoretical and practical understandings that go into making them. These understandings as contained in this work, are the essentials from which all of these rules have been derived.*

EXAMPLE 7.2

Given the circuit shown in Figure 7.17, design a lumped element matching network at 1 GHz that would transform
$Z_L = 10 + j10 \ \Omega$ *into a 50 Ω transmission line.*

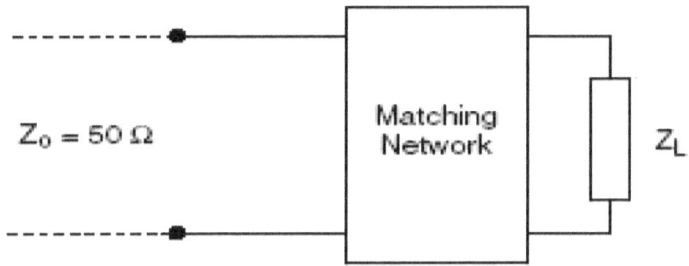

Figure 7.17 Circuit for Example 7.2.

Solution:

a. $Z_O = 50 \ \Omega$, $Y_O = 0.02$ S
$(Z_L)_N = (10 + j10)/50 = 0.2 + j0.2$

b. Locate $(Z_L)_N$ on the smith chart as shown in Figure 7.18.

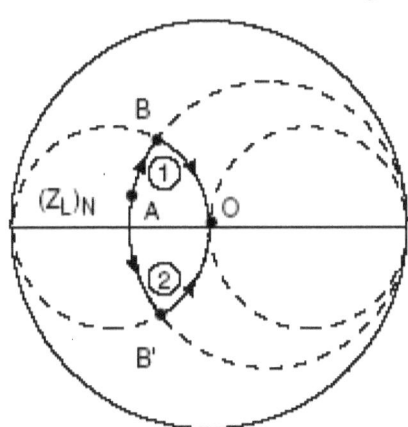

Figure 7.18 Smith Chart Solutions.

c. Since $(Z_L)_N$ is inside the unity conductance circle, this would correspond to case II and has two possible solutions.

SOLUTION (1):
Start from the load on a constant-R circle and move up from point "A" to "B". This Yields Series L:
$j\omega L = j0.2 \times 50 \Rightarrow L = 1.59$ nH
-Now starting from point "B", a motion downward on the unity conductance circle yields a shunt C:
$j\omega C = j2.0 \times 0.2 \Rightarrow C = 6.37$ pF
The final circuit schematic is shown in Figure 7.19.

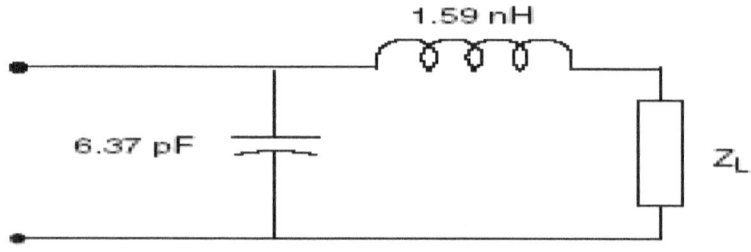

Figure 7.19 Circuit for solution 1.

SOLUTION (2):
Starting from the load, move downward on a constant-R circle (series C) to point "B'" and the upward (shunt L) to arrive at "C" s follows:
Series $C \Rightarrow 1/j\omega C = -j0.6 \times 50 \Rightarrow C = 5.3$ pF
Shunt $L \Rightarrow 1/j\omega L = -j2 \times 0.02 \Rightarrow L = 3.98$ nH
The schematic for solution (2) is shown in Figure 7.20.

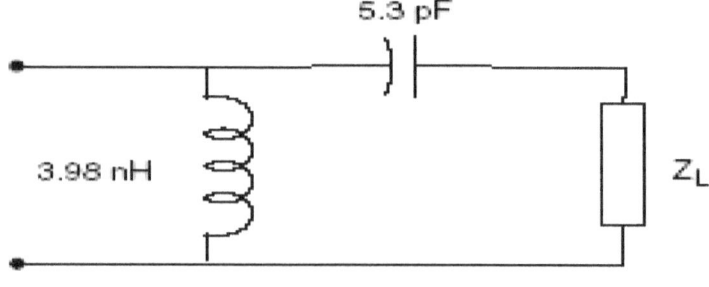

Figure 7.20 Circuit for solution 2.

7.6 MATCHING NETWORK DESIGN USING DISTRIBUTED ELEMENTS

At higher frequencies where the component or circuit size is comparable with wavelength, distributed components may be used to match the load to the transmission line.

The most common technique in this type of design is the use of a single open-circuited or short-circuited length of a transmission line (called a stub) connected either in parallel or in series with the transmission feed line at a certain distance from the load as shown in Figure 7.21.

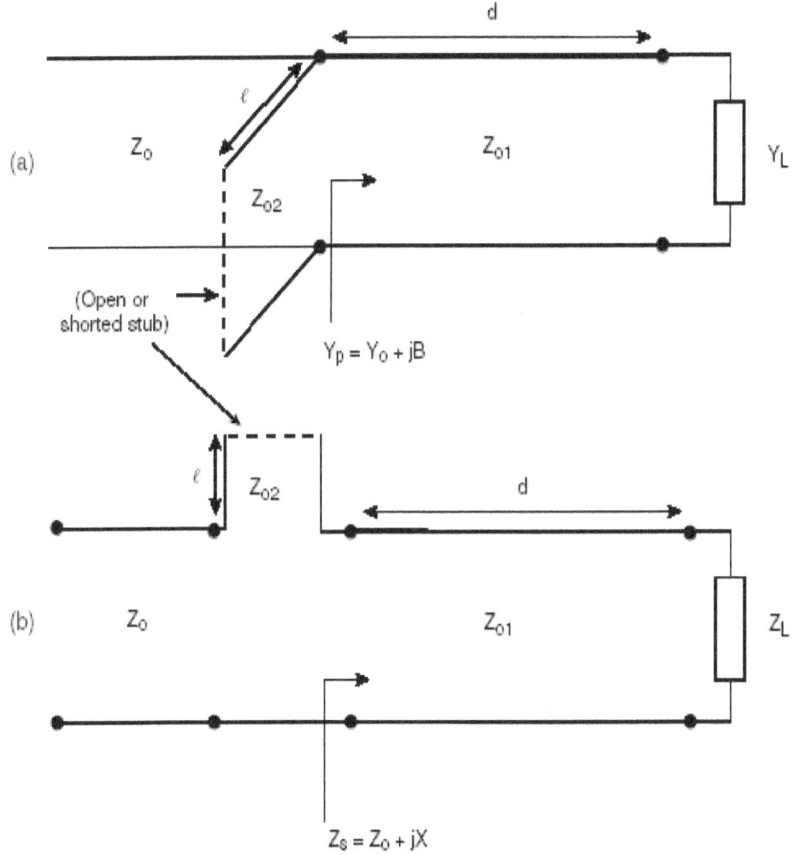

Figure 7.21 A stub located a distance "d" from the load.

Rather than using a two-wire transmission line schematic, alternate microstrip line schematic for short- and open-stubs can be drawn more effectively as shown in Figure 7.22 (a),(b),(c) and (d).

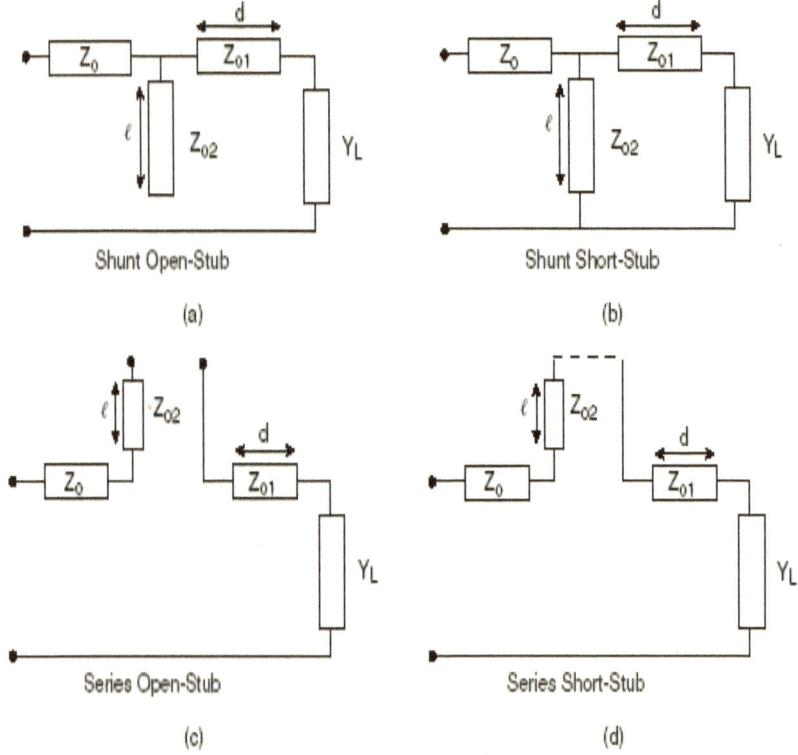

Figure 7.22 Microstrip-line schematics for shunt and series stubs.

Such a matching network is easy to build using microstrip or stripline technology. In single-stub matching networks, the two variable parameters are the distance "d" (from load to stub) and the length "ℓ" (stub's length) which provides the value of stub susceptance or reactance.

Selection of distance "d" is crucial for both shunt and series stub as explained below:

a. For the shunt stub case, "d" should be chosen such that the input admittance Y_P (seen looking into the line before adding the stub) is of the form $Y_P = Y_O + jB$ with the stub susceptance selected as (-jB) resulting in a matched condition.

b. for the series stub case, "d" should be chosen such that the impedance Z_S (seen looking into the line before the addition of the stub) is of the form $Z_S = Z_O + jX$ with the stub reactance selected as -jX, resulting in a matched condition.

7.6.1 Choice of Short- or Open-Circuited Stubs

With a $\lambda/4$ difference in length between the two, a short or open transmission line with proper length, can provide any value of reactance or susceptance needed for the design.

Structural considerations behind the choice of a short- versus an open-stub are as follows:

a. OPEN STUBS: For microstrip and stripline technology use of open stubs is preferred. the use of short stubs requires a via-hole through the substrate to the ground plane, which adds extra work and can be eliminated through the use of open circuits.

b. SHORT STUBS: For coaxial line or waveguide as a transmission line media,   the use of short stubs is preferred because the open stubs may radiate causing power losses thus making the stub no longer a purely reactive element.

7.6.2 Stub Realization Using Microstrip Lines

Series transmission lines and shunt stubs (short or open) can easily be realized using design steps for microstrip line technology (as outlined in Chapter 7).

Given a dielectric constant (ε_r), its height (h) and a certain characteristic impedance value (Z_O), the width of the microstrip line (W) can be calculated.

For example, a series transmission line and an open shunt stub schematic as well as its microstrip realization is shown in Figure 7.23.

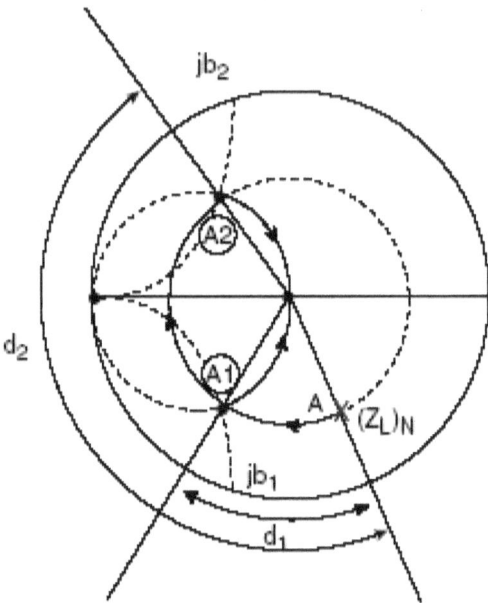

Figure 7.23 Smith Chart solution.

To minimize the microstrip transition interaction and improve the input VSWR, many designers use a balanced approach for shunt stubs rather than a single stub. Using balanced shunt stubs technique, two stubs of the same length (as the single stub) but twice the characteristic impedance are placed in parallel as shown in Figure 7.24. That is,

$$\ell_2' = \ell_2 \tag{7.11}$$

$$Z_{02}' = 2Z_{02} \tag{7.12}$$

The reason we use twice the characteristic impedance for each open shunt stub is due to the fact that each half of the balanced stub (Y_{stub}) must provide half the total admittance (Y_{tot}), i.e.,

$$Y_{stub} = jY_{02}' \tan\beta\ell_2' = \frac{1}{2} Y_{tot} \tag{7.13}$$

Where

$$Y_{tot} = j Y_{02} \tan\beta\ell_2 \tag{7.14}$$

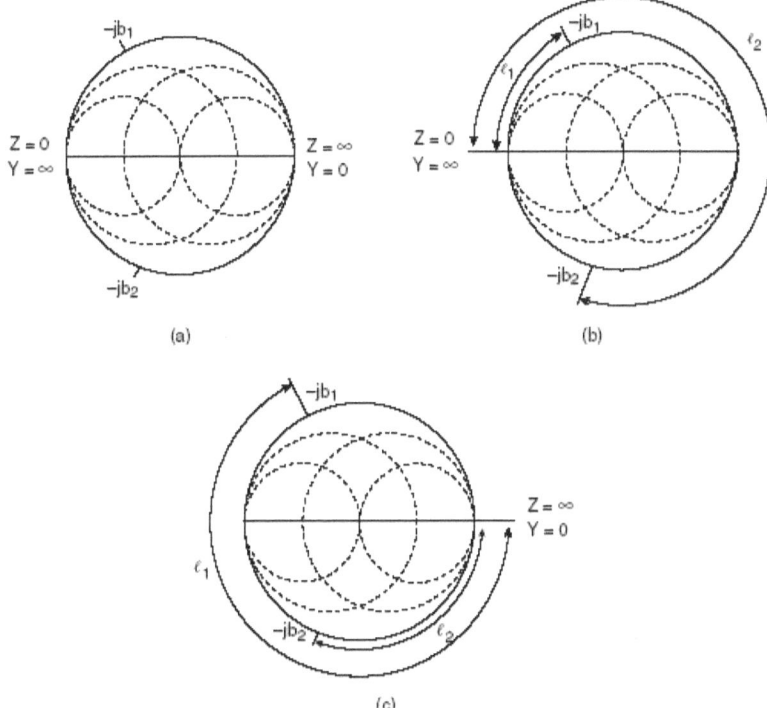

Figure 7.24 Plotting stub values on the Smith Chart.

Thus substituting Equation (7.14) in (7.13), we have:

$$Y_{O2}' \tan\beta\ell_2' = \frac{1}{2} Y_{O2} \tan\beta\ell_2 \qquad (7.15)$$

Choosing $\ell_2' = \ell_2$ yields:

$$Y_{O2}' = \frac{1}{2} Y_{O2} \qquad (7.16)$$

Or,

$$Z_{O2}' = 2Z_{O2} \qquad (7.17)$$

A similar discussion applies to a short shunt stub with the same conclusions.

7.6.3 Design Steps for Single Stub Matching (Using The Same Characteristic Impedance)

The Smith chart can be used effectively to design the distance "d" of the stub to the load and the length "ℓ" of the stub to create the proper value of susceptance or reactance (for more details see application #8, Chapter 6). There are two circuit configurations where the stub is either in parallel or in series.   Each case needs to be treated separately as follows:

a. PARALLEL STUB DESIGN
This is shown in Figure 7.23. The design process has the following steps:

Step 1. Plot $(Y_L)_N$ on the ZY chart
(Please note that a single Y-chart could also be used as well but the load impedance has to be inverted first)

Step 2. Draw the appropriate VSWR circle which goes through $(Z_L)_N$

Step 3. On the VSWR circle, move (toward generator) to intersect the $(1+jb)$ conductance unity circle at two solutions located at points "A1" and "A2" as shown in Figure 7.23:

- **Solution #1: $Y_1 = 1+ jb_1$ (distance d_1)**
- **Solution #2: $Y_2 = 1+ jb_2$ (distance d_2)**

Step 4. Now add a shunt susceptance of either $-jb_1$ (solution #1) or $-jb_2$ (solution #2) to arrive at the center of the chart.

Step 5. To determine lengths ℓ_1 and ℓ_2, we first locate $-jb_1$ and $-jb_2$ on the Smith chart as shown in Figure 7.24a.

Then starting from $Z = \infty$ (for open stubs) or $Z = 0$ (for short stubs), we travel along the outer edge of the chart "toward generator" to arrive at $-jb_1$ (for open) or $-jb_2$ (for short) stubs. The lengths can be read off on the circular scale on the outer edge of the chart. These are shown in Figures 7.24b and 7.24c.

EXAMPLE 7.3 (PARALLEL STUB DESIGN)
Design a matching network using a single shunt open stub as a tuning element to match a load impedance $Z_L=15+j10$ Ω to a 50 Ω transmission line(see Figure 7.25)

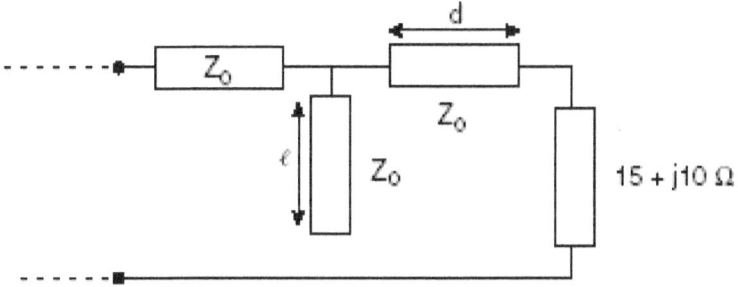

Figure 7.25 Circuit for example 7.3.

Solution:
a. Plot $(Z_L)_N=(15+j10)/50=0.3+j0.2$ on the ZY-chart
b. Draw the VSWR circle through $(Z_L)_N$ as shown in Figure 7.26.

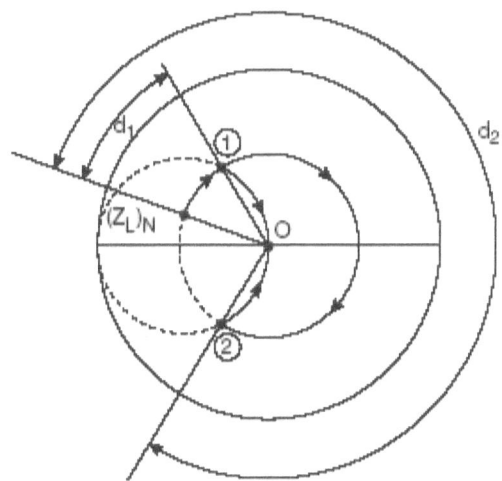

Figure 7.26 Smith Chart solution.

c. Move toward generator to meet the $(1+jb)$ circle at two points, giving two solutions:
$(Y_1)_N=1-j1.33$ with $d_1=0.044\lambda$ and $-jb_1=j1.33$
$(Y_2)_N=1+j1.33$ with $d_2=0.387\lambda$ and $-jb_2=-j1.33$
d. From Figure 7.27 we can read off ℓ_1 and ℓ_2 as:
$\ell_1=0.147\lambda$
$\ell_2=0.353\lambda$

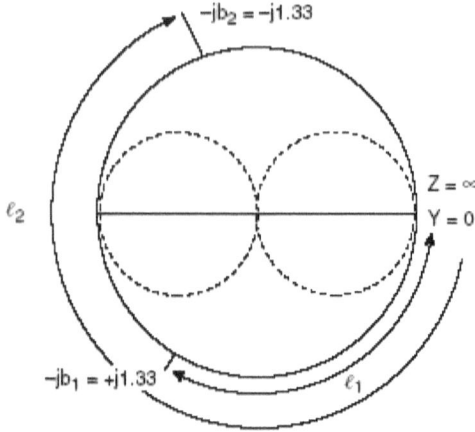

Figure 7.27 Stubs on the Smith Chart.

The two possible design schematics are shown in Figure 7.28.

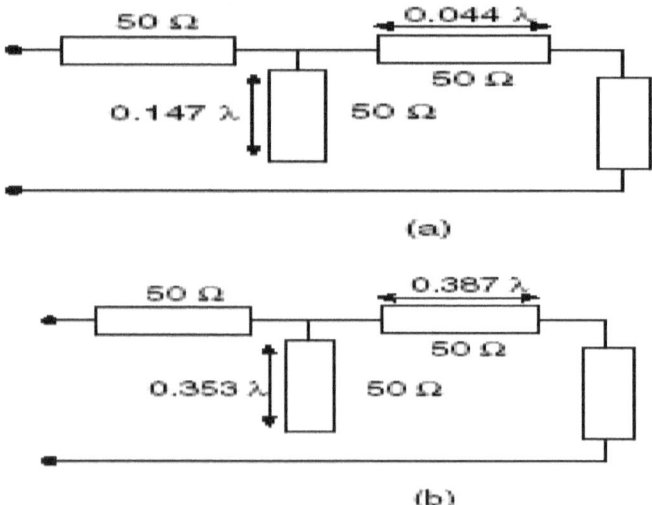

Figure 7.28 Two circuit design schematics.

NOTE: *Design #1 is more desirable Since it is usually preferred to keep the matching stub as close as possible to the load (i.e. smaller "d") in order to improve:*

❀ *The sensitivity of the matching network to frequency and thus providing a larger bandwidth,*
❀ *The standing wave losses occurring on the line between the stub and the load possibly due to a high VSWR.*

b. SERIES STUB DESIGN
From Figure 7.29, the design steps are as follows:
Step 1. Plot $(Z_L)_N$ on the ZY-chart.

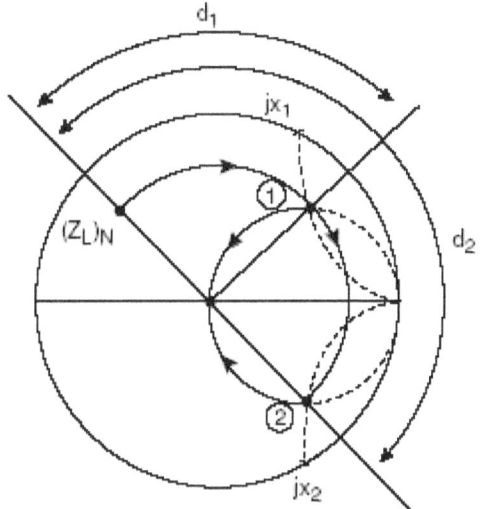

Figure 7.29 Smith Chart solution.

Step 2. Draw the appropriate VSWR circle.

Step 3. Move "toward generator" to intersect the $(1+jx)$, unity resistance circle, at two points (1) and (2) as shown in Figure 7.29:

- Solution #1: $Z_1 = 1 + jx_1$ (distance d_1), $Z_{1stub} = -jx_1$
- Solution #2: $Z_2 = 1 + jx_2$ (distance d_2) $Z_{2stub} = -jx_2$

Step 4. Now add a series reactance of either $-jx_1$ for solution#1 (or $-jx_2$ for solution #2) to cancel the existing reactance.

Step 5. stub lengths ℓ_1 (or ℓ_2) is now calculated by first locating reactance $-jx_1$ (or $-jx_2$) on the Smith chart (see Figures 7.30a,b).

For a series open stub, start from $Z = \infty$ and travel on the outer edge of the chart "Toward generator" to arrive at $-jx_1$ for solution #1 (or $-jx_2$ for solution #2). On the other hand, for a series short stub, repeat the above procedure except, start from $Z = 0$ as shown in Figure 7.30.

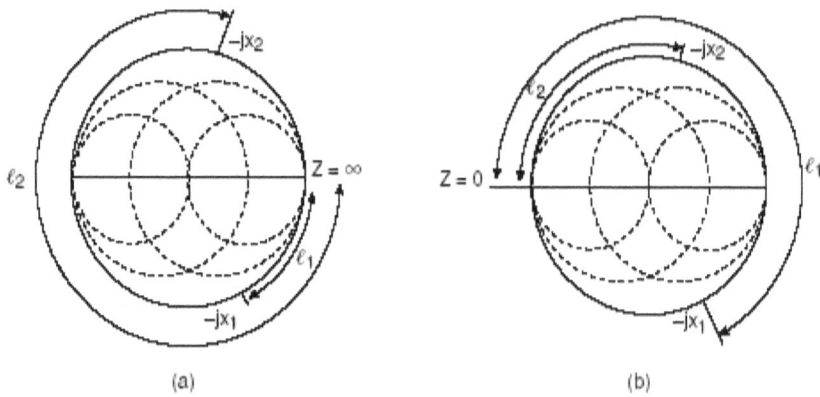

(a)   (b)

Figure 7.30 Smith Chart plot of stubs.

EXAMPLE 7.4 (SERIES STUB DESIGN)
Using a single series open stub, design a matching network that will transform a load impedance $Z_L = 100 + j80$ Ω to a 50 Ω Feed transmission line as shown in Figure, 7.31.

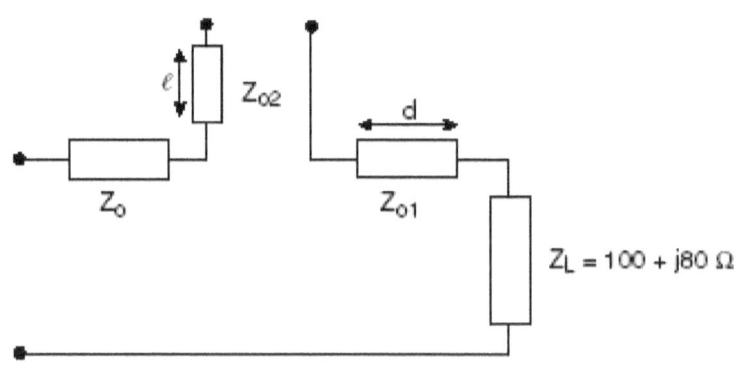

Figure 7.31 Circuit for example 7.4.

Solution:

a. Plot $(Z_L)_N = (100 + j80)/50 = 2 + j1.3$ on the smith chart.

b. Draw the VSWR circle (see Figure 7.32)

c. Move "toward generator" to intersect (1+jx) circle at two points (1) and (2) giving:

$Z_1=1-j1.33$ with $d_1=0.120\lambda$ and $Z_{1stub}=j1.33$
$Z_2=1+j1.33$ with $d_2=0.463\lambda$ and $Z_{2stub}=-j1.33$

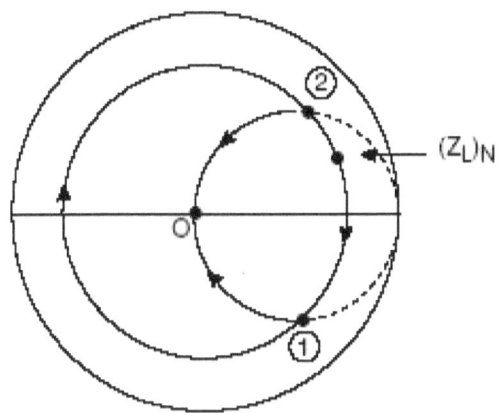

Figure 7.32 Smith Chart solution.

d. Lengths ℓ_1 and ℓ_2 can be read off from Figure 7.33 as:
$\ell_1=0.397\lambda$
$\ell_2=0.103\lambda$

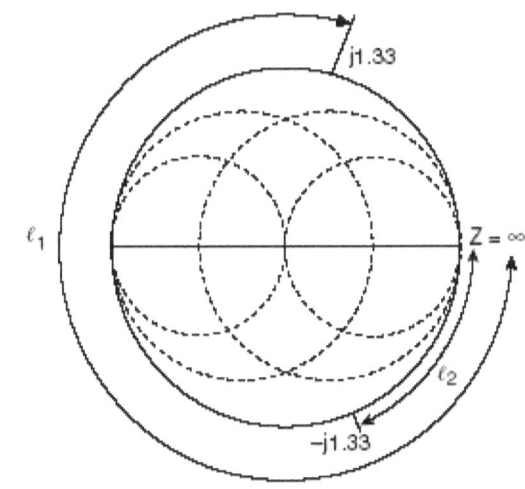

Figure 7.33 Stubs on the Smith Chart.

The two possible design schematics are shown in Figures 7.34a,b.

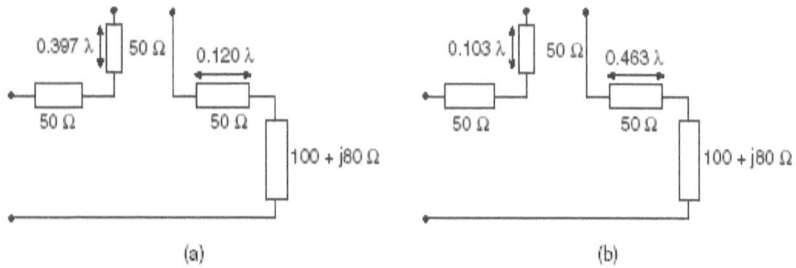

Figure 7.34 Final design schematics.

7.6.4 Design of Single Stub Matching (using a different characteristic impedance)

As can be seen from the previous two examples, the stub, the feed transmission line, and the line between the load and the stub all have the same characteristic impedance (Z_O).

However, if a characteristic impedance other than Z_O is desired, the location and length of the stub will now be different. The following procedure can be used effectively to solve for this situation:

A. Add a stub at the load such that it will reduce the load to a purely resistive value as shown in Figure 7.35. The following two types of stubs (each with the possibility of being short- or open-circuited) should be considered:

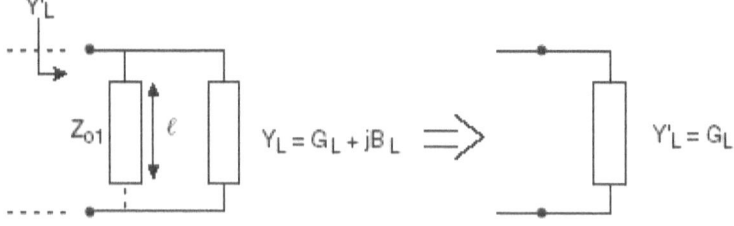

Figure 7.35 Conversion of the load to a purely resistive load.

A1. Shunt Stub (see Figure 7.35a):
Load: $\mathbf{Y_L} = \mathbf{G_L} + \mathbf{jB_L}$
Stub: $-jB_L$
Giving: $\mathbf{Y_L'} = \mathbf{G_L}$

The new load (Y_L') is purely resistive. To determine the length of the stub, we have to specify the stub termination as being short or open:

-SHORT SHUNT STUB(Z_{O1})

$Y_{SC} = -jY_{O1}/\tan\beta\ell = -jB_L$

$\Rightarrow Y_{O1} = B_L\tan\beta\ell,$

$Z_{O1} = 1/Y_{O1} = 1/B_L\tan\beta\ell$

By proper choice of stub length(ℓ), any positive value of stub characteristic impedance can be realized; for example:

$\ell = \lambda/8 \Rightarrow Z_{O1} = 1/B_L$ (for $B_L > 0$)

$\ell = 3\lambda/8 \Rightarrow Z_{O1} = -1/B_L$ (for $B_L < 0$)

-OPEN SHUNT STUB(Z_{O1})

$Y_{OC} = jY_{O1}\tan\beta\ell = -jB_L$

$\Rightarrow Y_{O1} = -B_L/\tan\beta\ell,$

$Z_{O1} = 1/Y_{O1} = -\tan\beta\ell/B_L$

By proper choice of stub length(ℓ), any positive value of stub characteristic impedance can be realized; for example:

$\ell = \lambda/8 \Rightarrow Z_{O1} = -1/B_L$ (for $B_L < 0$)

$\ell = 3\lambda/8 \Rightarrow Z_{O1} = 1/B_L$ (for $B_L > 0$)

A2. Series stub (see Figure 7.35b):

Load: $Z_L = R_L + jX_L$

Stub: $-jX_L$

Giving: $Z_L' = R_L$

The new load (Z_L') is purely resistive. To determine the length of the stub, we have to specify the stub termination as being short or open:

-SHORT SERIES STUB(Z_{O1})

$Z_{SC} = -jZ_{O1}\tan\beta\ell = -jX_L$

$\Rightarrow Z_{O1} = X_L/\tan\beta\ell,$

By proper choice of stub length(ℓ), any positive value of stub characteristic impedance can be realized; for example:

$\ell = \lambda/8 \Rightarrow Z_{O1} = X_L$ (for $X_L > 0$)

$\ell = 3\lambda/8 \Rightarrow Z_{O1} = -X_L$ (for $X_L < 0$)

-Open Series Stub(Z_{O1})

$Z_{OC}=jZ_{O1}/\tan\beta\ell=-jX_L$

$\Rightarrow Z_{O1}=-X_L\tan\beta\ell,$

By proper choice of stub length (ℓ), any positive value of stub characteristic impedance can be realized; for example:

$\ell=\lambda/8 \Rightarrow Z_{O1}=-X_L$ (for $X_L<0$)

$\ell=3\lambda/8 \Rightarrow Z_{O1}=X_L$ (for $X_L>0$)

B. Considering that the new load (Y_L' or Z_L') is now purely resistive (G_L or R_L), a simple quarter-wave ($\lambda/4$) transformer (Z_O') can be used efficiently to transform (Y_L' or Z_L') to Z_O. The $\lambda/4$ transformer has a characteristic impedance of:

Shunt stub: $\mathbf{Z_O'=(Z_O/Y_L')^{1/2}=(Z_O/G_L)^{1/2}}$

Or,

Series stub: $\mathbf{Z_O'=(Z_L'Z_O)^{1/2}=(R_LZ_O)^{1/2}}$

Where Z_O is the characteristic impedance of the feed line (assumed to be lossless), and is a positive real number. This method is depicted on the Smith chart, as shown in Figure 7.36.

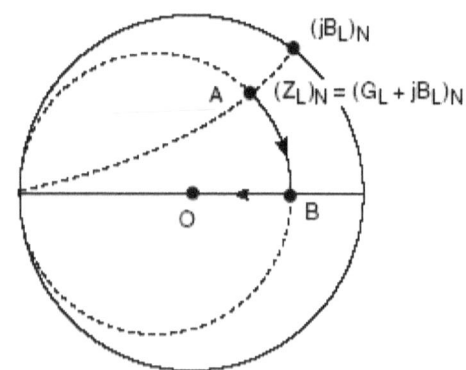

Figure 7.36 Smith Chart solution.

Starting from point "A", we travel on a constant conductance circle to point "B" (located on the purely resistive axis). A quarter-wave transformer would then take it from point "B" to "O", the center of the chart.

c) The final circuit schematic for the matching network is shown below in Figure 7.37.

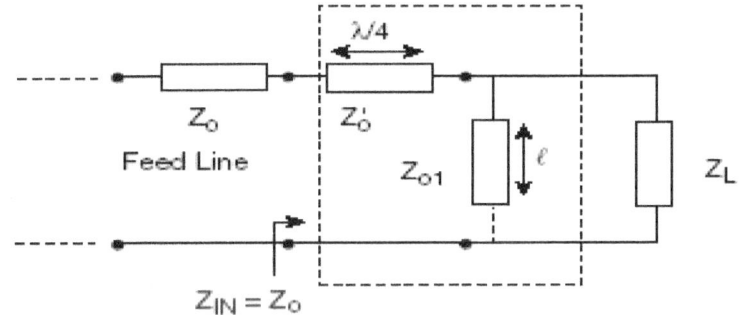

Figure 7.37 Final circuit schematic.

Example 7.5

Design a matching network using a quarter-wave transformer that would transform a 50 Ω load to $\Gamma_{in}=0.68\angle 97°$as shown in Figure 7.38.

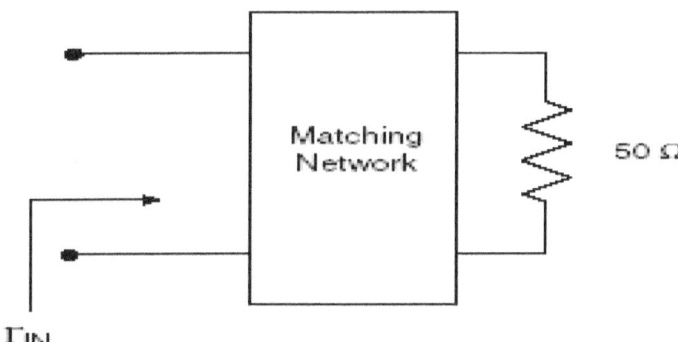

Figure 7.38 Circuit for example 7.5.

Solution:

To design the matching network we need first transform the load to the resistive part of Γ_{in} and then add a reactance equal to the reactance of Γ_{in} as follows:

a) Plot Γ_{in} =0.68∠97° on the smith chart as shown in Figure 7.38 and read off $(Y_{in})_N$:

$(Y_{in})_N=(0.4-j1.05) \Rightarrow Y_{in}=Y_O(Y_{in})_N =0.008-j0.021$ S

b) Starting from the center of the chart (i.e. load at point "O"), A quarter-wave transformer (Z_{O1}) is now added to transform 50 Ω to ($R_{in}=1/0.008=125$ Ω) (point "A" in Figure 7.39) :

$Z_{O1}=(50/0.008)^{1/2}= 79$ Ω

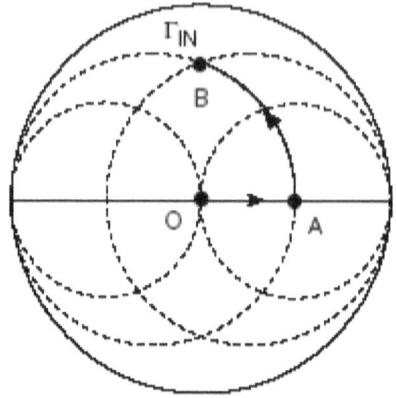

Figure 7.39 Smith Chart solution.

c) The reactive part (B_L=-j0.021) can now be synthesized by using an open-circuited shunt stub of length ℓ=3λ/8 to arrive at point "B". The characteristic impedance is given by:

Z_{O2}=1/0.021=47.6 Ω

(Note: Y_{OC}=jY_{O2}tan$\beta\ell$=-j/47.6 since tan$\beta\ell$=-1)

d) The final circuit schematic is shown in Figure 7.40.

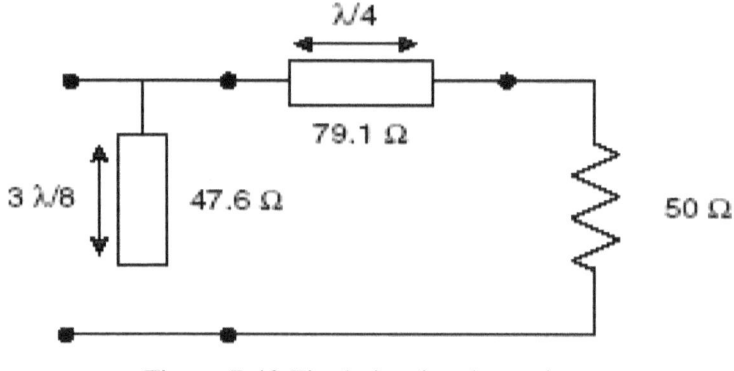

Figure 7.40 Final circuit schematic.

7.6.5 Generalized impedance matching network for a non-zero reflection coefficient

When the input and the output reflection coefficients are both non-zero, then the matching network would transform the load impedance to a point (other than the center of the chart) corresponding to the desired input reflection coefficient.

This case is a very generalized concept of a matching network and is in contrast with the case where either the input or the output reflection coefficient was at the center of the chart.

This type of matching network could occur for example in an intermediate matching stage of a two-stage amplifier. Figure 7.41 shows the generalized concept of a matching network along with the plot of the input and output reflection coefficients in the Smith chart.

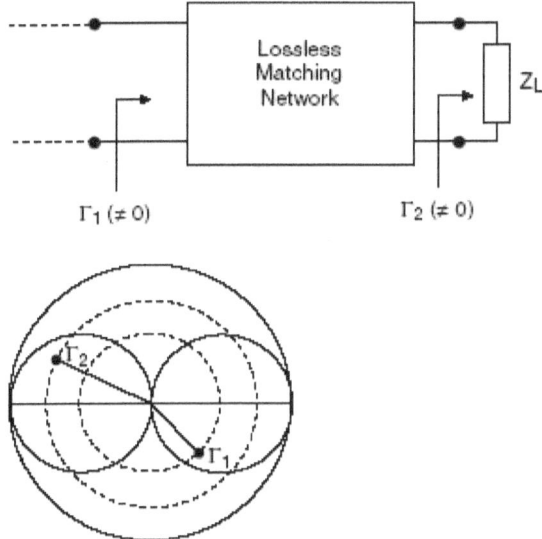

Figure 7.41 The generalized concept of matching.

There are a number of possible solutions that would lead to the desired matching network.

a. SOLUTION #1:
Assuming the input and output reflection coefficients to be Γ_1 and Γ_2, respectively, an obvious solution is to convert the load (Γ_2) to 50 Ω first and then match Γ_1 to 50 Ω, which is located at the center of the chart as shown in Figure 7.42.

To realize the matching circuit, we start from the load at point A and travel "toward generator" on the constant VSWR circle to arrive at point B. Adding an opposite shunt stub

at this point will move this point to the center of the Smith chart ($\Gamma_2'=0$) which is the new load at point "O" as shown.

Now starting from point "O", add a series stub (to arrive at point "C") and then a series transmission line to end up at Γ_1 which is the input reflection coefficient. The resulting distributed circuit is shown in Figure 7.42.

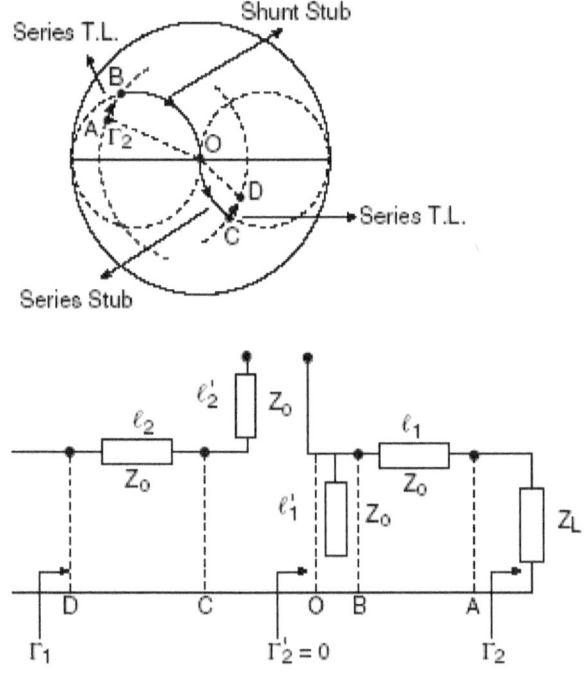

Figure 7.42 Smith Chart solution and the matching circuit.

NOTE 1: *As described earlier, it is best to keep the matching philosophy of moving from the load impedance toward the center of the chart which means to move from the output (load) and progress backward to the input end.*

NOTE 2: *The matching network above could have been alternately realized with lumped elements using the constant conductance or resistance circle (as described in the "Lumped element design" section) rather than the distributed element design.*

b. SOLUTION #2

First we draw the constant VSWR circle for (Γ_1) as shown in Figure 7.43. Then starting from (Γ_2) at point "A", we travel on a constant conductance circle to intersect the (Γ_1) VSWR circle at point B. This corresponds to a shunt stub element.

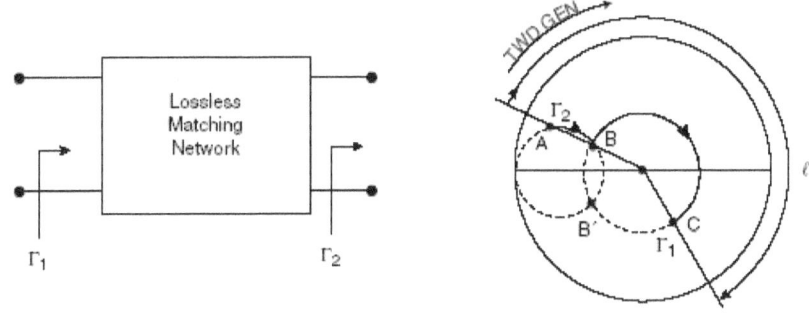

(a) Circuit block diagram   (b) Smith chart solution

Figure 7.43 The block diagram and Smith Chart solution.

Next, we travel on the constant VSWR circle "toward generator " (clockwise) to arrive at point "C". This would correspond to a series transmission line of length "ℓ", as shown in the schematic in Figure 7.44.

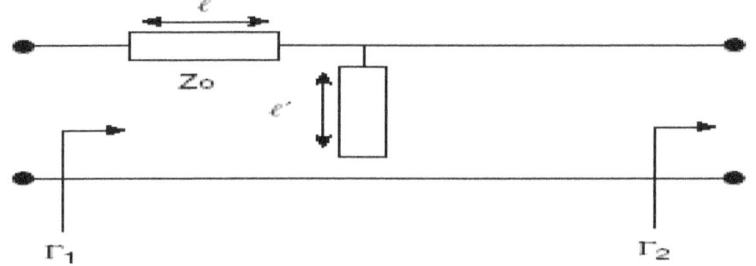

Figure 7.44 Smith Chart solution.

NOTE 1: *There is a second solution that uses the second intersection of the constant-G circle with the constant VSWR circle at point B' leading to a similar design procedure as described above.*

NOTE 2: *Solutions #2 provide only two elements (compared to four elements given by solution #1 which usually results in the best bandwidth for the interstage design.*

POINT OF INTEREST: *In some cases it is more convenient to work with the equivalent problem of conjugate reflection coefficient which functionally yields an equivalent circuit as shown in Figures 7.45 and 7.46.*

In Figure 7.47, since one is matching Γ_1^* to Γ_2^*, we have to start from Γ_1^* (as the load) and progress backwards to Γ_2^* at the other end. Traveling from point "A" to "B" gives a series transmission line, followed by going from "B" to "C" (producing a shunt stub), which is identical to what was obtained earlier.

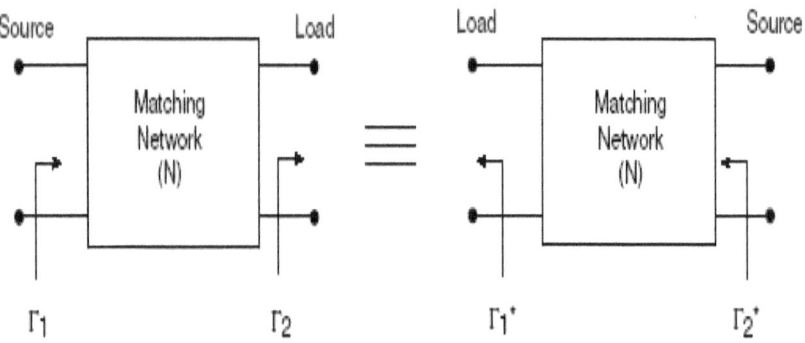

(a)   (b)

Figure 7.45 An equivalent representation.

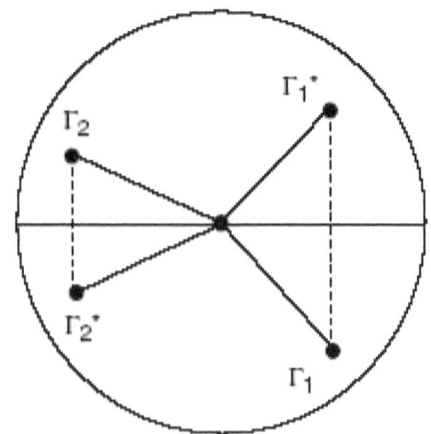

Figure 7.46 Plot of conjugate reflection coefficients.

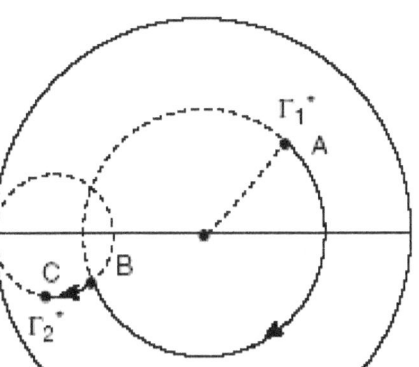

Figure 7.47 Smith Chart solution.

Chapter 7- Symbol list

A symbol will not be repeated again, once it has been identified and defined in an earlier chapter, with its definition remaining unchanged.

P_{av} - Average power

$(P_{av})_{max}$ - Maximum average power

$(Z_L)_{opt}$ - Optimum load impedance

CHAPTER-7 PROBLEMS

7.1) Design a single stub matching circuit (see Figure P7.1) that transforms a load ($Z_L=30+j50$ Ω) to a transmission line as follows (use a 100 Ω system) :

    a) Assume $Z_{O1}=Z_{O2}=100$ Ω.

    b) Assume $Z_{O1}=100$ Ω and $Z_{O2}=200$ Ω

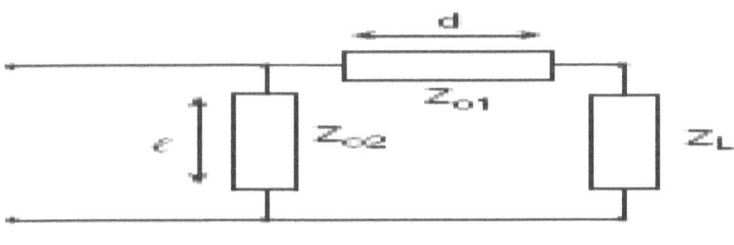

Figure P7.1

7.2) Design a lumped matching network to match the load $Y_L=(4-j6)\times10^{-3}$ S to a transmission line ($Z_O=100$ Ω). Find the element values at 10 GHz.

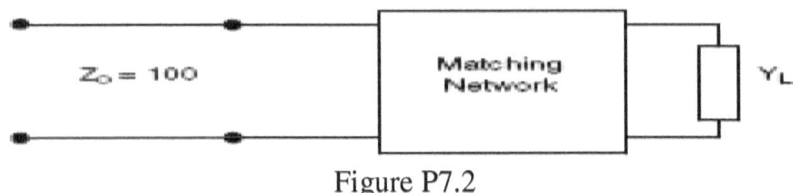

Figure P7.2

7.3) Design the matching network shown in Figure P7.3 to match the load, $Z_L=100+j100$ Ω to a 50 Ω transmission line at f=1 GHz ($Z_{O1}=100$ Ω).

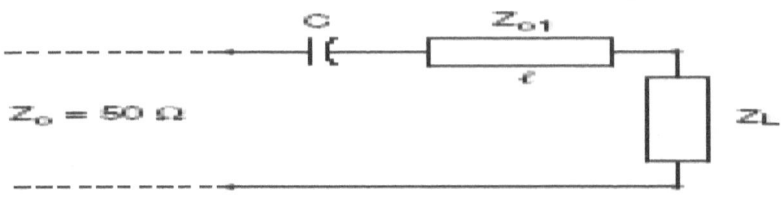

Figure P7.3

7.4) Design a matching network to transform a load impedance ($Z_L=50+j50$Ω) to the input impedance ($Z_{in}=25-j25$ Ω) at 1 GHz as shown in Figure P7.4 for the following two cases ($Z_O=50$ Ω):
   a. Lumped element design,
   b. Distributed element design.

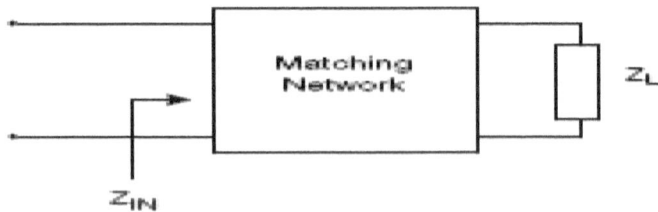

Figure P7.4

7.5) Design a matching network that will match a 50 Ω load to an input reflection coefficient of $\Gamma_{in}=0.5\angle150°$ in a 50 Ω system as shown in Figure P7.5. The matching network should use a quarter-wave transformer.

$$\Gamma_{IN} = 0.5\angle+150°$$

Figure P7.5

7.6) In the circuit shown below, a load $Z_L=90+j60$ Ω is to be matched to a line as shown in Figure P7.6. Determine Z_1, Z_{O1} and ℓ.

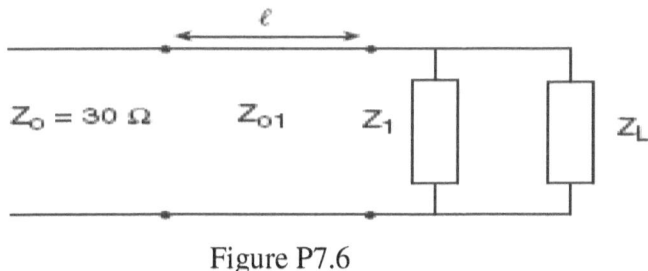

Figure P7.6

7.7) A certain microwave device has $Z_d=50-j50$ Ω. Design a matching network to match the device impedance to a 25 Ω system for:

    a) Lumped element circuit design.

    b) Distributed element circuit design.

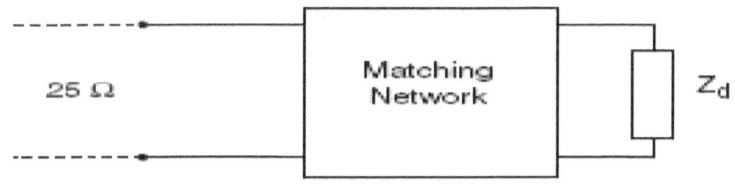

Figure P7.7

7.8) A lossless transmission line ($Z_O=50$ Ω) is to be matched to a load, $Z_L=5.5-j10.5$, by means of a short-circuited stub ($Z_{O1}=100$ Ω) as shown in Figure P7.8. Determine the position and length of the stub.

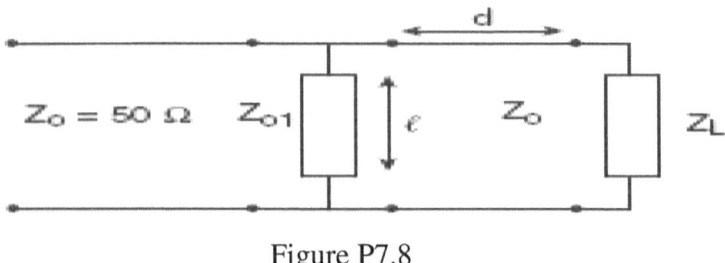

Figure P7.8

REFERENCES

[7.1] Anderson, E. M. *Electric Transmission Line Fundamentals.* Upper Saddle River: Prentice Hall, 1985.

[7.2] Cheng, D. K. *Fundamentals of Engineering Electromagnetics.* Reading: Addison Wesley, 1993.

[7.3] Cheung, W. S. and F. H. Levien. *Microwave Made Simple, Principles and Applications.* Norwood: Artech House, 1985.

[7.4] Gonzalez, G. *Microwave Transistor Amplifiers, Analysis and Design,* 2nd ed. Upper Saddle River: Prentice Hall, 1997.

[7.5] Liao, S. Y. *Microwave Circuit Analysis and Amplifier Design.* Upper Saddle River: Prentice Hall, 1987.

[7.6] Pozar, D. M. *Microwave Engineering,* 2nd ed. New York: John Wiley & Sons, 1998.

[7.7] Schwarz, S. E. *Electromagnetics for Engineers.* Orlando: Sanders College Publishing, 1990.

CHAPTER 8

Stability in Active Networks

8.1 INTRODUCTION

In any amplifier design, one of the very important considerations is the stability of the circuit under different source or load conditions. This term needs to be defined at this point:

DEFINITION- STABILITY: *Is defined to be in general, the ability of an amplifier to maintain effectiveness in its nominal operating characteristics in spite of large changes in the environment such as physical temperature, signal frequency, source or load conditions, etc.*

In this chapter, the stability requirements for an amplifier circuit or a more general two-Port network with known S-parameters will be discussed (see Figure 8.1). Furthermore, we will limit our stability considerations primarily to source and load conditions and will develop exact criteria for unconditional stability as well as conditional stability (also called potentially unstable condition).

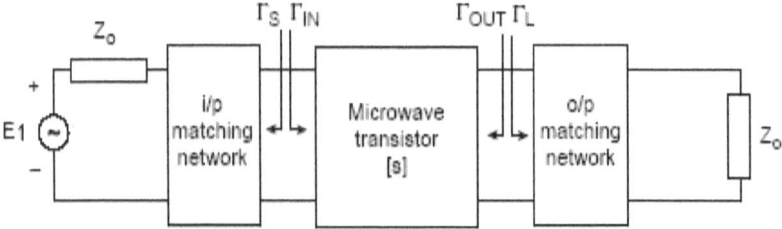

Figure 8.1 A single-stage amplifier.

When an amplifier becomes unstable (i.e. not able to maintain its nominal characteristics), it no longer acts as an amplifier but an oscillator.

From Figure 8.1, we can observe that since Γ_{in} and Γ_{out} depend on the source and load matching networks, therefore the stability of the amplifier circuit depends on Γ_S and Γ_L.

It should be noted that since the S-parameters of a two-Port network is frequency dependent, thus the stability condition of a circuit depends on the frequency of operation. Thus it is possible to have a circuit functioning well as an amplifier at an intended frequency while possibly oscillating for out of band frequencies.

8.1.1 ANALYSIS

In a two-Port network (see Figure 8.2), when the input impedance or output impedance presents a negative resistance, i.e.,

$$\textbf{Re}(\textbf{Z}_{in})\textbf{<0} \tag{8.1}$$

Or,

$$\textbf{Re}(\textbf{Z}_{out})\textbf{<0} \tag{8.2}$$

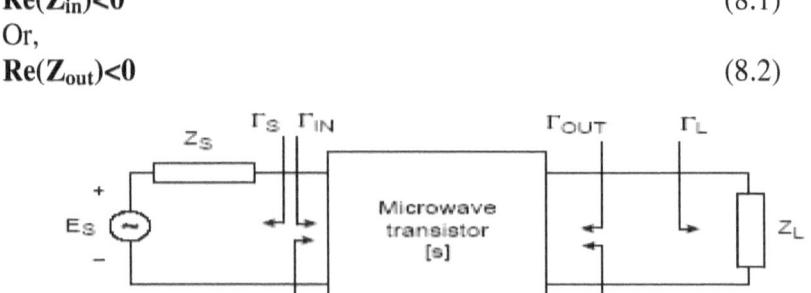

Figure 8.2 A General two-port network.

This negative resistance condition at the input (or the output) Port means that the reflected signal from the input (or output) port has higher power than the incident signal thus making $|\Gamma_{in}|$ or $|\Gamma_{out}|$ larger than unity, i.e.,

$$|\Gamma_{in}| > 1 \tag{8.3}$$

Or,

$$|\Gamma_{out}| > 1 \tag{8.4}$$

Thus considering a passive source or load impedance where:
$|\Gamma_S| < 1$, and $|\Gamma_L| < 1$, two types of "stability conditions" can be defined as follows:

1. Unconditional stability: a network is said to be "Unconditionally stable" in a frequency range if, and only if:

$$|\Gamma_{in}| < 1, \tag{8.5a}$$

and

$$|\Gamma_{out}| < 1 \qquad \textbf{for all } |\Gamma_S| < 1 \textbf{ and } |\Gamma_L| < 1 \tag{8.5b}$$

2. Conditional stability: a network is set to be "conditionally stable" or "Potentially unstable", in a frequency range, if

$$|\Gamma_{in}| < 1 \tag{8.6a}$$

and,

$$|\Gamma_{out}| < 1 \tag{8.6b}$$

only for a limited range of values of passive source and load impedances (or $|\Gamma_S|$ and $|\Gamma_L|$), but not for all values.

Using signal flow graphs from earlier chapters, Γ_{in} and Γ_{out} are derived and are given by:

$$\Gamma_{in} = S_{11} + \frac{S_{12}S_{21}\Gamma_L}{1 - S_{22}\Gamma_L} \tag{8.7}$$

$$\Gamma_{out} = S_{22} + \frac{S_{12}S_{21}\Gamma_S}{1 - S_{11}\Gamma_S} \tag{8.8}$$

A SPECIAL CASE--THE UNILATERAL TRANSISTOR
If the transistor is unilateral, i.e. $S_{12} = 0$, then the equations for unconditional stability simplifies to the following:

$$|S_{11}| < 1 \tag{8.9a}$$
$$|S_{22}| < 1 \tag{8.9b}$$

If the above condition can not be met by the transistor, then the amplifier circuit is considered to be conditionally stable for a certain range of Γ_S and Γ_L which will be discussed in more detail shortly.

8. 2 STABILITY CIRCLES

As discussed earlier, the unconditional stability for a general two-Port network having passive source/load impedances (as shown in Figure 8.3) requires the conditions as given by equations (8.5a,b).

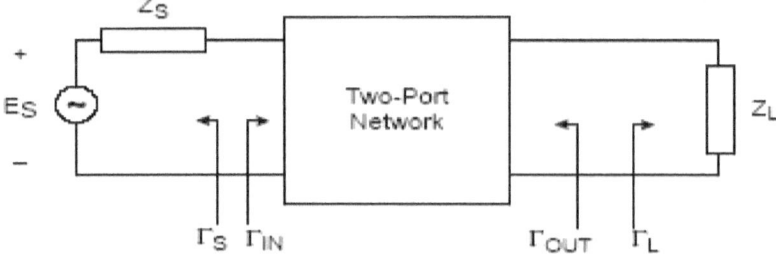

Figure 8.3 Stability of a two-port network.

However, the boundaries between stable and potentially unstable regions of Γ_S and Γ_L are determined by replacing the inequality signs in (8.5) with equality signs, giving:

$$\left|\Gamma_{in}\right| = \left|S_{11} + \frac{S_{12}S_{21}\Gamma_L}{1-S_{22}\Gamma_L}\right| = 1 \qquad (8.10a)$$

$$\left|\Gamma_{out}\right| = \left|S_{22} + \frac{S_{12}S_{21}\Gamma_S}{1-S_{11}\Gamma_S}\right| = 1 \qquad (8.10b)$$

$|\Gamma_S|<1 \qquad\qquad\qquad\qquad\qquad\qquad (8.10c)$

$|\Gamma_L|<1 \qquad\qquad\qquad\qquad\qquad\qquad (8.10d)$

8.3 GRAPHICAL SOLUTION OF STABILITY CRITERIA

Considering the Γ_L and Γ_S planes, the loci of points for which $|\Gamma_{in}|=1$ and $|\Gamma_{out}|=1$ are found to be two circles:

a. Input Stability Circle,
b. Output Stability Circle.

These two stability circles define the boundaries between "stable" and "unstable" regions for different values of Γ_S and Γ_L. For passive matching networks these values lie inside the standard Smith chart ($|\Gamma_S| \leq 1$, $|\Gamma_L| \leq 1$). Thus the intersection of the two stability circles with the standard smith chart provides the stable and unstable regions.

The equation for the output stability circle ($|\Gamma_{in}|=1$) drawn in the Γ_L plane can be written as (see Equation 8.11):

$$|\Gamma_L - C_L| = R_L \tag{8.12}$$

where:

$$C_L = \frac{\left(S_{22} - \Delta S_{11}^{*}\right)^{*}}{D_L} \tag{8.13a}$$

$$R_L = |S_{12}S_{21}/D_L| \tag{8.13b}$$
$$D_L = |S_{22}|^2 - |\Delta|^2 \tag{8.13c}$$
$$\Delta = S_{11}S_{22} - S_{12}S_{21} \tag{8.13d}$$

Similarly, the input stability circle ($|\Gamma_{out}|=1$) drawn in the Γ_S-plane is obtained by interchanging S_{11} for S_{22} in Equations (8.13) and is given by:

$$|\Gamma_S - C_S| = R_S \tag{8.14}$$

where:

$$C_S = \frac{\left(S_{11} - \Delta S_{22}^{*}\right)^{*}}{D_S} \tag{8.15a}$$

$$R_S = |S_{12}S_{21}/D_S| \tag{8.15b}$$
$$D_S = |S_{11}|^2 - |\Delta|^2 \tag{8.15c}$$
$$\Delta = S_{11}S_{22} - S_{12}S_{21} \tag{8.15d}$$

NOTE: *The general equation of a circle in the Γ plane can be written as:*

$$|\Gamma - C| = R \tag{8.15e}$$

Where "C" is a complex number representing the center of the circle and "R" is a real positive number representing the circle's radius.

Figures 8.4 and 8.5 illustrate the graphical plot of the input and output stability circles where the circles divide their respective planes (Γ_S-plane or Γ_L-plane) into two regions:

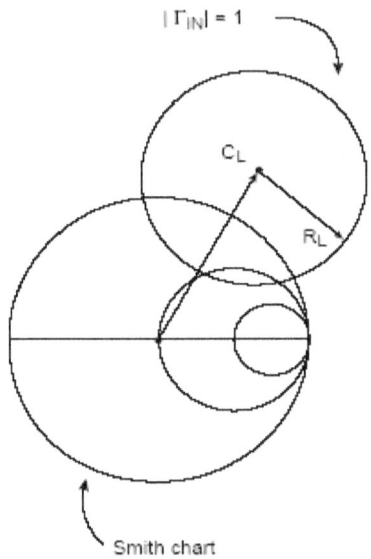

Figure 8.4 Output stability circle.

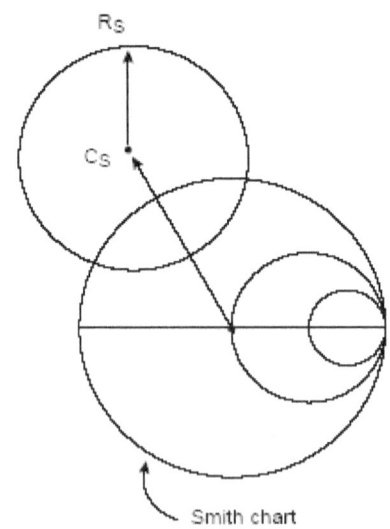

Figure 8.5 Input stability circle.

a.   The first region is the stable region and is characterized by $|\Gamma_{out}|<1$ for input stability circle in the Γ_S-plane (or $|\Gamma_{in}|<1$ for output stability circle in the Γ_L-plane), and

b.   The second region is the unstable region and is characterized by $|\Gamma_{out}|>1$ for input stability circle (or $|\Gamma_{in}|>1$ for output stability circle)

We now need a method to determine which region corresponds to the stable region that can be subsequently used for amplifier design.

OBSERVATION: *Because the S-parameters are frequency dependent, therefore the size and position of the input and output stability circles would change as the frequency is varied. This is because the center and radius of each circle are expressed in terms of the S-parameters.*

8.3.1 OUTPUT STABILITY CIRCLE
The output stability circles are plotted in the Γ_L plane as shown in Figure 8.4. If we set $\Gamma_L=0$ (i.e. $Z_L=Z_O$, the center of the chart), then from (8.7) we have:

$$|\Gamma_{in}|=|S_{11}| \tag{8.16}$$

Now if the $|S_{11}|$ of the device is less than unity (i.e. $|S_{11}|<1$), then that region of the output stability circle which includes the center of the smith chart is the "stable region" and the second region would be the unstable area.

For example, if the stability circle does not include the center of the smith chart as shown in Figure 8.4, then Γ_L values located outside of the stability circle are in the stable region and Γ_L values inside the stability circle are in the unstable region.

Vice versa, if $|S_{11}|$ of the device is more than unity (i.e. $|S_{11}|>1$), then the center of the smith chart is in the unstable region. For example, in Figure 8.4, the Stable region would be inside of the stability circle and the outside would be the unstable region.

8.3.2 Input Stability Circle

Similar to the output stability circles, the input stability circles are plotted in the Γ_S plane. If we set $\Gamma_S=0$ (i.e. the center of the smith chart), then from (8.8) we have:

$$|\Gamma_{out}|=|S_{22}| \tag{8.17}$$

In this case, if $|S_{22}|$ of the device is less than unity (i.e. $|S_{22}|<1$) then one region of the circle containing the center of the smith chart is the stable region and the other region is unstable. For example, the input stability circle as shown in Figure 8.5 does not include the center of the Smith chart. Thus outside of the circle is stable and the inside unstable. Vice versa, for $|S_{22}|>1$ the outside of the input stability circle is unstable and the inside stable

8.3.3 Special Case: Unconditional Stability

For $|S_{11}|<1$ and $|S_{22}|<1$, the amplifier circuit will be unconditionally stable when either of the following two conditions holds true:

a.   Both stability circles fall completely outside the Smith Chart, or
b.   Both stability circles completely enclose the smith chart.

Therefore for all passive source and load impedances, the unconditional stability can be concisely stated in mathematical form as:

$$||C_L|-R_L|>1 \quad \text{for } |S_{11}|<1 \tag{8.18a}$$
$$||C_S|-R_S|>1 \quad \text{for } |S_{22}|<1 \tag{8.18b}$$

Equations (8.18) in essence state that the distance between the center of the Smith Chart and the center of the stability circle must be larger than the stability circle's radius by an amount equal to the Smith Chart's radius. An example of the input stability circle in the Γ_S plane is shown in Figure 8.6.

NOTE: *If $|S_{11}|>1$ or $|S_{22}|>1$, then the amplifier can not be unconditionally stable since at the center of the Smith chart (where $\Gamma_s=0$, $\Gamma_L=0$), we have $|\Gamma_{in}|=|S_{22}|>1$ and $|\Gamma_{out}|=|S_{11}|>1$ which obviously is an unstable region.*

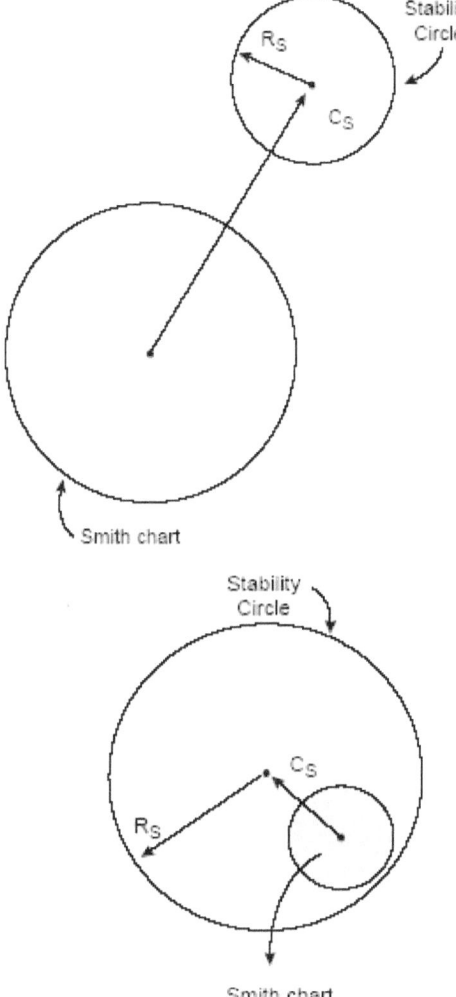

Figure 8.6 Two possibilities of input stability circles relative to the Smith Chart ($|S_{22}|<1$).

8.4 ANALYTICAL   SOLUTION   OF   STABILITY CRITERIA

The necessary and sufficient conditions for an amplifier to be unconditionally stable can be mathematically derived from Equations (8.10). The result of this derivation is a set of mathematical conditions which can be concisely referred to as "two-or three-parameter test criteria", as follows:

Let us define the determinant of the S-matrix (Δ), factors K and B_1 as:

$$\Delta = S_{11}S_{22} - S_{12}S_{21}, \tag{8.19}$$

$$K = \frac{1 - |S_{11}|^2 - |S_{22}|^2 + |\Delta|^2}{2|S_{12}S_{21}|} \tag{8.20}$$

and,

$$B_1 = 1 + |S_{11}|^2 - |S_{22}|^2 - |\Delta|^2 \tag{8.21}$$

Then based on these definitions, a two-port network will be unconditionally stable if, and only if, either one of the following mathematically equivalent criteria is satisfied:

CRITERION #1: THREE-PARAMETER TEST CRITERION

$$K > 1, \tag{8.22}$$

and

$$\frac{1 - |S_{11}|^2}{|S_{12}S_{21}|} > 1, \tag{8.23}$$

and

$$\frac{1 - |S_{22}|^2}{|S_{12}S_{21}|} > 1 \tag{8.24}$$

We can shrink the three-parameter test into a two-parameter test as the "K-Δ Test" below.

CRITERION #2: TWO-PARAMETER TEST CRITERION: (K-Δ TEST)

$$K > 1 \tag{8.25}$$
$$|\Delta| < 1 \tag{8.26}$$

This is often referred to as "K-Δ Test".

CRITERION #3: TWO-PARAMETER TEST CRITERION: (K-B_1 TEST)

$$K > 1 \tag{8.27}$$
$$B_1 > 0 \tag{8.28}$$

This may also be called "K-B_1 Test".

These three criteria are mathematically equivalent and if a device satisfies any one of the three criteria, the other two are automatically satisfied.

Thus a two-Port network will be unconditionally stable if and only if any one of the above three criteria are satisfied.

NOTE 1: *The two-parameter test criteria (#2 and #3) are more popular and more often used than criterion #1, primarily due to its simplicity and ease of calculations.*

NOTE 2: *For a **unilateral transistor** we have:*
$$S_{12}=0 \Rightarrow K=\infty >1 \qquad \qquad (8.29)$$
and,
$$|\Delta|=|S_{11}S_{22}| \qquad \qquad (8.30)$$

Since K>1 has already been satisfied, therefore in order to satisfy the condition for unconditional stability we desire $|\Delta|<1$, which requires:
a. $|S_{11}|<1$ $\qquad \qquad$ (8.31)
b. $|S_{22}|<1$ $\qquad \qquad$ (8.32)
for all passive values of Z_S and Z_L.
This conclusion is in agreement with the earlier discussion.

EXAMPLE 8.1
Determine the stability of a GaAs FET that has the following S-parameters at 2 GHz in a 50 Ω system both graphically and mathematically:
$S_{11}=0.89\angle-60°$
$S_{21}=3.1\angle123°$
$S_{12}=0.02\angle62°$
$S_{22}=0.78\angle-27°$

Solution:
a. Graphical method- we calculate the following values:
$C_L=1.36\angle47°$, $R_L=0.5$
$C_S=1.13\angle68°$, $R_S=0.2$
Input and output stability circles are plotted in Figure 8.7.

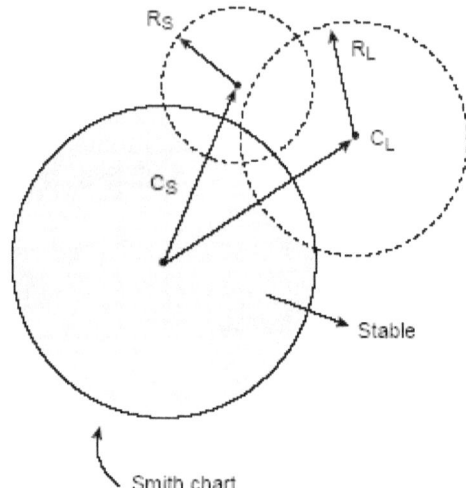

Figure 8.7 Input and output stability circles ($|S_{11}|<1$, $|S_{22}|<1$).

From this Figure we can see that the GaAs FET is "potentially unstable" and the center of the Smith Chart ($\Gamma_s=0$, $\Gamma_L=0$), being outside of the circles, represents stable regions since at this point we have:

$|\Gamma_{in}|=|S_{11}|=0.89<1$

$|\Gamma_{out}|=|S_{22}|=0.78<1$

b. Mathematical method- we calculate the following values:

K=0.6

$\Delta=0.7\angle-83°$

Since K<1, the transistor is potentially unstable which is in agreement with the graphical method.

Criterion 4: Single-Parameter Test (μ-Parameter Test)

Considering the more "popular two-parameter test", we can see that the "two-parameter test" (in particular the "K-Δ test") described above, provides a set of mathematical conditions on two parameters for unconditional stability, and only indicates whether a device is stable or not. However, due to the fact that certain constraints are imposed on the two parameters, the "two-parameter" test can not be used to show the degree of stability of one device relative to other similar devices.

To determine the unconditional stability of a device as well as its degree of stability relative to other devices, a new criterion has been derived that combines the "K-Δ parameters" into a single-parameter test and is often referred to as the "**μ-parameter test**". The parameter "μ" is defined as:

$$\mu = \frac{1 - |S_{11}|^2}{|S_{22} - S_{11}^* \Delta| + |S_{21} S_{12}|}$$

For unconditional stability, the following must be satisfied:
$$\mu > 1 \tag{8.33}$$

Furthermore, if "device A" has a parameter "μ_A" which is greater than "μ_B" corresponding to "device B" i.e.,
$$\mu_A > \mu_B \tag{8.34}$$
Then "device A" is said to be more stable than "device B".

Equation (8.34) indicates that a device with a larger value of "μ" is more desirable for an amplifier design since it implies a greater degree of stability.

8.5 POTENTIALLY UNSTABLE CASE
Sometimes when Γ_s and Γ_L are chosen such that:
$$|\Gamma_{in}| > 1 \text{ or } |\Gamma_{out}| > 1 \tag{8.35}$$
then the amplifier circuit becomes potentially unstable. In these situations, the device could be made unconditionally stable if the total input and output loop resistance is made to be positive, i.e.,
$$\text{Re}(Z_S + Z_{in}) > 0 \tag{8.36}$$
$$\text{Re}(Z_L + Z_{out}) > 0 \tag{8.37}$$

To achieve a positive loop resistance and thus making a potentially unstable transistor into a conditionally stable one, two methods normally are employed:
a. Resistively loading the transistor, or

b. Adding negative feedback (not commonly done)

The use of these techniques brings about a reduction in the gain, an increase in the noise Figure and a degradation of the amplifier power output.

These two techniques are useful in broadband potentially unstable amplifiers where the wide frequency range increases the probability of unstability. First the resistive loading is used to stabilize the transistor and then negative feedback is used to provide a relatively constant gain in low input and output VSWR.

In narrowband amplifiers, the use of these techniques is not recommended instead, careful selection of Γ_s and Γ_L in the early stages of the design is necessary to ensure a stable amplifier.

EXAMPLE 8.2

A BJT has the following S-parameters:

$S_{11}=0.65\angle-95°$

$S_{21}=5.0\angle115°$

$S_{12}=0.035\angle40°$

$S_{22}=0.8\angle-35°$

Is this transistor unconditionally stable? If not, use resistive loading to make the transistor conditionally stable. What are the resistor values?

Solution:

Simple calculations give us:

$K=0.547$

$\Delta=0.504\angle-110.4°$

Since $K<1$, therefore the transistor is potentially unstable!

To draw the output stability circles, we find the output stability circle C_L and R_L from (8.13) as follows:

$$C_L = \frac{\left(S_{22} - \Delta S_{11}^{*}\right)^{*}}{D_L},$$

$R_L = |S_{12}S_{21}/D_L|$

$D_L = |S_{22}|^2 - |\Delta|^2$

$\Delta = S_{11}S_{22} - S_{12}S_{21}$

$\Rightarrow C_L = 1.3\angle48°, \ R_L = 0.45$

Similarly, for the input stability circle, we calculate C_S and R_S from (8.15) as follows:

$$C_S = \frac{\left(S_{11} - \Delta S_{22}^{*}\right)^{*}}{D_S}$$

$R_S = |S_{12}S_{21}/D_S|$
$D_S = |S_{11}|^2 - |\Delta|^2$
$\Rightarrow C_S = 1.79\angle 122°, \ R_S = 1.04$

These two circles are drawn in Figure 8.8.

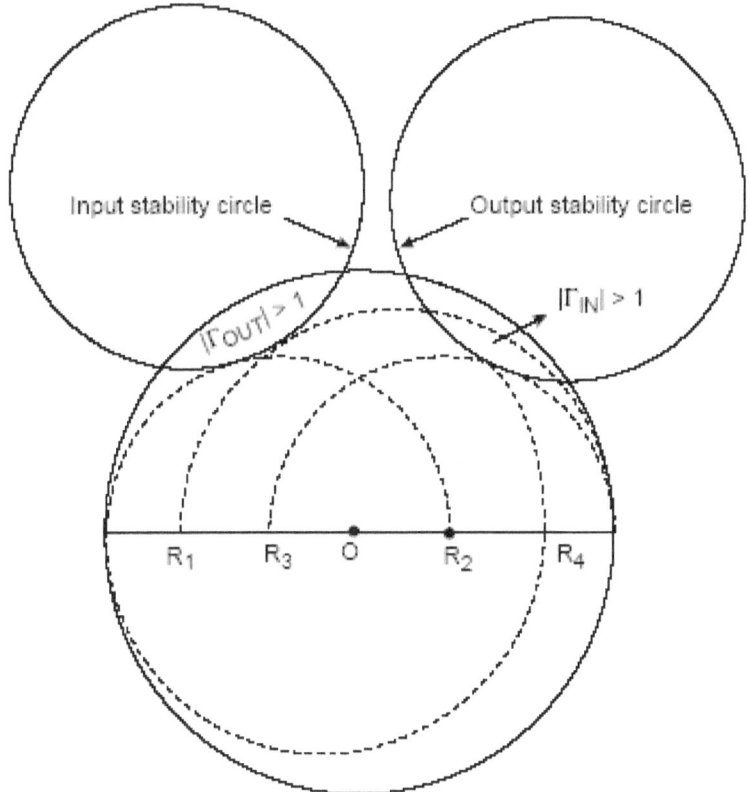

Figure 8.8 Input and output stability circles.

There are four types of resistive loading possible to improve stability as shown in Figure 8.9.

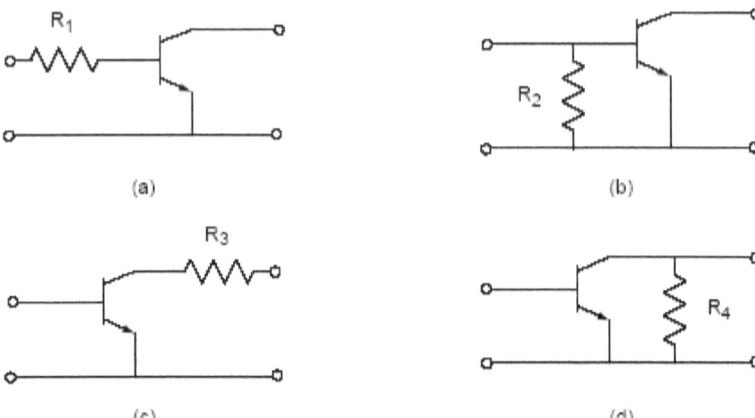

Figure 8.9 Four types of resistive loading.

Using the input stability circle, it can be seen that:
a.   A series resistor of R_1=9 Ω (Figure 8.9a) or,
b.   a shunt resistor of R_2=71 Ω (Figure 8.9b) at the input of the transistor will restore the stability.

On the other hand, using the output stability circle, we can see that:
c.   A series resistor of R_3=43 Ω (Figure 8.9c) or,
d.   A shunt resistor of R_4=500 Ω (Figure 8.9d) at the output of the transistor will assure stability.

It should be noted that in most cases, stabilizing either the input or the output port will restore stability to the transistor. Thus any one of the four types of resistive loading should be sufficient to create a stable amplifier.

POINT OF CAUTION: *Use of resistive loading at the input of the transistor (see Figures 8.9a, b) is not recommended due to an increase in the input loss which translates into a higher noise Figure at the output of the amplifier. Any resistive loading is preferred to take place at the output of the amplifier to minimize the increase in the amplifier's Noise Figure. This effect will be studied in more depth in Chapter 14.*

Chapter 8- Symbol List

A symbol will not be repeated again, once it has been identified and defined in an earlier chapter, with its definition remaining unchanged.

K - Stability factor used to evaluate the stability of a two-port network (K-Δ Test)

Δ - Determinant of the S-matrix used to evaluate stability (K-Δ Test)

μ - A parameter used to evaluate the stability of a network, The test is called the μ- parameter test.

CHAPTER-8 PROBLEMS

8.1) In each of the stability circle drawings shown in Figure P8.1, clearly indicate the possible locations for a stable source reflection coefficient.

8.2) Output stability circles are shown in Figure P8.2. Determine the stable region for the load reflection coefficient.

8.3) The scattering parameters for three different transistors are given below. Determine the stability in each case and in a potentially unstable case, draw the input and output stability circles:

a.
$$S = \begin{bmatrix} 0.67\angle-67° & 0.075\angle6.2° \\ 1.74\angle36.4° & 0.6\angle-92.6° \end{bmatrix}$$

b.
$$S = \begin{bmatrix} 0.385\angle-55° & 0.045\angle90° \\ 2.7\angle78° & 0.89\angle-26.5° \end{bmatrix}$$

c.
$$S = \begin{bmatrix} 0.7\angle-50° & 0.27\angle75° \\ 5\angle120° & 0.6\angle80° \end{bmatrix}$$

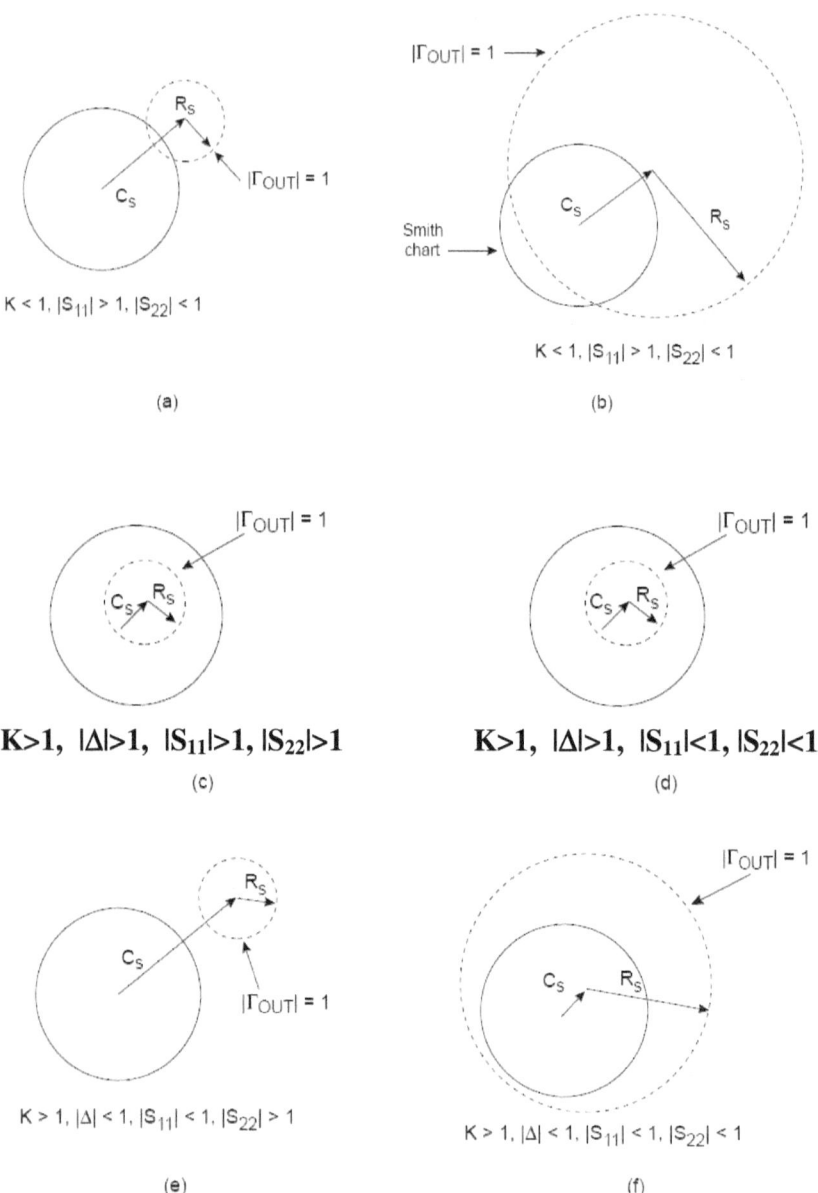

Figure P8.1

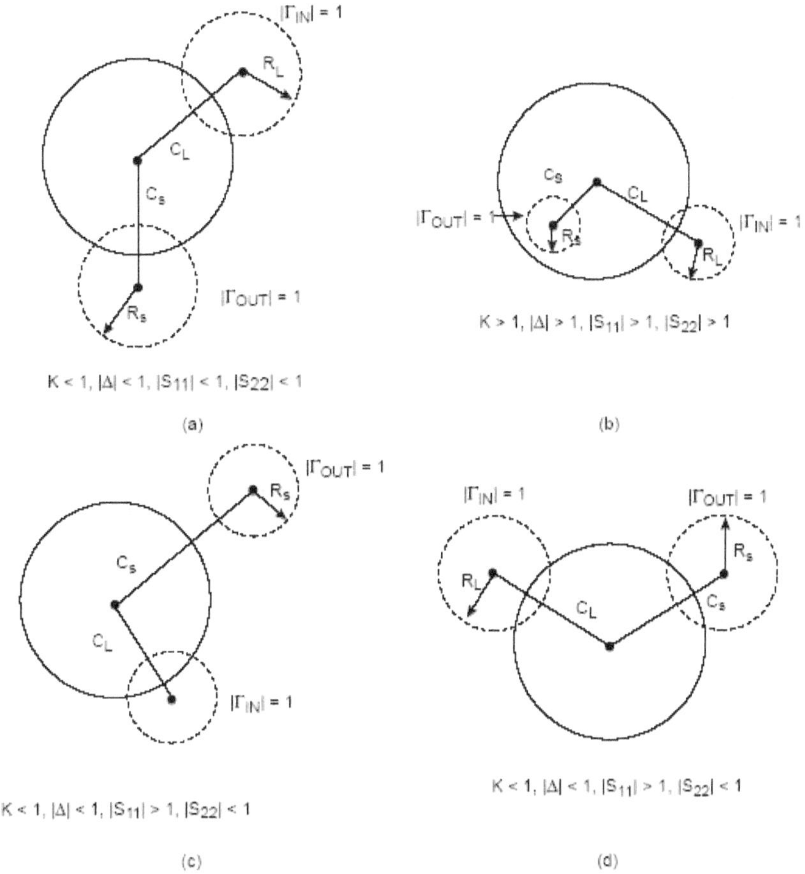

(a)

K < 1, |Δ| < 1, |S₁₁| < 1, |S₂₂| < 1

(b)

K > 1, |Δ| > 1, |S₁₁| > 1, |S₂₂| > 1

(c)

K < 1, |Δ| < 1, |S₁₁| > 1, |S₂₂| < 1

(d)

K < 1, |Δ| < 1, |S₁₁| > 1, |S₂₂| < 1

Figure P8.2

8.4) The S parameters of a GaAs FET at a certain Q-point are given in the following table:

f(GHz)	S_{11}	S_{12}	S_{21}	S_{22}
4	$0.9\angle-67°$	$0.076\angle43°$	$2.3\angle-118°$	$0.68\angle-39°$
6	$0.84\angle-97°$	$0.112\angle24°$	$2.06\angle87°$	$0.6\angle-58°$
8	$0.73\angle-140°$	$0.135\angle-5°$	$2.04\angle53°$	$0.47\angle-85°$
10	$0.67\angle-178°$	$0.146\angle-27°$	$1.81\angle18°$	$0.42\angle-120°$
12	$0.63\angle115°$	$0.133\angle-66°$	$1.42\angle-38°$	$0.36\angle-172°$

Draw the input stability circles (at each frequency) in a smith chart and the output stability circles. Indicate the unstable regions.

8.5) The S parameters of several two-Port networks are given by:

a. $S = \begin{bmatrix} 0.7\angle 0° & 0.7\angle 180° \\ 0.7\angle 0° & 0.7\angle 0° \end{bmatrix}$

b. $S = \begin{bmatrix} 0.7 & 1.7 \\ 1.7 & 0.7 \end{bmatrix}$

c. $S = \begin{bmatrix} 1 & 0.7 \\ 0.7 & 1 \end{bmatrix}$

Determine K and $|\Delta|$. Draw the input and output stability circles for each case as well.

8.6) a. show that in the limit as S_{12} approaches zero, the center and radius of the stability circles are given by:
$C_S \approx 1/S_{22}$, $r_S \approx 0$, $C_L \approx 1/S_{11}$ and $r_L \approx 0$.
b. The S parameters of a two-Port network are:
$$S = \begin{bmatrix} 2\angle 90° & 0 \\ 2 & 0.1\angle 45° \end{bmatrix}$$
Draw the stability circles and show the unstable regions.

8.7) Show how resistive loading can stabilize a resistor whose S-parameters at f=750 MHz are:
$$S = \begin{bmatrix} 0.69\angle 78° & 0.033\angle 41.4° \\ 5.67\angle 123° & 0.84\angle 25° \end{bmatrix}$$
Consider all four types of resistive loading for this problem.

8.8) A microwave GaAs FET has the following S parameters measured at V_{DS}=3 V and I_D=10 mA at 4 GHz:
$$S = \begin{bmatrix} 0.89\angle -50° & 0.06\angle 66° \\ 3.26\angle 141° & 0.58\angle -24° \end{bmatrix}$$
a. Calculate the delta factor (Δ).
b. The stability factor (K).

c. Find the center and radius of the input stability circle and plot the circle.

d. Determine the center and radius of the output stability circle and plot the circle.

8.9) Considering the diagram shown in Figure P8.9, indicate the possible locations for a stable source impedance and stable load impedance. The solid circle is the smith chart.

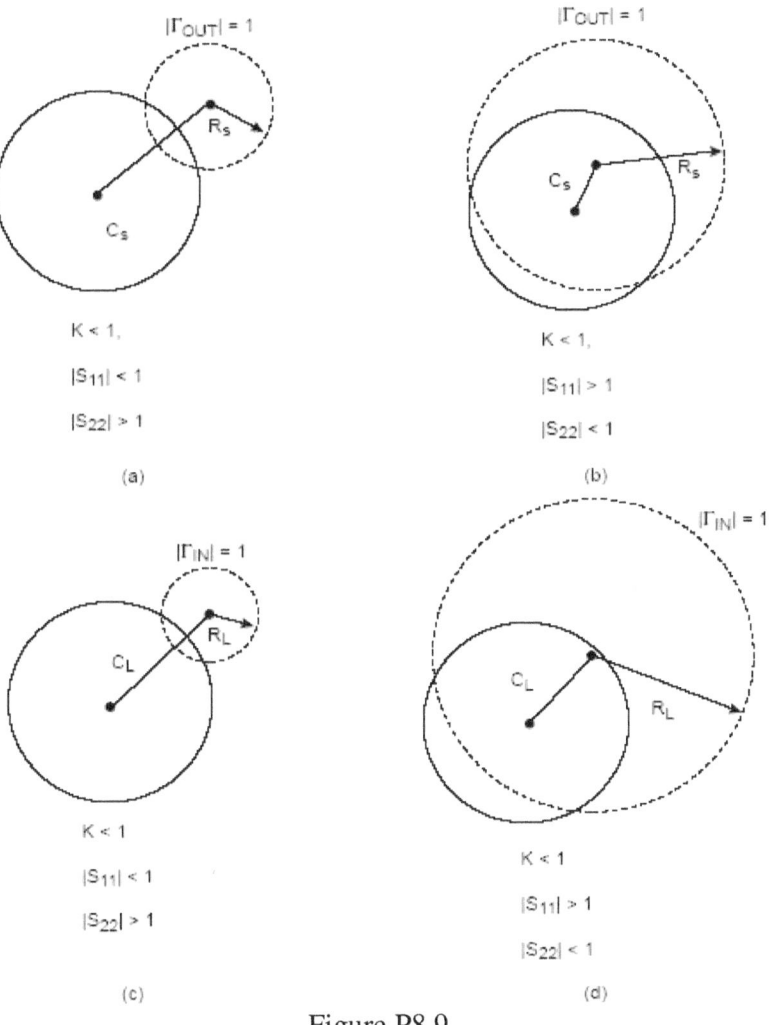

Figure P8.9

8.10) A microwave transistor has the following S parameters:

$$S = \begin{bmatrix} 0.8\angle -170° & 5.1\angle 80° \\ 0.3\angle 70° & 0.62\angle -40° \end{bmatrix}$$

Determine the stability, and plot the stability circles if the device is potentially unstable.

REFERENCES

[8.1] Cheung, W. S. and F. H. Levien. *Microwave Made Simple, Principles and Applications.* Norwood: Artech House, 1985.

[8.2] Gonzalez, G. *Microwave Transistor Amplifiers, Analysis and Design,* 2nd ed. Upper Saddle River: Prentice Hall, 1997.

[8.3] Liao, S. Y. *Microwave Circuit Analysis and Amplifier Design.* Upper Saddle River: Prentice Hall, 1987.

[8.4] Pozar, D. M. *Microwave Engineering,* 2nd ed. New York: John Wiley & Sons, 1998.

[8.5] Schwarz, S. E. *Electromagnetics for Engineers.* Orlando: Saunders College Publishing, 1990.

[8.6] Vendelin, George D. *Design of Amplifiers and Oscillators by the S-Parameter Method.* New York: John Wiley & Sons, 1982.

[8.7] Vendelin, George D., Anthony M. Pavio, and Ulrich L. Rhode. *Microwave Circuit Design, Using Linear and Non-Linear Techniques.* New York: John Wiley & Sons, 1990.

[8.8] Woods, D. *Reappraisal of the Unconditional Stability Criteria.* IEEE Transactions on Circuits and Systems, Feb. 1976.

CHAPTER 9

Gain Concepts in Amplifiers

9.1 INTRODUCTION

Gain consideration in an amplifier plays an important role in the design process. As discussed earlier, the primary consideration in an amplifier is its stability with its power gain following very closely as second in importance.

9.2 POWER GAIN CONCEPTS

Consider the single-stage microwave transistor amplifier with the transistor straddled by two matching networks on either side as shown in Figure 9.1.

There are several power gain concepts that are commonly used in the amplifier design process. Each one of these power gains has a specific name with a specific definition.

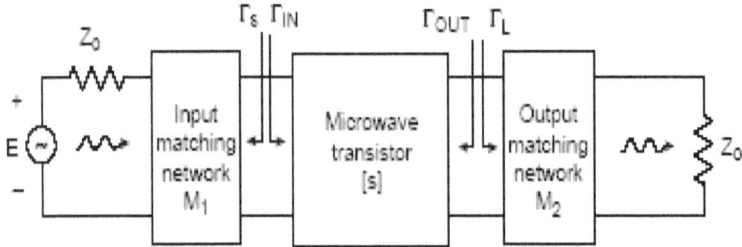

Figure 9.1 A general block diagram for a transistor amplifier.

Therefore at the outset let us first define the various power levels existing in the circuit as shown in Figure the 9.2.

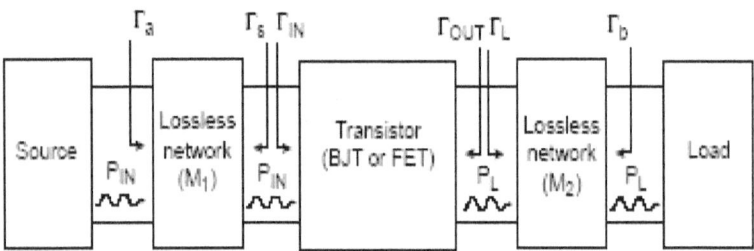

Figure 9.2 A transistor amplifier with matching networks.

$P_{IN} \equiv$ *Power input to the transistor or to the input matching network.*

$P_{AVS} \equiv$ *Power available from the source under matched condition. This is a special case of P_{IN} when $\Gamma_{IN} = \Gamma_S^*$.*

$P_L \equiv$ *Power delivered to the load or the output matching network.*

$P_{AVN} \equiv$ *Power available from the transistor under matched condition; a special case of P_L when $\Gamma_L = \Gamma_{OUT}^*$.*

Based on these definitions of power, we can now define the following power gain equations:

1. $G_T \equiv \dfrac{P_L}{P_{AVS}}$ (**Transducer power gain**)

2. $G_P \equiv \dfrac{P_L}{P_{IN}}$ (**Operating power gain, also called power gain**)

3. $G_A \equiv \dfrac{P_{AVN}}{P_{AVS}}$ (Available power gain)

Using signal flow graphs and Mason's rule, the three power ratios as defined above, are derived and the results are stated as follows

1. Transducer Gain (G_T)

$$G_T = \frac{1-|\Gamma_S|^2}{|1-\Gamma_{IN}\Gamma_S|^2}|S_{21}|^2\frac{1-|\Gamma_L|^2}{|1-S_{22}\Gamma_L|^2} \qquad (9.1)$$

Equation (9.1) can be written as :
$G_T = G_S.G_O.G_L$ \qquad (9.2)
Where

$$G_S = \frac{1-|\Gamma_S|^2}{|1-\Gamma_{IN}\Gamma_S|^2} \qquad (9.3)$$

$$G_O = |S_{21}|^2, \qquad (9.4)$$

$$G_L = \frac{1-|\Gamma_L|^2}{|1-S_{22}\Gamma_L|^2} \qquad (9.5)$$

From Equation (9.2) we may attribute G_O to the gain of the transistor while G_S and G_L are attributable to the effective gains of the input and output matching networks.

Alternately, Equation (9.1) can also be written in terms of Γ_{OUT} (rather than Γ_{IN}) as:

$$G_T = \frac{1-|\Gamma_S|^2}{|1-S_{11}\Gamma_S|^2}|S_{21}|^2\frac{1-|\Gamma_L|^2}{|1-\Gamma_{OUT}\Gamma_L|^2} \qquad (9.6)$$

2. Operating power gain (G_P)

$$G_P = \frac{1}{1-|\Gamma_{IN}|^2}|S_{21}|^2\frac{1-|\Gamma_L|^2}{|1-S_{22}\Gamma_L|^2} \qquad (9.7)$$

3. Available Power gain (G_A)

$$G_A = \frac{1-|\Gamma_S|^2}{|1-S_{11}\Gamma_S|^2}|S_{21}|^2\frac{1}{1-|\Gamma_{OUT}|^2} \qquad (9.8)$$

Where

$$\Gamma_{IN} = S_{11} + \frac{S_{12}S_{21}\Gamma_L}{1 - S_{22}\Gamma_L} \tag{9.9}$$

$$\Gamma_{OUT} = S_{22} + \frac{S_{12}S_{21}\Gamma_S}{1 - S_{11}\Gamma_S} \tag{9.10}$$

NOTE: *From Equation (9.1), the terms represent input and output matching network's degree of matching to the transistor at its input or its output. The matching networks are made up of passive components and have no inherent gain, thus they are incapable of generating power. Nevertheless, since input and output matching networks are capable of increasing the degree of match in the circuit as the signal flows through, they can be considered to have a positive gain in a relative manner. Thus we can write Equation (9.2) in dB as:*

$$G_T(dB) = G_S(dB) + G_O(dB) + G_L(dB) \tag{9.11}$$

9.3 A SPECIAL CASE-UNILATERAL TRANSISTOR

If the transistor is unilateral, i.e. $S_{12}=0$, then G_S and G_L gain blocks, as well as Γ_{IN} and Γ_{OUT}, simplify into:

$$G_{TU} = G_{SU} \cdot G_O \cdot G_{LU} \tag{9.12}$$

Where:

$$\Gamma_{IN} = S_{11}, \tag{9.13}$$

$$\Gamma_{OUT} = S_{22}, \tag{9.14}$$

$$G_{SU} = \frac{1 - |\Gamma_S|^2}{|1 - S_{11}\Gamma_S|^2} \tag{9.15}$$

$$G_{LU} = \frac{1 - |\Gamma_L|^2}{|1 - S_{22}\Gamma_L|^2} \tag{9.16}$$

9.4 THE MISMATCH FACTOR

The source mismatch factor (M_S), can be defined as:

$$M_S \equiv \frac{P_{IN}}{P_{AVS}} \qquad (9.17)$$

This ratio is always less than or at best equal to unity, i.e. $M_S \leq 1$, since P_{IN} is always less than or equal to P_{AVS}.

By observation we can see that the equation for M_S can be obtained simply by taking the ratio of G_T/G_P which yields:

$$M_S = \frac{(1-|\Gamma_S|^2)(1-|\Gamma_{IN}|^2)}{|1-\Gamma_S\Gamma_{IN}|^2} \qquad (9.18)$$

Source mismatch factor (M_S) is used to quantify the portion of P_{AVS} that is delivered to the input of the transistor.

If the input port is matched (i.e., $\Gamma_{IN} = \Gamma_S^*$) making $P_{IN}=P_{AVS}$, then we can see that $M_S=1$. This means that all of the available power from the source is delivered to the transistor and no mismatch exists at the input port, i.e.,

$$P_{IN} = P_{AVS}|_{\Gamma_{IN}=\Gamma_S^*} \Rightarrow M_S = 1 \qquad (9.19)$$

Similarly, The load mismatch factor (M_L), can be defined as:

$$M_L \equiv \frac{P_L}{P_{AVN}} \qquad (9.20)$$

This ratio is always less than or at best equal to unity, i.e. $M_L \leq 1$, since P_L is always less than or equal to P_{AVN}.

By observation we can see that the equation for M_L can be obtained simply by taking the ratio of G_T/G_A which yields:

$$M_L = \frac{(1-|\Gamma_L|^2)(1-|\Gamma_{OUT}|^2)}{|1-\Gamma_{OUT}\Gamma_L|^2} \qquad (9.21)$$

The load mismatch factor (M_L) is used to quantify the portion of P_{AVN} that is delivered to the load.

If the output port is matched (i.e., $\Gamma_{OUT} = \Gamma_L^*$) making $P_L=P_{AVN}$, then we can see that $M_L=1$. This means that all of the available power from the transistor is delivered to the load and no mismatch exists at the output port, i.e.,

$$\mathbf{P_{OUT}} = \mathbf{P_{AVN}}\big|_{\Gamma_{OUT}=\Gamma_L^*} \Rightarrow \mathbf{M_L} = 1 \tag{9.22}$$

Note: The "mismatch factor" is also called "mismatch loss" which (in dB) signifies the amount of power loss due to mismatch. From Equations (9.17) and (9.20) we can write the following:

$$\mathbf{M_S(dB)}= \mathbf{P_{IN}(dBm)} - \mathbf{P_{AVS}(dBm)}, \quad \mathbf{M_S}<0 \tag{9.23}$$
$$\mathbf{M_L(dB)}= \mathbf{P_L(dBm)} - \mathbf{P_{AVN}(dBm)}, \quad \mathbf{M_L}<0 \tag{9.24}$$

9.4.1 Constancy of Mismatch Factor

For a "lossless network", since the output power equals the input power, It can be shown mathematically that mismatch factor always remains constant.

For example, the mismatch factor (M_S) at the input of the lossless matching network (M_1), where the source is connected, has the same value as at its output where the transistor input is connected, i.e.,

At the input of M1: $\mathbf{P_{IN}=M_SP_{AVS}}$ (power into M1)
At the output of M1: $\mathbf{P_{OUT}=M_S'P_{AVS}}$ (power into the transistor)
Lossless network: $\mathbf{P_{OUT}=P_{IN}} \Rightarrow \mathbf{M_S=M_S'}$

Similarly, mismatch factor (M_L) remains unchanged at the input and output of the lossless matching network (M2), i.e.,
At the output of M2: $\mathbf{P_L=M_LP_{AVN}}$ (power into the load)
At the input of M2: $\mathbf{P_{OUT}=M_L'P_{AVN}}$ (power into M2)
Lossless network: $\mathbf{P_{OUT}=P_{IN}} \Rightarrow \mathbf{M_L=M_L'}$

In summary, we can state the following conclusion:
CONCLUSION: *For a lossless network, the Mismatch factor is an "invariant quantity".*

9.5 INPUT AND OUTPUT VSWR

In many cases the microwave amplifier's specification is in terms of the input VSWR and the output VSWR. Therefore we would like to obtain a relationship between the mismatch factor and VSWR.

9.5.1 Input-Port VSWR

From Figure 9.2, we can express the input power (P_{IN}) entering the input port of the matching network (M1), in terms of the input reflection coefficient (Γ_a) as follows:

$$P_{IN}=P_{AVS}(1-|\Gamma_a|^2) \Rightarrow P_{IN}=M_S P_{AVS} \qquad (9.25a)$$
$$M_S=1-|\Gamma_a|^2 \qquad (9.25b)$$

Where

$$\Gamma_a = \frac{Z_a - Z_O}{Z_a + Z_O} \qquad (9.25c)$$

Thus from Equation (9.25b), we can write:

$$|\Gamma_a| = \sqrt{1-M_S} \qquad (9.25d)$$

Therefore at the input of the lossless matching network (M1), the input VSWR is given by:

$$(VSWR)_{IN} = \frac{1+|\Gamma_a|}{1-|\Gamma_a|} = \frac{1+\sqrt{1-M_S}}{1-\sqrt{1-M_S}} \qquad (9.25e)$$

Thus $(VSWR)_{IN}$ can be calculated simply by knowing the input mismatch factor (M_S).

9.5.2 Output-Port VSWR

Similarly, From Figure 9.2 we can express the output power (P_L) exiting the output port of the matching network (M2). This power can be expressed in terms of the output reflection coefficient (Γ_b) as follows:

$$P_L=P_{AVN}(1-|\Gamma_b|^2) \Rightarrow P_L=M_L P_{AVN} \qquad (9.26a)$$
$$M_L=1-|\Gamma_b|^2 \qquad (9.26b)$$

Where

$$\Gamma_b = \frac{Z_b - Z_O}{Z_b + Z_O} \qquad (9.27)$$

Thus from Equation (9.26b), we can write:

$$|\Gamma_b| = \sqrt{1-M_L} \qquad (9.28)$$

Therefore at the output of the lossless matching network (M2), the output VSWR is given by:

$$(VSWR)_{OUT} = \frac{1+|\Gamma_b|}{1-|\Gamma_b|} = \frac{1+\sqrt{1-M_L}}{1-\sqrt{1-M_L}} \qquad (9.29)$$

Thus $(VSWR)_{OUT}$ can be calculated simply by knowing the input mismatch factor (M_L).

NOTE: *In all of the above equations the normalizing factor, Z_O (usually the source impedance: $Z_S = Z_O$), is assumed to be a real positive number. In cases where this assumption can not be upheld (i.e., when the source impedance is a complex number), the above equations no longer apply and have to be modified appropriately to account for this fact. Such a modification is outside the scope of this work and thus will not be discussed here.*

Example 9.1:
Given the amplifier circuit shown in Figure 9.3, having:
$\Gamma_S = 0.5\angle120°$,
$\Gamma_L = 0.4\angle90°$,
$S_{11}=0.6\angle-160°$, $S_{12}=0.045\angle16°$,
$S_{21}=2.5\angle30°$, $S_{22}=0.5\angle-90°$
determine:
a. G_T, G_A and G_P.
b. The power levels P_L, P_{IN}, P_{AVS} and P_{AVN}, and
c. The mismatch loss (in dB) at the input and output of the transistor.
d. The input and output reflection coefficient magnitude ($|\Gamma_a|, |\Gamma_b|$)
e. $(VSWR)_{IN}$ and $(VSWR)_{OUT}$

Solution:
a. The following can be calculated to be:
$\Gamma_{IN} = 0.627\angle-165°$,
$\Gamma_{OUT} = 0.47\angle-98°$,
$G_T=9.4=9.75$ dB
$G_P=13.5=11.3$ dB
$G_A=9.6=9.8$ dB

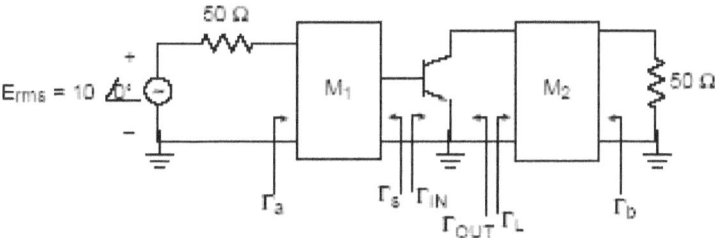

Figure 9.3 A block diagram for example 9.1.

b. To find the power levels, we note that:
$G_T=P_L/P_{AVS}=9.4$
$G_P=P_L/P_{IN}=13.5$
$\Rightarrow P_{AVS}/P_{IN}=13.5/9.4 = 1.43$
P_{AVS} is the input power under matched conditions as shown in Figure 13.4:

$$P_{AVS} = \frac{1}{2}(\frac{E_{rms}}{2Z_0})^2 = 0.5 \text{ W}$$

Thus:
$P_{IN} = P_{AVS}/1.43 = 0.35 \text{ W}$
$P_L = G_P P_{IN} = 4.72 \text{ W}$
$P_{AVN} = G_A P_{AVS} = 4.8 \text{ W}$

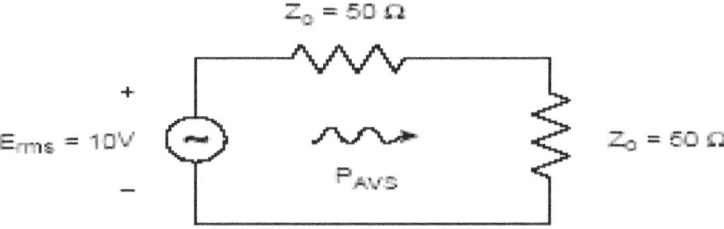

Figure 9.4 P_{AVS}.

c. The mismatch factor (being an invariant quantity) at the input of the transistor has the same value as at the input of the matching network (M1):
$M_S=P_{IN}/P_{AVS}=1/1.43=0.70= -1.55 \text{ dB}$
Similarly, the mismatch factor at the output of the transistor is the same value as at the load:
$M_L=P_L/P_{AVN}=4.73/4.8 = 0.99= -0.044 \text{ dB}$

d. $|\Gamma_a| = (1\text{-}M_S)^{1/2} = (1\text{-}0.7)^{1/2} = 0.55$

    $|\Gamma_b| = (1\text{-}M_L)^{1/2} = (1\text{-}0.99)^{1/2} = 0.1$

e. $(VSWR)_{IN} = \dfrac{1+0.55}{1-0.55} = 3.44$

    $(VSWR)_{OUT} = \dfrac{1+0.1}{1-0.1} = 1.22$

9.6 MAXIMUM GAIN DESIGN

From (9.4) we can observe that since G_O is fixed for any given transistor, the overall gain of the amplifier is controlled by the gain blocks G_S and G_L corresponding to the input and output matching networks, respectively.

Therefore in order to obtain the maximum possible gain from the amplifier circuit, we must maximize G_S and G_L values, which effectively implies that the input and output matching sections must provide a conjugate match at the transistor's input and output port.

Furthermore, under this conjugate matched condition at the input and the output of the transistor, maximum power will be transferred into the input port and out of the output port of the transistor as shown in Figure 9.5.

Based on conjugate impedance matching concepts, maximum power transfer a) from the input matching network to the transistor and, b) from the transistor to the output matching network, will occur when:

$$\Gamma_{IN} = \Gamma_S^* \tag{9.30}$$
$$\Gamma_{OUT} = \Gamma_L^* \tag{9.31}$$

NOTE: *Due to the inherent mismatch between the transistor and the matching networks (M1, M2) and the fact that the conjugate match and maximum power transfer will occur theoretically only at one particular frequency, this type of a circuit design is considered to be a "narrow-band design".*

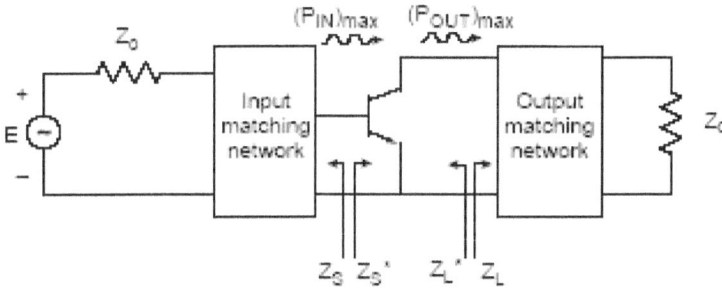

Figure 9.5 Maximum gain under conjugate matched condition.

9.7 UNILATERAL CASE (MAXIMUM GAIN)

When the transistor is unilateral (i.e. $S_{12}=0$) then Equations (9.3) through (9.5) simplify into the following:

$$\Gamma_{IN} = S_{11} \tag{9.32}$$
$$\Gamma_{OUT} = S_{22} \tag{9.33}$$
$$G_{TU} = G_S.G_O.G_L \tag{9.34}$$

Where

$$G_S = \frac{1-|\Gamma_S|^2}{|1-\Gamma_{IN}\Gamma_S|^2} \tag{9.35}$$

$$G_O = |S_{21}|^2, \tag{9.36}$$

$$G_L = \frac{1-|\Gamma_L|^2}{|1-S_{22}\Gamma_L|^2} \tag{9.37}$$

Under conjugately matched (maximum gain) conditions:

$$\Gamma_S = S_{11}^* \tag{9.38}$$
$$\Gamma_L = S_{22}^* \tag{9.39}$$
$$G_{TU,max} = G_{S,max}.G_O.G_{L,max} \tag{9.40}$$

Where

$$G_{S,max} = \frac{1}{1-|S_{11}|^2} \tag{9.41}$$

$$G_O = |S_{21}|^2, \tag{9.42}$$

$$G_{L,max} = \frac{1}{1-|S_{22}|^2} \tag{9.43}$$

Thus we can Equation (9.40) write as:

$$G_{TU,max} = G_{S,max}G_O G_{L,max} = \frac{1}{1-|S_{11}|^2}|S_{21}|^2\frac{1}{1-|S_{22}|^2}$$

NOTE: *Each gain block (G_S or G_L) is bound at the lower end by a gain of zero and at the upper end by the maximum gain ($G_{S,max}$ or $G_{L,max}$), as follows:*

$$0 \leq G_S \leq G_{S,max} \tag{9.44}$$
$$0 \leq G_L \leq G_{L,max} \tag{9.45}$$

We can normalize these two equations to obtain:

$$0 \leq g_S \leq 1 \tag{9.46}$$
$$0 \leq g_L \leq 1 \tag{9.47}$$

where the normalized gain factors (g_S, g_L) are defined as:

$$g_S = \frac{G_S}{G_{S,max}} = \frac{1 - |\Gamma_S|^2}{|1 - S_{11}\Gamma_S|^2}(1 - |S_{11}|^2) \tag{9.48}$$

$$g_L = \frac{G_L}{G_{L,max}} = \frac{1 - |\Gamma_L|^2}{|1 - S_{22}\Gamma_L|^2}(1 - |S_{22}|^2) \tag{9.49}$$

9.8 CONSTANT GAIN CIRCLES (UNILATERAL CASE)

Considering Equations (9.35) and (9.37), it can be shown (see the next section) that the values of Γ_S and Γ_L that produce a constant gain (or normalized gain) lie in a circle in the Smith chart. These circles are called constant G_S or G_L circles, respectively.

To obtain the equations for these circles, we start with Equations (9.48) and (9.49). It is shown that the values of Γ_S (or Γ_L) that produce a constant value of g_S (or g_L) lie in a circle described by the following equations:

$$|\Gamma_S - C_{gS}| = R_{gS}, \tag{9.50a}$$
$$|\Gamma_L - C_{gL}| = R_{gL} \tag{9.50b}$$

Where the center and radius (C_S, R_S) and (C_L, R_L) for each of the two circles are given by:

$$C_{gS} = \frac{g_S S_{11}{}^*}{1 - |S_{11}|^2 (1 - g_S)} \tag{9.51a}$$

$$R_{gS} = \frac{\sqrt{(1 - g_S)(1 - |S_{11}|^2)}}{1 - |S_{11}|^2 (1 - g_S)} \tag{9.51b}$$

and

$$C_{gL} = \frac{g_L S_{22}{}^*}{1 - |S_{22}|^2 (1 - g_L)}$$ (9.52a)

$$R_{gL} = \frac{\sqrt{(1 - g_L)(1 - |S_{22}|^2)}}{1 - |S_{22}|^2 (1 - g_L)}$$ (9.52b)

The two equations in (9.50) represent equations of two families of circles where the centers of each family of circles lie along the straight line given by the angle of $S_{11}{}^*$ and $S_{22}{}^*$ as shown in Figure 9.6.

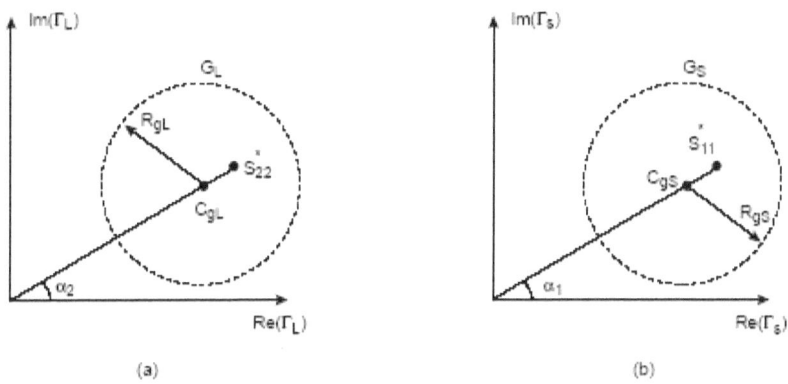

(a)   (b)

Figure 9.6 Families of constant gain circles.

OBSERVATIONS:

a. From Equations (9.51) and (3.52) we note that when g_S or $g_L = 1$ (i.e. maximum gain condition), we have:

$R_{gS} = 0$,   (9.53a)
$R_{gL} = 0$,   (9.53b)
$C_{gS} = S_{11}{}^*$,   (9.53c)

And

$C_{gL} = S_{22}{}^*$.   (9.53d)

This indicates that the maximum gain occurs only at one point located at $S_{11}{}^*$ and $S_{22}{}^*$ in the Γ_S or Γ_L plane, respectively. This observation is in agreement with our earlier results as expressed by Equations (9.38) and (9.39).

b. The 0-dB circles (i.e., $G_S=1$, $G_L=1$ circles) will always pass through the origin (i.e. $\Gamma_S=0$, $\Gamma_L=0$ points). This can be shown by noting that g_S from (9.48) can be written as:

$G_S=1$ and $\Gamma_S=0 \Rightarrow g_S=1-|S_{11}|^2$

$$|C_{gs}| = R_{gs} = \frac{S_{11}^{*}}{1+|S_{11}|^2} \qquad (9.53e)$$

Similarly for g_L:

$G_L=1$ and $\Gamma_L=0 \Rightarrow g_L=1-|S_{22}|^2$

These results show that the radius and distance from the origin to the center of the 0-dB constant G_S or G_L circle are identical and thus prove our observation.

c.   At the outer edge of the smith chart:

$|\Gamma_S|=1 \Rightarrow G_S=0=-\infty$ **dB**

$|\Gamma_L|=1 \Rightarrow G_L=0=-\infty$ **dB**

Since this gain value is impossible to achieve, therefore the gain circles never intersect the outer edge of the Smith chart.

d.   For a particular gain value, there are an infinite number of points on the constant gain circle that provide the same gain. Thus the choice of Γ_S and Γ_L, along the constant gain circles, is not unique, but in order to minimize mismatch loss and maximize bandwidth it is best to choose points close to the center of the smith chart.

"Choosing points close to the center of the smith chart" is true only for cases where noise is not of importance. In fact for low-noise amplifier design, (as we will see in the next chapter) one needs to use mismatch at the input matching network in order to obtain minimum noise from the amplifier.

9.8.1 Derivation of the Constant Gain Circles

The unilateral gain equations are given by:

$$g_S = \frac{G_S}{G_{S,max}} = \frac{1-|\Gamma_S|^2}{|1-S_{11}\Gamma_S|^2}(1-|S_{11}|^2) \qquad (9.54a)$$

$$g_L = \frac{G_L}{G_{L,max}} = \frac{1-|\Gamma_L|^2}{|1-S_{22}\Gamma_L|^2}(1-|S_{22}|^2) \qquad (9.54b)$$

We will derive the constant gain circle equation in the Γ_S plane using Equation (M.1). A similar procedure could be utilized for Equation (M.2) in the Γ_L plane.

We will present the general equation of a circle first, and then proceed to obtain the desired derivation from Equation (M.1).

a. General Equation of a Circle
The general equation for a circle in the Γ_S plane, with radius (R_{gs}) and a center at (C_{gs}), is given by:
$$|\Gamma_S - C_{gs}| = R_{gs} \qquad (9.54c)$$

Squaring both sides of Equation (9.54c), we have:
$$|\Gamma_S - C_{gs}|^2 = R_{gs}^2 \qquad (9.54d)$$

Equation (9.54d) can be rewritten as:
$$(\Gamma_S - C_{gs})(\Gamma_S - C_{gs})^* = R_{gs}^2$$

$$\Rightarrow (\Gamma_S - C_{gs})(\Gamma_S - C_{gs})^* = R_{gs}^2 \qquad (9.54e)$$

Applying the conjugate operation and Multiplying the terms in Equation (9.54e) yields the desired equation form for a circle as:
$$|\Gamma_S|^2 - C_{gs}^*\Gamma_S - C_{gs}\Gamma_S^* = R_{gs}^2 - |C_{gs}|^2 \qquad (9.54f)$$

b. CONSTANT GAIN CIRCLES
Multiplying the terms, Equation (9.54a) can be written as:
$$g_s(1+|S_{11}\Gamma_S|^2 - S_{11}\Gamma_S - S_{11}^*\Gamma_S^*) = 1-|\Gamma_S|^2 - |S_{11}|^2 + |\Gamma_S|^2|S_{11}|^2 \qquad (9.54g)$$

Separating the various terms, we can write Equation (9.54g) as:
$$|\Gamma_S|^2(1-|S_{11}|^2 + g_s|S_{11}|^2) - g_s S_{11}\Gamma_S - g_s S_{11}^*\Gamma_S^* = 1- g_s - |S_{11}|^2 \qquad (9.54h)$$

Dividing both sides by the $|\Gamma_S|^2$ coefficient, Equation (9.54h) can be recast as:

$$\left|\Gamma_{s}\right|^{2}-\left(\frac{g_{s}S_{11}}{1-\left|S_{11}\right|^{2}(1-g_{s})}\right)\Gamma_{s}-\left(\frac{g_{s}S_{11}^{*}}{1-\left|S_{11}\right|^{2}(1-g_{s})}\right)\Gamma_{s}^{*}=\frac{1-g_{s}-\left|S_{11}\right|^{2}}{1-\left|S_{11}\right|^{2}(1-g_{s})}$$

$$(9.54i)$$

Comparing Equation (9.54f) with (9.54i) yields:

$$C_{gs}=\frac{g_{s}S_{11}^{*}}{1-\left|S_{11}\right|^{2}(1-g_{s})} \tag{9.54j}$$

$$R_{gs}^{2}-\left|C_{gs}\right|^{2}=\frac{1-g_{s}-\left|S_{11}\right|^{2}}{1-\left|S_{11}\right|^{2}(1-g_{s})} \tag{9.54k}$$

Substituting for C_{gs} from Equation (9.54j) in (9.54k), we obtain R_{gs} as:

$$R_{gs}=\frac{\sqrt{(1-g_{s})(1-\left|S_{11}\right|^{2})}}{1-\left|S_{11}\right|^{2}(1-g_{s})} \tag{9.54\ell}$$

Equations (9.54j) and (9.54ℓ) provide the equations for C_{gs} and R_{gs} as exactly given earlier.

9.9 UNILATERAL FIGURE OF MERIT

We already noticed that under the unilateral assumption, power gain analysis greatly simplifies. However, in most cases $S_{12}\neq0$. Thus if we still wish to use the unilateral assumption and the simplified unilateral gain equations for the amplifier design (when $S_{12}\neq0$), we need to determine the error involved in our analysis.

The error involved lies in the magnitude ratio of G_T/G_{TU} which is obtained by dividing Equation (9.1) by (9.12):

$$\frac{G_T}{G_{TU}}=\frac{1}{\left|1-X\right|^{2}} \tag{9.55}$$

Where

$$X=\frac{S_{12}S_{21}\Gamma_{s}\Gamma_{L}}{(1-S_{11}\Gamma_{s})(1-S_{22}\Gamma_{L})} \tag{9.56}$$

It can be shown that the ratio of G_T/G_{TU} is bounded by:

$$\frac{1}{(1+|X|)^2} < \frac{G_T}{G_{TU}} < \frac{1}{(1-|X|)^2} \tag{9.57}$$

When $\Gamma_S = S_{11}{}^*$ and $\Gamma_L = S_{22}{}^*$, GTU achieves its maximum value, $G_{TU,max}$. The maximum error introduced using the unilateral assumption (i.e. using G_{TU} instead of G_T) is bounded by:

$$\frac{1}{(1+U)^2} < \frac{G_T}{G_{TU,max}} < \frac{1}{(1-U)^2} \tag{9.58}$$

where

$$U = \frac{|S_{12}\| S_{21} \| S_{11} \| S_{22}|}{(1-|S_{11}|^2)(1-|S_{22}|^2)} \tag{9.59}$$

Where U is defined to be the "unilateral Figure of merit" which varies with frequency due to its S-parameter dependence. Thus "U" needs to be calculated at each frequency in order to obtain the limits of the error involved due to unilateral assumption.

From table 9.1, which lists various values of "U" vs. G_T/G_{TU}, one can determine if the calculated value of "U" gives a tolerable error value for G_T/G_{TU}.

U		$R = G_T/G_{TU,max}$	
(ratio)	(dB)	(ratio)	(dB)
0.010	−20.0	0.980 < R < 1.020	−0.086 < R < 0.087
0.020	−17.0	0.961 < R < 1.041	−0.170 < R < 0.180
0.030	−15.2	0.943 < R < 1.063	−0.26 < R < 0.26
0.040	−14.0	0.925 < R < 1.085	−0.34 < R < 0.36
0.050	−13.0	0.907 < R < 1.108	−0.42 < R < 0.45
0.060	−12.2	0.890 < R < 1.132	−0.51 < R < 0.54
0.070	−11.5	0.873 < R < 1.156	−0.59 < R < 0.63
0.080	−11.0	0.857 < R < 1.181	−0.67 < R < 0.72
0.090	−10.5	0.842 < R < 1.208	−0.75 < R < 0.82
0.10	−10.0	0.826 < R < 1.235	−0.83 < R < 0.92
0.11	−9.6	0.812 < R < 1.262	−0.91 < R < 1.01
0.12	−9.2	0.797 < R < 1.291	−0.98 < R < 1.11
0.13	−8.9	0.783 < R < 1.321	−1.06 < R < 1.21
0.14	−8.5	0.769 < R < 1.352	−1.13 < R < 1.31
0.15	−8.2	0.756 < R < 1.384	−1.21 < R < 1.41

Table 9.1 Tabulation of values of U vs. R.

Usually an error of a few tenths of a dB in the $G_T/G_{TU,max}$ ratio is justifiable when using the unilateral assumption. The following example illustrates this point further:

EXAMPLE 9.2
Assume U=0.05 at 1 GHz for a microwave amplifier where $S_{12} \neq 0$.
a. Find the maximum error if we use unilateral gain equations for this transistor;
b. If the transistor is used in an amplifier design with a gain of 15 dB, can unilateral assumption be used in this case?

Solution:
a. From table 9.1 we can see that for U=0.05, the $G_T/G_{TU,max}$ is bound by:
$0.907 < G_T/G_{TU,max} < 1.108$
Or in dB, we have:
-0.42 dB< $G_T/G_{TU,max}$ <0.45 dB

b. Since the error due to unilateral assumption is bound between -0.42 dB and +0.45 dB, it is small enough (compared to 15 dB) to justify the unilateral assumption.

EXAMPLE 9.3
Find the maximum error range for the transducer gain value if we use unilateral gain equations for a transistor that has the following S-parameters:

$S_{11}=0.6\angle-160°, S_{12}=0.045\angle16°,$
$S_{21}=2.5\angle30°, S_{22}=0.5\angle-90°$

Solution:
From Equation (9.59) we have:
$$U = \frac{|S_{12}|\,||S_{21}||\,||S_{11}||\,||S_{22}|}{(1-|S_{11}|^2)(1-|S_{22}|^2)} = \frac{0.045 \times 2.5 \times 0.6 \times 0.5}{(1-0.6^2)(1-0.5^2)} = 0.070$$

The lower limit is:

$$\frac{1}{(1+u)^2} = \frac{1}{(1+0.070)^2} = 0.87 = -0.59 \text{ dB}$$

The upper limit is:

$$\frac{1}{(1-u)^2} = \frac{1}{(1-0.070)^2} = 1.156 = 0.63 \text{ dB}$$

Thus the error range for G_T is given by:
-0.59<G_T<0.63 dB

9.10 BILATERAL CASE

When $S_{12}\neq0$ and unilateral Figure of merit causes an unjustifiably high error in the gain equations, we are faced with the bilateral case where S_{12} can no longer be ignored.

We know that from Equations (9.30) and (9.31) the maximum power transfer occurs when:

$$\Gamma_{IN} = \Gamma_S{}^* = S_{11} + \frac{S_{12}S_{21}\Gamma_L}{1-S_{22}\Gamma_L} \tag{9.60}$$

$$\Gamma_{OUT} = \Gamma_L{}^* = S_{22} + \frac{S_{12}S_{21}\Gamma_S}{1-S_{11}\Gamma_S} \tag{9.61}$$

Under these conditions, the overall maximum gain using lossless matching networks is given by:

$$G_T = \frac{1}{1-|\Gamma_S|^2} |S_{21}|^2 \frac{1-|\Gamma_L|^2}{|1-S_{22}\Gamma_L|^2} \tag{9.62}$$

From Equation (9.60), we note that for a bilateral transistor Γ_S depends on Γ_L and vice versa, from (9.61) Γ_L depends on Γ_S. This means that these two equations are cross-coupled and must be solved simultaneously to obtain the simultaneous conjugate match values of Γ_S and Γ_L.

Solving Equations (9.60) and (9.61) simultaneously, we obtain the simultaneous conjugate match values of Γ_S and Γ_L (referred to as Γ_{MS} and Γ_{ML}):

$$\Gamma_{MS} = \frac{B_1 \pm \sqrt{B_1^2 - 4|C_1|^2}}{2C_1} \qquad (9.63a)$$

$$\Gamma_{ML} = \frac{B_2 \pm \sqrt{B_2^2 - 4|C_2|^2}}{2C_2} \qquad (9.63b)$$

where
$$B_1 = 1 + |S_{11}|^2 - |S_{22}|^2 - |\Delta|^2 \qquad (9.64a)$$
$$B_2 = 1 + |S_{22}|^2 - |S_{11}|^2 - |\Delta|^2 \qquad (9.64b)$$
$$C_1 = S_{11} - \Delta S_{22}^* \qquad (9.65a)$$
$$C_2 = S_{22} - \Delta S_{11}^* \qquad (9.65b)$$

NOTE: *It can be shown that for an unconditionally stable two-Port network ($K>1$, $|\Delta|<1$), the solutions from (9.63a,b) with a minus sign (-) should be considered in order to obtain meaningful values for Γ_{MS} and Γ_{ML} (i.e., $|\Gamma_{MS}|<1$ and $|\Gamma_{ML}|<1$).*

Under simultaneous conjugate matched conditions, $G_{T,max}$ from Equation (9.62) is obtained to be:
$$\Gamma_S = \Gamma_{IN}^* = \Gamma_{MS} \qquad (9.66a)$$
$$\Gamma_L = \Gamma_{OUT}^* = \Gamma_{ML} \qquad (9.66b)$$

$$G_{T,max} = \frac{1}{1 - |\Gamma_{MS}|^2} |S_{21}|^2 \frac{1 - |\Gamma_{ML}|^2}{|1 - S_{22}\Gamma_{ML}|^2} \qquad (9.67)$$

Substituting for Γ_{MS} and Γ_{ML} from Equations (9.63a,b) into (9.67), we obtain:

$$G_{T,max} = \frac{|S_{21}|}{|S_{12}|}(K - \sqrt{K^2 - 1}) \qquad (9.68)$$

When $K = 1$, we obtain the maximum stable gain G_{MSG} from (9.68):

$$G_{MSG}=G_{T,max}|_{K=1} =|S_{21}|/|S_{12}|$$

G_{MSG} is a figure of merit showing the maximum value that $G_{T,max}$ can achieve. Thus simply by looking at a transistor forward (S_{21}) and reverse (S_{12}) transmission coefficients, one can decide if the transistor is useful in providing the needed gain for a particular amplifier design or not.

9.11 SUMMARY
Before embarking upon the task of designing a functional amplifier we need to consider one more aspect, namely "Noise Figure" which for a sensitive receiver plays a very important role in the design process.

The concept of "Noise Figure" will be considered in detail in the next chapter and will complete the major designconsiderations (i.e., gain, stability, and noise) that accompany the design process of an amplifier. This is done so that when we start discussing the actual design process for several types of amplifiers, oscillators, mixers, etc., a full knowledge of the major design consideration has already been fully explored and understood by the designer.

Chapter 9 - List of Symbols
A symbol will not be repeated again, once it has been identified and defined in an earlier chapter, with its definition remaining unchanged.

G_A - Available power gain
G_0 - Transistor power gain
G_L - Output matching network power gain
G_{LU} - Output matching network power gain for unilateral case
G_{MSG} - Maximum stable gain
G_P - Operating power gain or input matching network power gain
G_S - Available power gain
G_{SU} - Available power gain for special case-unilateral transistor
G_T - Transducer power gain
$G_{T,max}$ - Maximum transducer power gain
G_{TU} - Transducer power gain for special case-unilateral transistor
M_L - Load mismatch factor

M_S - Source mismatch factor
P_{AVS} - Power available from the source under matched condition
P_{AVN} - Power available from the transistor under matched condition
P_{IN} - Input power to the transistor or the input matching network
P_L - Output power to the load or the output matching network
P_{OUT} - Output power from network under consideration
Γ_{ML} - Reflection coefficient for conjugate match value of Γ_L
Γ_{MS} - Reflection coefficient for conjugate match value of Γ_S

CHAPTER-9 PROBLEMS

9.1) The S parameters of a transistor are:

$$S = \begin{bmatrix} 0.7\angle 30° & 0 \\ 4\angle 90° & 0.5\angle 0° \end{bmatrix}$$

The transistor is used in the amplifier shown in Figure P9.1, where the output matching network produces $\Gamma_L=0.5\angle 90°$. Determine the values of G_T, G_P and G_A.

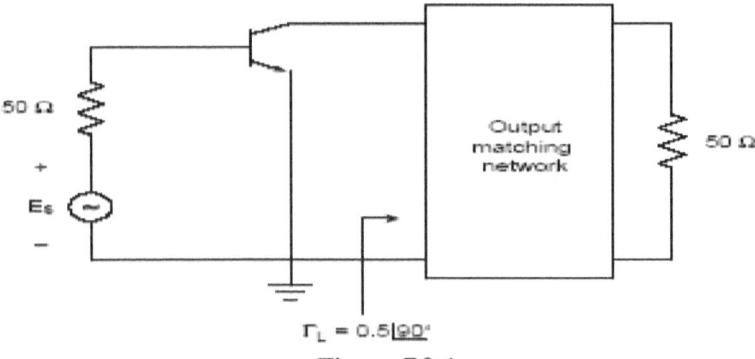

$\Gamma_L = 0.5\angle 90°$

Figure P9.1

9.2) a. Determine G_T, G_P and G_A in a microwave amplifier as shown in Figure P9.2, if $\Gamma_S=0.49\angle -150°$, $\Gamma_L=0.56\angle 90°$ and the S-parameters of the transistor are:

$$S = \begin{bmatrix} 0.54\angle 165° & 0.09\angle 20° \\ 2\angle 30° & 0.5\angle -80° \end{bmatrix}$$

b. Calculate P_{AVS}, P_{IN}, P_{AVN} and P_L if $E_1=10\angle 30°$, $Z_1=50\ \Omega$ and $Z_2=50\ \Omega$.

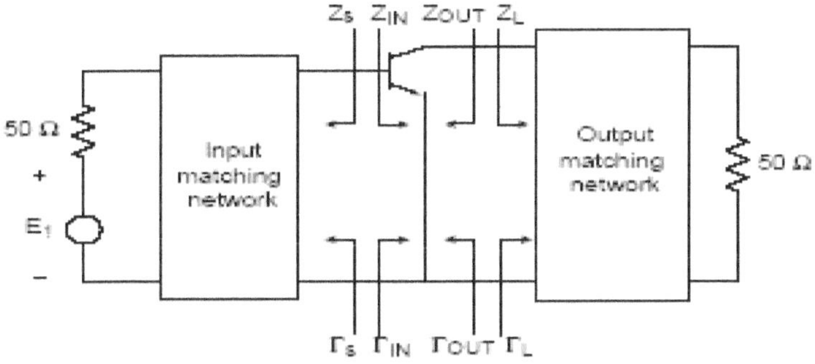

Figure P9.2

9.3) Prove that the maximum unilateral transducer power gain is obtained when $\Gamma_S = S_{11}^*$ and $\Gamma_L = S_{22}^*$.

9.4) a. Design a microwave transistor amplifier for $G_{TU,max}$ using a BJT whose S parameters in a 50 Ω system at $V_{CE}=10$ V, $I_C=10$ mA and f=1 GHz are given by:

$$S = \begin{bmatrix} 0.7\angle -160° & 0 \\ 5\angle 85° & 0.5\angle -20° \end{bmatrix}$$

9.5) A microwave amplifier is to be designed for $G_{TU,max}$ using a transistor with:

$$S = \begin{bmatrix} 0.5\angle 140° & 0 \\ 5\angle 45° & 0.6\angle -95° \end{bmatrix}$$

The S parameters were measured in a 50 Ω system at f=900 MHz, $V_{CE}=15$ V and $I_C=20$ mA.

a. Determine $G_{TU,max}$.

b. Draw the constant gain circle for $G_L=1$ dB.

c. If the S parameters at 1 GHz are:

$$S = \begin{bmatrix} 0.48\angle 137° & 0 \\ 4.6\angle 48° & 0.57\angle -99° \end{bmatrix}$$

Calculate the gain G_T at 1 GHz if $\Gamma_S=0.49\angle -150°$ and $\Gamma_L=0.56\angle 90°$.

9.6) An GaAs FET amplifier has the following S parameters for the active device:

$$S = \begin{bmatrix} 0.5\angle 180° & 0 \\ 4\angle 90° & 0.5\angle -45° \end{bmatrix}$$

a. Is the amplifier stable?
b. What is the maximum gain in dB?
c. What is the input impedance Z_{in}, in a 50 Ω system?
d. What is the load impedance Z_L for the maximum-gain case?

9.7) An amplifier is operating at 10 GHz using a GaAs FET with the following S parameters (see Figure P9.7):

$$S = \begin{bmatrix} 0.45\angle -50° & 0.01 \\ 4\angle 30° & 0.8\angle -150° \end{bmatrix}$$

a. Is the transistor stable? if not, how would you stabilize it?
b. Calculate the unilateral figure of merit and determine the error in dB for a unilateral design.
c. What is the maximum gain in dB?

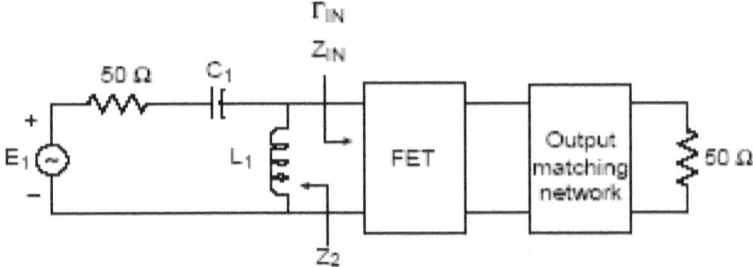

Figure P9.7

9.8) An FET device has the following S parameters at 3GHz:

$$S = \begin{bmatrix} 0.3\angle -60° & 0 \\ 2\angle 45° & 0.8\angle -30° \end{bmatrix}$$

Design an amplifier for maximum gain using this transistor and 50 Ω input and output transmission lines:
a. Check the stability of the device.
b. What is the maximum gain in dB?
c. Design an input-matching network for maximum gain using series L and shunt C elements to match a 50 Ω line to Z_S.

9.9) The FET amplifier shown in Figure P9.9 has the following S parameters in a 50 Ω system:

$$S = \begin{bmatrix} 0.5\angle 180° & 0 \\ 3.0\angle 90° & 0.5\angle -90° \end{bmatrix}$$

The circuit is terminated by R_S= 50 Ω and RL=100 Ω. Find:
a. Z_{in} and Z_{out},
b. The unilateral power gain, in dB.
c. If matching networks are used at the input and output ports such that maximum power transfer occurs, find the maximum unilateral power gain in dB.

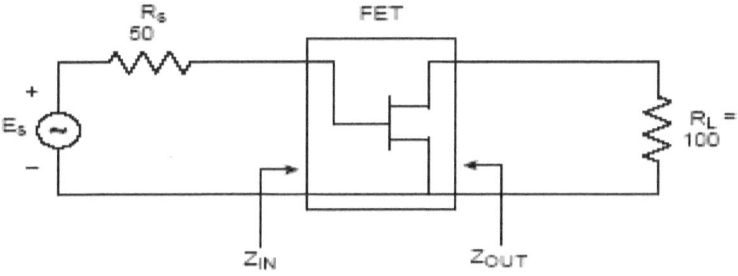

Figure P9.9

9.10) Consider the microwave network shown in Figure P9.10, consisting of a 50 Ω source, a 50 Ω-3 dB matched attenuator and a 50 Ω load:

   a. Compute the available power gain, the transducer power gain, and the actual power gain.
   b. How do these gains change if the load is changed to 25 Ω? How do these gains change if the source impedance is changed to 25Ω?

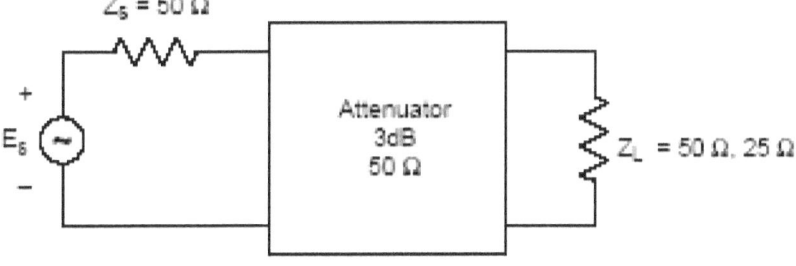

Figure P9.10

9.11) Use the new μ-parameter test to determine which of the following devices are unconditionally stable, and of these, which has the greatest stability:

Device	S_{11}	S_{12}	S_{21}	S_{22}
A	$0.34\angle-170°$	$0.06\angle70°$	$4.3\angle80°$	$0.45\angle-250°$
B	$0.75\angle-60°$	$0.2\angle70°$	$5.0\angle90°$	$0.51\angle60°$
C	$0.65\angle-140°$	$0.04\angle60°$	$2.4\angle50°$	$0.70\angle-65°$

9.12) Show that for a unilateral device where $S_{12}=0$, the μ-parameter test implies that:

$|S_{11}|<1$ and $|S_{22}|<1$ (for unconditional stability).

REFERENCES

[9.1] Bahl, I. and P. Bhartia. *Microwave Solid State Circuit Design.* New York: Wiley Interscience, 1988.

[9.2] Gonzalez, G. *Microwave Transistor Amplifiers, Analysis and Design,* 2nd ed. Upper Saddle River: Prentice Hall, 1997.

[9.3] Ha, T. T. *Solid State Microwave Amplifier Design.* New York: John Wiley & Sons, 1987.

[9.4] Liao, S. Y. *Microwave Circuit Analysis and Amplifier Design.* Upper Saddle River: Prentice Hall, 1987.

[9.5] Pozar, D. M. *Microwave Engineering,* 2nd ed. New York: John Wiley & Sons, 1998.

[9.6] Vendelin, George D. *Design of Amplifiers and Oscillators by the S-Parameter Method.* New York: John Wiley & Sons, 1982.

[9.7] Vendelin, George D., Anthony M. Pavio, and Ulrich L. Rhode. *Microwave Circuit Design, Using Linear and Non-Linear Techniques.* New York: John Wiley & Sons, 1990.

CHAPTER 10

Noise in Active Networks

10.1 INTRODUCTION
Having done a stability check and having met the gain requirements of an amplifier, the next important point to consider is noise. In an RF/microwave amplifier, the existence of a noise signal plays an important role in the overall design procedure and needs to be grasped before a meaningful design process can be developed.

Noise power results from random processes that exist in nature. These random processes can be classified in several important classes each generating a certain type of noise which will be characterized shortly.

Some of the most important types of random processes are:
1. *Thermal vibrations of atoms, electrons, and molecules in a component at any temperature above $0\,^\circ K$.*
2. *Flow of charges (electrons and/or holes) in a wire or a device.*
3. *Emission of charges (electrons or ions) from a surface such as the cathode of a diode or an electron tube, etc.*
4. *Wave propagation through the atmosphere or any other gas*

10.2 IMPORTANCE OF NOISE
Noise is passed into a microwave component or system either from an external source or is generated within the unit itself. Regardless

of the manner of entrance of the noise signal, the noise level of a system greatly affects the performance of the system by setting the minimum detectable signal in the presence of noise. Therefore it is often desirable to reduce the influence of external noise signals and minimize the generation of noise signals within the unit, in order to achieve the best performance.

10.3 NOISE DEFINITION

Since noise considerations are of important consequences, we need to define it first:

DEFINITION- ELECTRICAL NOISE (OR NOISE): *Is defined to be any unwanted electrical disturbance or spurious signal. These unwanted signals are random in nature and are generated either internally in the electronic components or externally through impinging electromagnetic radiation.*

Since signals are totally random and uncorrelated in time, they are best analyzed through statistical methods. Their statistical properties can be briefly summarized as:

a.   The "Mean value" of the noise signal is zero, i.e.,

$$\overline{V_n} = Lim_{T\to\infty} 1/T \int_{t_1}^{t_1+T} V_n(t)dt = 0 \qquad (10.1)$$

Where $\overline{V_n}$ is the noise mean value, $V_n(t)$ is the instantaneous noise voltage, t_1 is any arbitrary point in time and T is any arbitrary period of time ideally a large one approaching ∞.

b.   The "mean-square-value" of the noise signal is a constant value, i.e.,

$$\overline{V_n^2} = Lim_{T\to\infty} 1/T \int_{t_1}^{t_1+T} [V_n(t)]^2 dt = Constant \qquad (10.2)$$

c.   The "root-mean-square" (rms) of a noise signal is given by:

$$V_{n,rms} = \sqrt{\overline{V_n^2}} \qquad (10.3)$$

or,

$$(V_n)^2_{rms} = \overline{V_n^{\,2}} \tag{10.4}$$

The concept of "root-mean-square value" of noise as given by Equation (10.3), is based on the fact that the "mean-square value", $\overline{V_n^{\,2}}$, is proportional to the "noise power". Thus if we take the square root of Equation (10.2), we obtain the "rms value" of the noise voltage which is the "effective value" of the noise voltage.

10.4 SOURCES OF NOISE
There are several types of noise which needs to be defined:

a. DEFINITION-THERMAL NOISE (ALSO CALLED JOHNSON NOISE OR NYQUIST NOISE): *is the most basic type of noise which is caused by thermal vibration of bound charges and thermal agitation of electrons in a conductive material. This is common to all passive or active devices.*

b. DEFINITION-SHOT NOISE (OR SCHOTTKY NOISE): *is caused by the random passage of discrete charge carriers (causing a current "I", due to motion of electrons or holes) in a solid state device while crossing a junction or other discontinuities. It is commonly found in a semiconductor device (e.g. in a pn junction of a diode or a transistor) and is proportional to $(I)^{1/2}$.*

c. DEFINITION-FLICKER NOISE (ALSO CALLED 1/f NOISE): *is small vibrations of a current due to the following factors:*

1. *Random injection or recombination of charge carriers at an interface, such as at a metal-semiconductor interface (in semiconductor devices).*

2. *Random charges in cathode emissions of electric charges such as at a cathode-air interface (in a thermionic tube).*

Flicker noise is a frequency-dependent noise, which distorts the signal by adding more noise to the lower part of the signal band than the upper part. It exists at lower frequencies, almost from DC extending down to approximately 500 kHz to 1 MHz at a rate of −10 dB per decade.

10.5 THERMAL NOISE ANALYSIS

To analyze noise, let us consider the circuit shown in Figure 10.1a where a noisy resistor is connected to the input port of a two-port network. Focusing primarily on thermal noise, we note that the available noise power (i.e. maximum power available under matched conditions) from any arbitrary resistor has been shown by Nyquist to be:

$$P_N = kTB \tag{10.5}$$

Where,

k= Boltzmann's constant (=1.374x10^{-23} J/K).
T= The resistor's physical temperature.
B= The 2-port network's bandwidth (i.e., B= f$_H$-f$_L$).

Since the noise power does not depend on the center frequency of operation but only on the bandwidth, it is called "white noise" as shown in Figure 10.1b.

There are a few observations about noise power (P_N) which is worth considering:

a. As bandwidth (B) is reduced, so does the noise power which means narrower bandwidth amplifiers are less noisy,

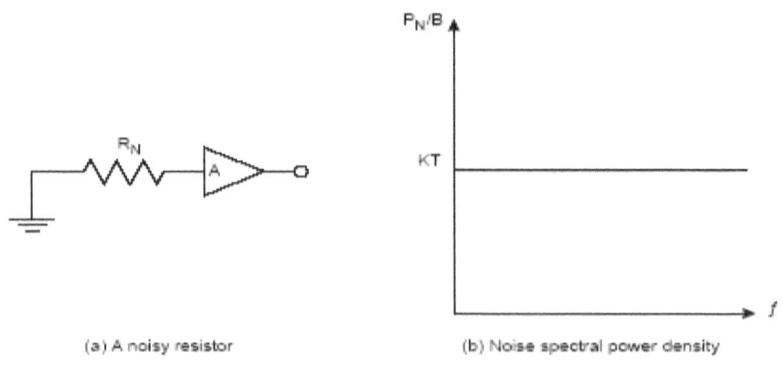

(a) A noisy resistor   (b) Noise spectral power density

Figure 10.1 White noise in an amplifier.

b. As temperature (T) is reduced, the noise power is also lessened which means cooler devices or amplifiers generate less noise power,

c. Increasing bandwidth to infinity causes an infinite noise power (called ultraviolet catastrophe) which is incorrect since (10.5) for noise power is only valid up to approximately 1000 GHz.

10.6 NOISE MODEL OF A NOISY RESISTOR

A noisy resistor (R_N) at a temperature (T) can be modeled by an ideal noiseless resistor (R_{NO}) at 0 °K in conjunction with a noise voltage source ($V_{n,rms}$) as shown Figure 10.2. If we assume that the resistor value is independent of temperature then $R_{NO}=R_N$.

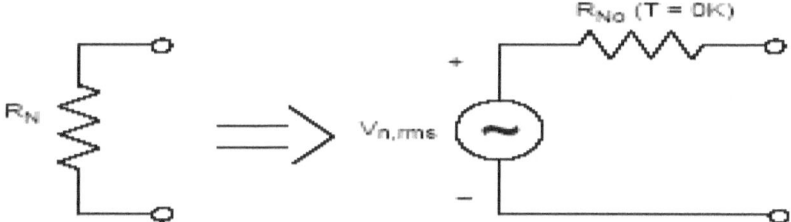

Figure 10.2 A simple model of a noisy resistor.

From this model, the available noise power to the load (under matched condition) is given by (see Figure 10.3):

$$P_N = \frac{V_{n,rms}^2}{4R_N} \qquad (10.6)$$

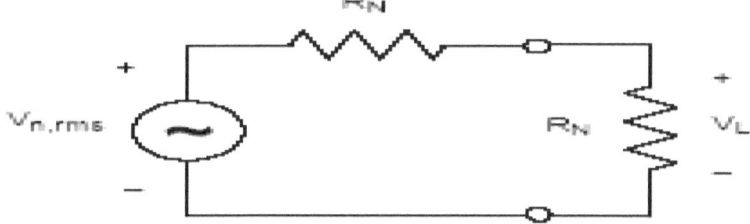

Figure 10.3 Available noise power.

Equation (10.17) provides the noise power available from a noisy resistor which equals Equation (10.5) for any arbitrary resistor. Thus:

$$P_N = kTB \qquad (10.7a)$$

$$V_{n,rms} = 2\sqrt{P_N R_N} = 2\sqrt{kTBR_N} \qquad (10.7b)$$

From Equation (10.7) we can observe that the noise voltage is proportional to ($R^{1/2}$). Thus higher-valued resistors have higher noise voltage even though they provide the same noise power level as the lower-valued resistors.

EXAMPLE 10.1

Calculate the noise power (in dBm) and rms noise voltage at T=290 °K for
a. R_N=1 Ω, B=1 Hz
b. R_N=2 MΩ, B=5 kHz.

Solution:
a. The noise power is given by:
$k=1.374\times10^{-23}$ J/°K
$P_N=kTB=1.374\times10^{-23}$ x290x1=3.985x10^{-21} W
Or in dBm, we have:
P_N(dBm)=10log(3.985x10^{-21}/10^{-3})=-174 dBm
This is the power per unit Hz. The corresponding noise voltage for a 1 Ω resistor is given by:

$$V_{n,rms} = 2\sqrt{P_N R_N} = 2\sqrt{3.985\times10^{-21}\,x1} = 12.6\times10^{-11}\,V$$
$$=12.6\times10^{-5}\ \mu V$$

b. For a 5 kHz bandwidth, we have
$P_N=kTB=3.985\times10^{-21}$x5000=19.925x10^{-18} W
The corresponding noise voltage for a 2 MΩ resistor is given by

$$V_{n,rms} = 2\sqrt{P_N R_N} = 2\sqrt{19.925\times10^{-18}\,x2x10^6} = 12.6\times10^{-6}\,V$$
$$=12.6\ \mu V$$

10.7 EQUIVALENT NOISE TEMPERATURE

Any type of noise, in general, has a power spectrum which can be plotted in the frequency domain. If the noise power spectrum is not a strong function of frequency (i.e., it is "White" noise) then it can be modeled as an equivalent thermal noise source characterized by an "equivalent noise temperature" (T_e).

To define "the equivalent noise temperature" (Te), we consider an arbitrary white noise source with an available power (P_S) having a noiseless source resistance (R_S) as shown in Figure 10.4a. This white noise source can be replaced by a noisy resistor with an equivalent noise temperature (T_e) defined by:

$$T_e = \frac{P_S}{kB} \tag{10.8}$$

Where B is the bandwidth of the system or the component under consideration.

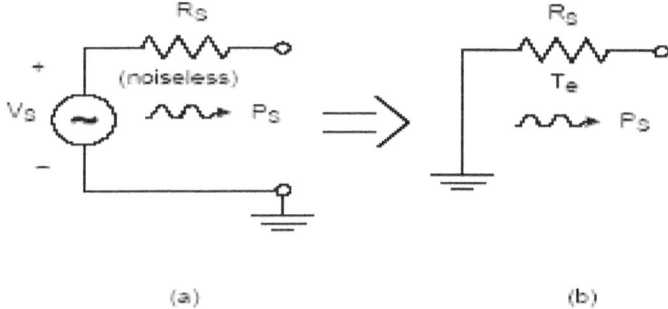

(a)   (b)

Figure 10.4 An arbitrary white noise source and its equivalent circuit.

EXAMPLE 10.2:
Consider a noisy amplifier with available power gain (G_A) and bandwidth (B) connected to a source and load resistance (R) both at $T=T_S$ as shown in Figure 10.5. Determine the overall noise temperature of the combination and the total output noise power if the amplifier all by itself creates an output noise power of P_n.

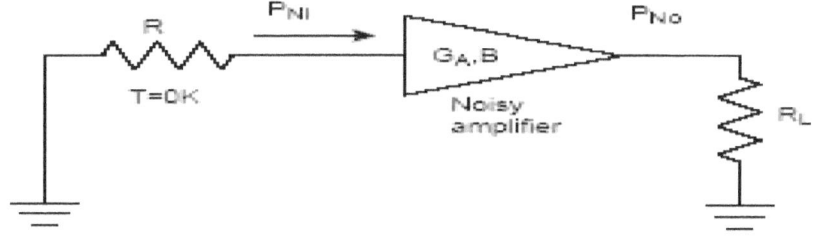

Figure 10.5 A noisy amplifier.

Solution:
To simplify the analysis, let us first assume that the source resistor is at $T=0°K$. This means that no noise enters the amplifier, i.e., $P_{Ni}=0$.

The noisy amplifier can be modeled by a noiseless amplifier with an input resistor at an equivalent noise temperature of:

$$T_e = \frac{P_n}{G_A kB}$$ (10.9)

T_e is called the equivalent noise temperature of the amplifier "referred to the input" as shown in Figure 10.6.

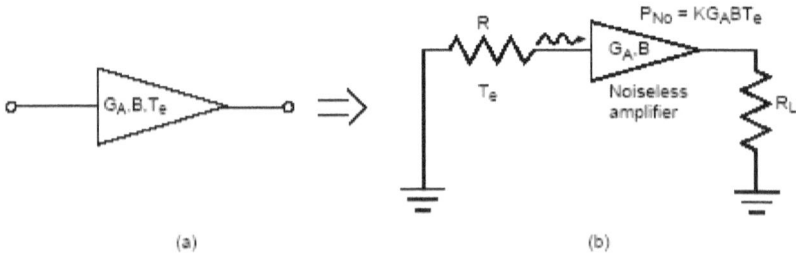

(a)   (b)

Figure 10.6 An equivalent model of a noisy amplifier.

Since source resistor (R) is at a physical temperature other than zero (T=T$_S$), then as a result the combined equivalent noise temperature (T_e') is the addition of the two noise temperatures:

$$T_e' = T_e + T_S$$ (10.10)

Assuming the noise power at the input terminals of the amplifier is P_{NI} (=kT_SB), the total output noise power due to the amplified input thermal noise power will be ($G_A P_{Ni}$) which adds to the amplifier's generated noise power (P_n) linearly by using the superposition theorem (see Figure 10.7), i.e.,

$$P_{No,tot} = G_A P_{Ni} + P_n = G_A kB(T_S + T_e)$$
$$P_{No,tot} = G_A kB T_e'$$ (10.11)

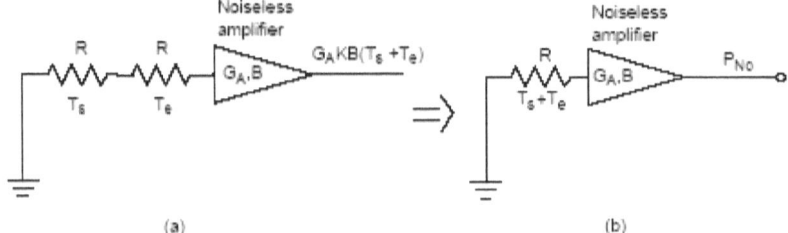

(a)   (b)

Figure 10.7 Total output noise power and its equivalent circuit.

NOTE: *It is important to note that from (10.11), the "equivalent noise temperature" (T_e') is defined by "referring" the total output noise power to the input port. Thus the same noise power is delivered to the load by driving a "noiseless amplifier" with a resistor at an equivalent temperature ($T_e'=T_e+T_S$).*

10.7.1 A Measurement Application: Y-Factor Method

The concept of equivalent noise temperature is commonly used in the measurement of noise temperature of an unknown amplifier using the "Y-factor method". In this method, the physical temperature of a matched resistor is changed to two distinct and known values:

a. One temperature (T_1) is at boiling water ($T_1=100°C$) or at room temperature ($T_1=290\ °K$),

b. The second temperature (T_2) is obtained by using either a noise source
    (hotter source than room temperature) or a load immersed in liquid nitrogen at $T=77\ °K$ (a colder source than room temperature) as shown in Figure 10.8.

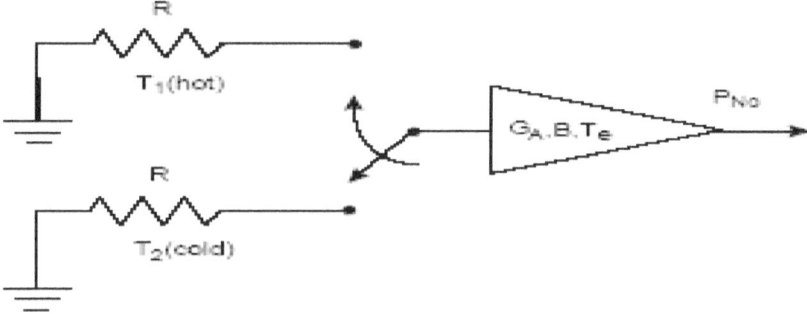

Figure 10.8 The Y-factor method.

The amplifier's unknown noise temperature (T_e) can be obtained as follows:

$$P_{No,1}=G_AkB(T_1+T_e) \tag{10.12}$$
$$P_{No,2}=G_AkB(T_2+T_e) \tag{10.13}$$

Now define:

$$Y \equiv \frac{P_{No,1}}{P_{No,2}}$$

Thus we can write:

$$Y = \frac{T_1 + T_e}{T_2 + T_e} \tag{10.14}$$

Or,

$$T_e = \frac{T_1 - YT_2}{Y - 1} \tag{10.15}$$

From a measurement of T_1, T_2, and Y the unknown amplifier's noise temperature (T_e) can be found.

POINT OF CAUTION: *To obtain an accurate value for Y, the two temperatures ideally must be far apart; otherwise $Y \approx 1$ and the denominator of Equation (10.15) will create relatively inaccurate results.*

NOTE: *A noise source "hotter" than room temperature, as used in the Y-factor measurement, would be a solid-state noise source (such as an IMPATT diode) or a noise tube. Such active sources, providing a calibrated and specific noise power output in a particular frequency range, are most commonly characterized by their "excess noise ratio" values vs. frequency. The term "excess noise ratio" or ENR is defined as:*

$$ENR(dB) = 10\log_{10}\left(\frac{P_N - P_O}{P_O}\right) = 10\log_{10}\left(\frac{T_N - T_O}{T_O}\right) \tag{10.16}$$

where P_N and T_N are the noise power and equivalent noise temperature of the active noise generator, P_O and T_O are the noise power and noise temperature of a passive device (such as a matched load, etc.), respectively.

10.8 DEFINITIONS OF NOISE FIGURE

As discussed earlier, a noisy amplifier can be characterized by an equivalent noise temperature (T_e). An alternate method to characterize a noisy amplifier, is through the use of the concept of noise Figure which we need to define first.

DEFINITION- NOISE FIGURE: *Is defined to be the ratio of the total available noise power at the output, $(P_O)_{tot,}$ to the output available noise power $(P_O)_i$ due to thermal noise coming only from the input resistor at the standard room temperature ($T_O=290$ °K).*

To formulate an equation for noise figure (F), let us transfer the noise generated inside the amplifier (P_n) to its input terminals and model it as a "noiseless" amplifier which is connected to a noisy resistor (R) at noise temperature (T_e) in series to another resistor (R) at $T=T_O$, both connected at the input terminals of the "noiseless" amplifier as shown in Figure 10.9. From this configuration we can write:

$$P_n=G_A kT_e B \tag{10.17a}$$
$$(P_O)_i=G_A P_{Ni}=G_A kBT_O \tag{10.17b}$$
$$(P_O)_{tot}=P_{NO}=Pn+(P_O)_i \tag{10.18}$$
$$F = \frac{(P_O)_{tot}}{(P_O)_i} = \frac{(P_O)_i + P_n}{(P_O)_i} = 1 + \frac{P_n}{G_A P_{Ni}} \tag{10.19a}$$

Or,

$$F = 1 + \frac{T_e}{T_O} \tag{10.19b}$$

Or, in dB we can write:

$$F = 10\log_{10}\left(1 + \frac{T_e}{T_O}\right) \tag{10.20}$$

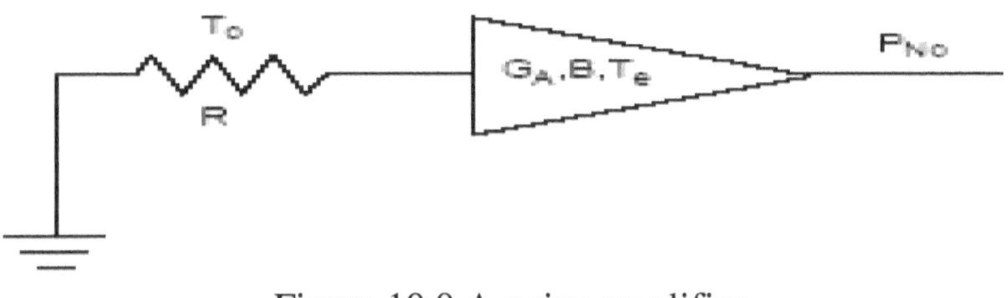

Figure 10.9 A noisy amplifier.

From (10.19) we can see that "F" is bounded by:
$$1 \le F \le \infty \tag{10.21}$$

The lower boundary (F=1) is the best-case scenario and is the Noise Figure of an ideal noiseless amplifier where T_e=0.

From Equation (10.19b), we can write:
$$T_e=(F-1)T_O \tag{10.22}$$

NOTE1: *Temperature (T_e) is the equivalent noise temperature of the amplifier referred to the input.*

NOTE 2: *Either F or T_e can interchangeably be used to describe the noise properties of a two-port network. However, For small noise Figure values (i.e., when F≈1), the use of T_e becomes preferable.*

POINT OF CAUTION: *It is interesting to note that the noise Figure is defined with reference to a matched input termination at room temperature (T_O=290 °K). Therefore if the physical temperature of the amplifier changes to some value other than T_O, we still use the room temperature (T_O=290 °K) to find the noise figure value.*

10.8.1 Alternate Definition of Noise Figure
From Equations (10.17) and (10.18), we can write:
$$P_{NO}=G_AP_{Ni}+P_n \tag{10.23}$$
$$(Po)_i=G_AP_{Ni} \tag{10.24}$$
Where $P_n=G_AkT_eB$ is the generated noise power inside the amplifier. The noise figure can now be written as:
$$F = \frac{P_{N_o}}{(P_o)_i} = \frac{P_{N_o}}{G_A P_{N_i}} = 1 + \frac{P_n}{G_A P_{N_i}} \tag{10.25}$$

The available power gain (G_A) by definition is given by:
$$G_A = \frac{P_{S_o}}{P_{S_i}}$$

where P_{S_o} and P_{S_i} are the available signal power at the output and the input, respectively. Thus Equation (10.25) can now be written as:
$$F = \frac{P_{S_i}/P_{N_i}}{P_{S_o}/P_{N_o}} = \frac{(SNR)_i}{(SNR)_o} \tag{10.26}$$

where (SNR)$_i$ and (SNR)$_o$ are the available signal-to-noise ratio at the input and output ports, respectively.

Equation (10.26) indicates that the noise figure can also be defined in terms of the ratio of available signal-to-noise power ratio to the available signal-to-noise power ratio at the output

10.8.2 Noise Figure of a Lossy Two-Port Network
This is an important case, where the two-port network considered earlier, is a lossy passive component such as an attenuator or a lossy transmission line, as shown in Figure 10.10.

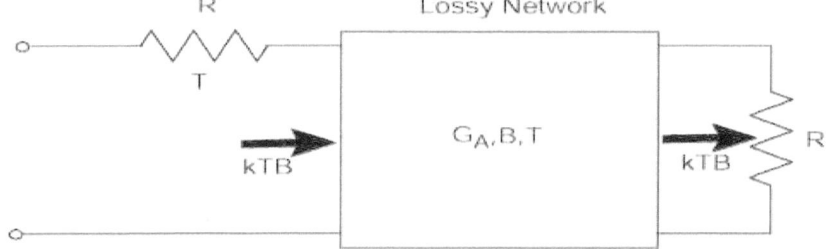

Figure 10.10 A lossy two-port network.

A lossy network has a gain ($\mathbf{G_A} = \dfrac{\mathbf{P_o}}{\mathbf{P_i}}$) less than unity, which can be expressed in terms of the loss factor or attenuation (L) as:

$$\mathbf{G_A} = \frac{1}{\mathbf{L}} \qquad (G_A < 1) \tag{10.27}$$

Since the gain of a lossy network is less than unity it follows that the loss or attenuation factor (L) is more than unity (i.e., L=P_i/P_o>1) for any lossy network or component.

Expressing the attenuation factor (L) in "dB" gives the following:

$$\mathbf{L(dB)} = -10\log_{10}\left(\frac{\mathbf{P_i}}{\mathbf{P_o}}\right) \tag{10.28}$$

For example, if the lossy component attenuates the input power by ten times, then we can write:

$$G_A = \frac{P_0}{P_i} = 0.1 \implies L = 1/G_A = 10 = 10 \text{ dB}$$

If the lossy network is held at a temperature (T), the total available output noise power according to Equation (10.5) is given by:

$$P_{NO} = kTB \tag{10.29}$$

On the other hand, from (10.23) the available output noise power is also given by the addition of the input noise power and the generated noise inside the circuit (P_n):

$$P_{NO} = G_A kTB + P_n = KTB/L + P_n \tag{10.30}$$

where P_n is the noise generated inside the two-port. Equating Equations (10.29) and (10.30), we obtain P_n as:

$$P_n = \left(\frac{L-1}{L}\right) kTB \tag{10.31a}$$

NOTE: *If we refer the noise generated inside the amplifier (P_n) to the input side (P_n)$_i$, from (10.31a) we have:*

$$(P_n)_i = P_n/G_A = LP_n = (L-1)kTB \tag{10.31b}$$

Using Equations (10.31) we can now define the equivalent noise temperature (T_e) of a lossy two-port, referred to the input terminals, as:

$$T_e = \frac{(P_n)_i}{kB} \implies T_e = (L-1)T \tag{10.32}$$

Thus the noise figure of a lossy network is given by:

$$F = 1 + \frac{T_e}{T_0} = 1 + (L-1)\frac{T}{T_0} \tag{10.33}$$

A SPECIAL CASE:
For a lossy network at room temperature, i.e., $T = T_0$, Equation (10.33) gives:

$$F = L \tag{10.34}$$

Equation (10.34) indicates that the noise figure of a lossy network at room temperature equals the attenuation factor (L). For example: if $G_A=1/5$ then $L=1/G_A=5$, giving F=5 or 7 dB for $T=T_o=290$ K.

Example 10.3
A wideband amplifier (2-4 GHz) has a gain of 10 dB, an output power of 10 dBm and a noise figure of 4dB at room temperature. Find the output noise power in dBm.

Solution:
B=2 GHz
GA=10 dB
F=4 dB
$F=P_{NO}/G_A P_{Ni}=P_{NO}/G_A kT_o B$
Thus:
$P_{NO}=FG_A kT_o B$
$10 \log_{10}P_{NO}=P_{NO}(dB)=F(dB)+G_A(dB)+10 \log_{10}(kT_o)+10 \log_{10}(B)$
$$=4+10-174+10 \log_{10}(2x10^9)= -67 \text{ dBm}$$

10.9 NOISE FIGURE OF CASCADED NETWORKS
A microwave system usually consists of several stages or networks connected in a cascade where each adds noise to the system thus degrading the overall signal-to-noise ratio.
If the noise figure (or noise temperature) of each stage is known, the overall noise figure (or noise temperature) can be determined.

10.9.1 Cascade of Two Stages
To analyze a two-stage amplifier, let us consider a cascade of two amplifiers each with its own gain, noise temperature, or noise figure as shown in Figure 10.11. The noise power of each stage is given as follows:
$$P_{NO1}=G_{A1}kB(T_o+T_{e1}) \qquad (10.35)$$
$$P_{NO2}=G_{A2}P_{NO1}+G_{A2}kT_{e2}B \qquad (10.36)$$

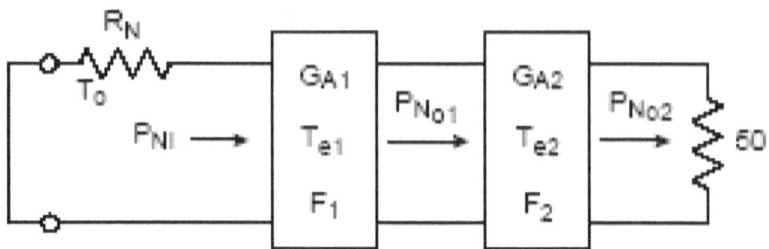

Figure 10.11 Cascade of two stages.

Combining Equations (10.35) and (10.36) we get:

$$P_{NO2}=G_{A1}G_{A2}kB(T_o+T_{e1}+T_{e2}/G_{A1}) \tag{10.37}$$

The two-stage amplifier as a whole has a total gain of $G_A=G_{A1}G_{A2}$, an overall equivalent noise temperature (T_e) and a total output noise power (P_{NO}) given by:

$$P_{NO}=G_AkB(T_o+T_e) \tag{10.38}$$

Comparing Equation (10.38) to (10.37) we have:

$$T_e=T_{e1}+T_{e2}/G_{A1} \tag{10.39}$$

The overall noise figure (F) for the two-stage amplifier is found by using (10.39):

$$F=1+T_e/T_o=1+(T_{e1}+T_{e2}/G_{A1})/T_o \tag{10.40}$$

By noting that:

$$F_1=1+T_{e1}/T_o, \tag{10.41}$$
$$F_2=1+T_{e2}/T_o \tag{10.42}$$

Equation (10.40) can be written as:

$$F = F_1 + \frac{F_2 - 1}{G_{A1}} \tag{10.43}$$

Equations (10.39) and (10.43) show that the first stage noise figure F_1 (or noise temperature, T_{e1}), and gain (G_{A1}) have a large influence on the overall noise figure (or noise temperature). This is because the second stage noise figure, F_2 (or noise temperature, T_{e2}) is reduced by the gain of the first stage (G_{A1}).

Thus the key to low overall noise figure is a primary focus on the first stage by reducing its noise and increasing its gain. Later stages have a greatly reduced effect on the overall noise figure.

NOISE MEASURE

In order to determine systematically the order or sequence in which two similar amplifiers need to be connected to produce the lowest possible noise figure, we first must define a quantity called "noise measure" as:

$$M = \frac{F-1}{1-1/G_A} \qquad (10.44)$$

If amplifier #1 (AMP1) has a noise measure (M_1) and amplifiers #2 (AMP2) a noise measure (M_2) then there are two possible cases which need to be addressed (in order to obtain the lowest possible noise from the cascade), as follows:

Case I: $M_1<M_2$ -- Then AMP1 should precede AMP2 since $F_{12}<F_{21}$

Case II: $M_2<M_1$ -- Then AMP2 should precede AMP1 since $F_{21}<F_{12}$

Where F_{12} is the overall noise figure of the two-stage amplifier when AMP1 precedes AMP2; and vice versa F_{21} is for the case when AMP2 precedes AMP1.

NOTE: *It can easily be shown mathematically that for example:*
If $M_1<M_2$ then $F_{12}<F_{21}$ $\qquad (10.45)$
where

$$F_{12} = F_1 + \frac{F_2-1}{G_{A1}} \qquad (10.46)$$

$$F_{21} = F_2 + \frac{F_1-1}{G_{A2}} \qquad (10.47)$$

And vice versa, if $M_2<M_1$ then $F_{21}<F_{12}$.

Example 10.4

An antenna is connected to an amplifier via a transmission line which has an attenuation of 3 dB (see Figure 10.12). The amplifier has the following specifications:
$G_A=20$ dB
$B=200$ MHz
$T_e=145$ K
Calculate the overall noise figure and gain of the cascade at 300 K.

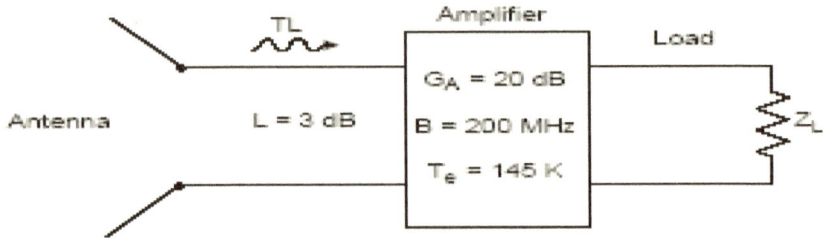

Figure 10.12 Circuit for example 10.4.

Solution:
a. For the transmission line we have:
Since $L=1/G_{TL} \Rightarrow L(dB)=-G_{TL}(dB)$
$L=3\ dB=2 \Rightarrow G_{TL}=-3\ dB=1/2$
$F_{TL}=1+(L-1)T/T_o=1+(2-1)300/290=2.03=3.1\ dB$
b. For the amplifier we have:
$F_{AMP}=1+T_e/T_o=1.5=1.8\ dB$
c. The overall noise figure and gain are calculated to be:
$F_{TOT}=F_{TL}+(F_{AMP}-1)/G_{TL}=2.03+(1.5-1)/0.5=3.03=4.8\ dB$
$G_{TOT}=G_{TL}+G_{AMP}=-3+20=17\ dB$

Therefore we can see that due to the addition of a lossy transmission line in front of the amplifier, we have three deleterious effects: 1) the overall noise figure has increased (from 1.8 dB to 4.2 dB) 2) the second stage noise contribution has been intensified since the transmission line has a gain less than unity ($G_{TL}<1$), and 3) the overall gain dropped by 3 dB which represents the third side effect.

___

10.9.2 Cascade of n Stages
For a cascade of "n" amplifiers, the overall noise figure is the generalization of equations for equivalent noise temperature ($T_{e,cas}$) and noise figure (F_{cas}) of a two-stage cascade as follows (see Figure 10.13):

$$T_{e,cas} = T_{e1} + \frac{T_{e2}}{G_{A1}} + \frac{T_{e3}}{G_{A1}G_{A2}} + \cdots + \frac{T_{en}}{G_{A1}G_{A2}\cdots G_{An-1}} \qquad (10.48a)$$

$$F_{cas} = F_1 + \frac{F_2 - 1}{G_{A1}} + \frac{F_3 - 1}{G_{A1}G_{A2}} + \cdots + \frac{F_n - 1}{G_{A1}G_{A2} \cdots G_{An-1}} \qquad (10.48b)$$

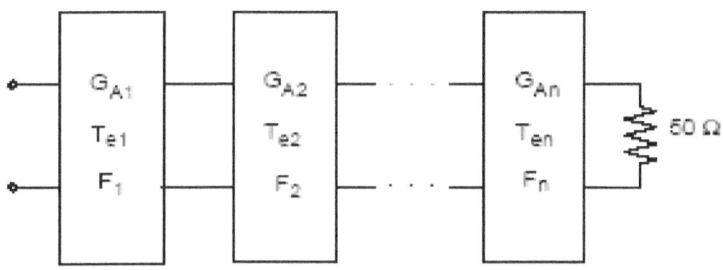

Figure 10.13 Cascade of "n" stages.

SPECIAL CASE: IDENTICAL STAGES
If all stages are identical, i.e.,

$$G_{A1} = G_{A2} = \ldots = G_{An} = G_A \qquad (10.49a)$$
$$T_{e1} = T_{e2} = \ldots = T_{en} = T_e \qquad (10.49b)$$
$$F_1 = F_2 = \ldots = F_n = F \qquad (10.49c)$$

Then Equations (10.48a,b) would greatly simplify as follows:

$$T_{e,cas} = T_e(1 + X + X^2 + \ldots + X^{n-1}), \qquad (10.50a)$$
$$F_{cas} = (F-1)(1 + X + X^2 + \ldots + X^{n-1}) + 1 \qquad (10.50b)$$

Where

$$X = \frac{1}{G_A}$$

Using the following identity for the geometric series:

$$1 + X + X^2 + \ldots + X^{n-1} = (1 - X^n)/(1 - X), \quad |X| < 1 \qquad (10.51)$$

We can write Equations (10.50a,b) as:

$$T_{e,cas} = T_e\left(\frac{1 - (1/G_A)^n}{1 - 1/G_A}\right) \qquad (10.52a)$$

$$F_{cas} = (F - 1)\left(\frac{1 - (1/G_A)^n}{1 - 1/G_A}\right) + 1 \qquad (10.52b)$$

AN INFINITE CHAIN OF IDENTICAL AMPLIFIERS
If n is very large (i.e., n→ ∞) then:

$$\text{Lim}_{n\to\infty}(X)^n = 0, \quad |X|<1 \tag{10.53a}$$

And the geometric series identity in Equation (10.51) further simplifies into:

$$1+X+X^2+.....+X^{n-1}+.... = \frac{1}{1-X}, \quad |X|<1 \tag{10.53b}$$

Using Equation (10.53b), we can see that Equations (10.52a,b) and for an infinite chain of amplifiers become:

$$T_{e,cas} = T_e\left(\frac{1}{1-1/G_A}\right) \tag{10.54a}$$

$$F_{cas} = (F-1)\left(\frac{1}{1-1/G_A}\right)+1 \tag{10.54b}$$

In terms of "noise measure", M, defined earlier as:

$$M = \frac{F-1}{1-1/G_A} \tag{10.55}$$

we can write (10.54) as:

$$T_{e,cas} = T_e\left(\frac{M}{F-1}\right) \tag{10.56a}$$

$$F_{cas} = M+1 \tag{10.56b}$$

NOTE 1: *For a "Minimum-noise amplifier", where each stage operates at minimum noise figure (i.e., $F_1=F_2=....=F_n=F_{min}$), we have:*

$$M_{min} = \left(\frac{F_{min}-1}{1-1/G_A}\right) \tag{10.57}$$

We can write (10.56) as:

$$T_{e,cas} = T_{e,min}\left(\frac{M_{min}}{F_{min}-1}\right) \tag{10.58a}$$

$$F_{cas} = M_{min}+1 \tag{10.58b}$$

NOTE 2: *If the gain of each stage is very large (i.e., $G_A\to\infty$), then Equation (10.56) becomes:*

$$G_A\to\infty \Rightarrow M=F-1 \tag{10.59}$$

$T_{e,cas} = T_e$ (10.60a)

$F_{cas} = F$ (10.60b)

This result indicates that a large cascade of very high-gain amplifiers will only result in the degradation of the signal by the first stage and the effect of all the many stages is null and void as far as the added noise is concerned.

This result is in agreement with the conclusion made earlier, in which it became apparent that the first stage's gain and noise figure value dominates and greatly affects the overall noise figure of the cascade.

10.10 CONSTANT NOISE FIGURE CIRCLES

As derived in reference [10.1], the Noise Figure of a two-Port amplifier is given by

$$F = F_{min} + \frac{r_n}{g_S} \left| Y_S - Y_{opt} \right|^2$$ (10.61)

Where

$r_n = \dfrac{R_n}{Z_O}$ (the equivalent normalized noise resistance of the two-port).

$Y_S = g_S + jb_S$ (the normalized source admittance corresponding to Γ_S as defined in Chapter 13)

$Y_{opt} = g_{opt} + jb_{opt}$ (the normalized source admittance for minimum noise figure, i.e. at $\Gamma_S = \Gamma_{opt} \Rightarrow F = F_{min}$)

Since Y_S and Y_{opt} are related to Γ_S and Γ_{opt} by the relations:

$$Y_S = \frac{1 - \Gamma_S}{1 + \Gamma_S}$$ (10.62)

$$Y_{opt} = \frac{1 - \Gamma_{opt}}{1 + \Gamma_{opt}}$$ (10.63)

Then using Γ_S and Γ_{opt} instead of Y_S and Y_{opt} in Equation (10.61), we can write:

$$F = F_{min} + \frac{4r_n \left| \Gamma_S - \Gamma_{opt} \right|^2}{\left(1 - \left| \Gamma_S \right|^2 \right)\left| 1 + \Gamma_{opt} \right|^2} \qquad (10.64)$$

We now define a parameter called the noise Figure parameter (N):

$$N = \frac{\left| \Gamma_S - \Gamma_{opt} \right|^2}{1 - \left| \Gamma_S \right|^2} \qquad (10.65)$$

Thus Equation (10.64) can be written as:

$$F = F_{min} + \frac{4r_n N}{\left| 1 + \Gamma_{opt} \right|^2} \qquad (10.66)$$

Parameters r_n, Γ_{opt}, and F_{min} are called the "**noise parameters**" of the transistor and are usually provided in the data sheets by the manufacturer.

NOTE 1: *Using Equation (10.66), we can write Equation (10.65) as:*

$$N = \frac{\left| \Gamma_S - \Gamma_{opt} \right|^2}{1 - \left| \Gamma_S \right|^2} = \frac{F - F_{min}}{4r_n} \left| 1 + \Gamma_{opt} \right|^2 \qquad (10.67)$$

From Equation (10.67) we can see that for a fixed (F), the parameter (N) is a positive real number.

NOTE 2: *Noise parameters may also be determined experimentally by the following procedure:*
a. *Vary Γ_S until a minimum noise figure occurs which is subsequently recorded (i.e., $F=F_{min}$).*
b. *Now using a vector network analyzer, measure Γ_S which provides the value for Γ_{opt}.*
c. *To find r_n, we set Γ_S to zero and measure the noise figure ($F_{\Gamma_S=0}$) at this point. Then using Equation (10.64) and the value of Γ_{opt} from step (b), we can obtain r_n as:*

$$r_n = \Delta F\left(\frac{\left|1+\Gamma_{opt}\right|^2}{4\left|\Gamma_{opt}\right|^2}\right)$$

(10.68)

Where

$$\Delta F = F_{\Gamma_S=0} - F_{min}$$

10.10.1 Analysis

By using Equation (10.67) and through rearranging terms and further mathematical manipulation of Equation (10.64), we obtain an equation of a circle in the Γ_S-plane as:

$|\Gamma_S - C_F| = R_F$ (10.69)

where C_F and R_F are the center and radius of noise figure circles given by:

$$C_F = \frac{\Gamma_{opt}}{N+1}$$

(10.70)

$$R_F = \frac{\sqrt{N^2 + N(1 - \left|\Gamma_{opt}\right|^2)}}{1+N}$$

(10.71)

Equation (10.69) represents a family of noise figure circles with the noise figure (F) value as a parameter.

OBSERVATIONS:
When $F=F_{min}$ then:

$\Gamma_S = \Gamma_{opt} \Rightarrow N=0$ (10.72)
$C_F = \Gamma_{opt}$ (10.73)
$R_F = 0$ (10.74)

Equation (10.73) and (10.74) indicate that F_{min} is a point uniquely located at Γ_{opt}. From Equation (10.67) we can see that since "N" is a real positive number, then all noise figure circles have centers located along Γ_{opt} vector in the Γ_S plane as shown in Figure 10.14.

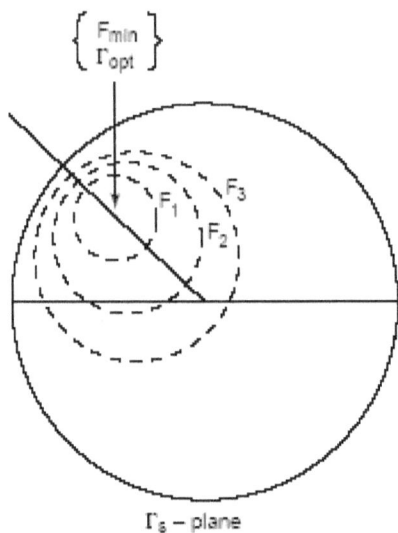

Figure 10.14 Family of Noise Figure circles, all centered at Γ_{opt}.

EXERCISE: Derive Equation (10.69) from Equations (10.64) through (10.67).

Hint: Use the following identity:

$$\left|\Gamma_S - \Gamma_{opt}\right|^2 = \left(\Gamma_S - \Gamma_{opt}\right)\left(\Gamma_S - \Gamma_{opt}\right)^*$$

to write (10.52) as:

$$\left|\Gamma_S\right|^2 - \frac{\left(\Gamma_S\Gamma_{opt}^* + \Gamma_S^*\Gamma_{opt}\right)}{1+N} + \frac{\left|\Gamma_{opt}\right|^2}{1+N} = \frac{N}{1+N}$$

Now add $\left|\Gamma_{opt}\right|^2/(N+1)^2$ to both sides of the above equation to obtain the desired relation for a circle as given by Equation (10.69).

10.10.2 Derivation of the Constant Noise Figure Circles

We will derive the general equation of a circle in the Γ_S-plane and then use it to derive the equation for the noise figure circles (Γ_S is a complex number).

a. GENERAL EQUATION OF A CIRCLE

The general equation for a circle in the Γ_S plane, with radius (R) and a center (C), is given by:

$$|\Gamma_S - C| = R \tag{10.75}$$

Squaring both sides of Equation (10.75), we have:

$$|\Gamma_S - C|^2 = R^2 \tag{10.76}$$

Equation (10.76) can be rewritten as:

$$(\Gamma_S - C)(\Gamma_S - C)^* = R^2$$

$$\Rightarrow (\Gamma_S - C)(\Gamma_S - C)^* = R^2 \tag{10.77}$$

Applying the conjugate operation and Multiplying the terms in Equation (10.77) yields the desired equation form for a circle as:

$$|\Gamma_S|^2 - C^* \Gamma_S - C \Gamma_S^* = R^2 - |C|^2 \tag{10.78}$$

NOTE: *The symbol "R," being a radius, is a real number, whereas "C," being the location of the center in the Γ_S plane, is a complex number.*

b. CONSTANT NOISE FIGURE CIRCLES

The general equation of a circle in the Γ_S-plane can be used successfully to derive the constant noise figure circle's center and radius (C_F, R_F). This is done by using the equation (10.65) as follows:

$$N = \frac{|\Gamma_S - \Gamma_{opt}|^2}{1 - |\Gamma_S|^2} \tag{10.79}$$

By expanding the terms of this equation and rearranging the terms, it can be rewritten as follows:

$$|\Gamma_S|^2 - \frac{\left(\Gamma_S \Gamma_{opt}^* + \Gamma_S^* \Gamma_{opt}\right)}{1+N} + \frac{|\Gamma_{opt}|^2}{1+N} = \frac{N}{1+N} \tag{10.80}$$

By comparing equations (10.78) and (10.80), we obtain:

$$C_F = \frac{\Gamma_{opt}}{N+1} \qquad\qquad (10.81)$$

$$R_F = \frac{\sqrt{N^2 + N(1 - |\Gamma_{opt}|^2)}}{1+N} \qquad\qquad (10.82)$$

Chapter 10- Symbol List

A symbol will not be repeated again, once it has been identified and defined in an earlier chapter, with its definition remaining unchanged.

B - Bandwidth
F - Noise Figure
k - Boltzmann's constant
M - Noise measure
N - Overall noise Figure
P_N - Noise power
P_{NI} - Input Noise power
$P_{N0,tot}$ - Total output noise power
R_N - Noisy resistor
R_{N0} - Noiseless resistor
T - Temperature
T_e - Equivalent noise temperature
T_0 - Standard room temperature (290° K)
T_S - Source and load temperature
$V_{n,rms}$ - Root-mean-square (rms) of noise
$\overline{V_n^2}$ - The mean-square value of noise

CHAPTER-10 PROBLEMS

10.1) The Y-factor method is to be used to measure the equivalent noise temperature of a component. A hot load of $T_1 = 300$ K and a cold load of $T_2 = 77$ K will be used. If the noise temperature of the amplifier is $T_e = 250$ K, what will be the ratio of power meter readings at the output of the component for the two loads?

10.2) A transmission line has a noise figure F=1 dB at a temperature T_O=290 K. Calculate and plot the noise figure of this line as its physical temperature ranges from T=0 K to 1000 K.

10.3) Assume that measurement error introduces an uncertainty of ΔY into the measurement of Y in a Y-factor measurement. Derive an expression for the normalized error of the equivalent noise temperature $(\Delta T_e/T_e)$ in terms of $(\Delta Y/Y)$ and the temperatures T_1, T_2 and T_e. Plot $(\Delta T_e/T_e)$ as a function T_e for two values of $(\Delta Y/Y)$: a) 0.1, and b) 0.20, and from these plots establish the minimum normalized error $(\Delta T_e/T_e)$ and the corresponding T_e for each case.

10.4) An amplifier with a bandwidth of 1 GHz has a gain of 15 dB and a noise temperature of 250 K. If it is used as a preamplifier in a cascade, preceding a microwave amplifier of 20 dB gain 5 dB noise figure, determine the overall noise temperature.

10.5) An amplifier with a gain of 12 dB, a bandwidth of 150 MHz and a noise figure of 4 dB feeds a receiver with a noise temperature of 900K. Find the noise figure of the overall system.

10.6) Consider the microwave system shown in Figure P10.6, where the bandwidth is 1 GHz centered at 20 GHz and the physical temperature of the system is T_O=300 K. What is the equivalent noise temperature of the source? What is the noise input of the amplifier in dB? When the noisy source is connected to the system what is the total noise power output of the amplifier in dBm?

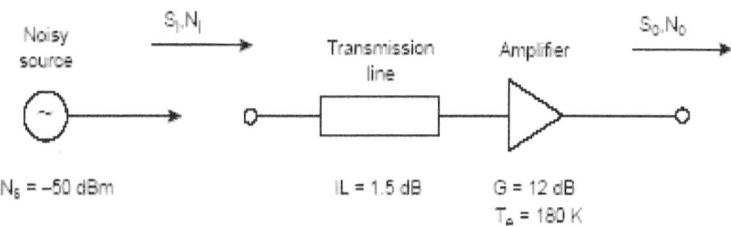

Figure P10.6

10.7) Consider the wireless local area network (WLAN) receiver front-end shown in Figure P10.7, where the bandwidth of the bandpass filter is 100 MHz centered at 2.4 GHz. If the system is at room temperature, find the noise figure of the overall system. What is the resulting signal to noise ratio at the output if the input signal power level is -90 dBm? Can the components be rearranged to give a better noise figure?

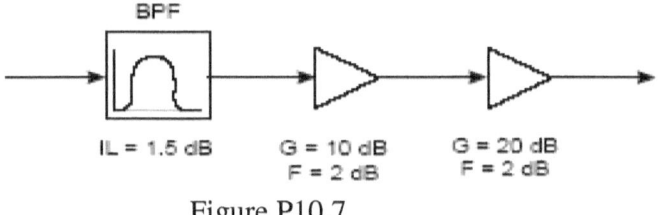

Figure P10.7

10.8) A two-way power divider has one output port terminated in a matched load as shown in Figure P10.8. Find the equivalent noise temperature of the resulting two-Port network if the divider is an equal-split two-way resistive divider.

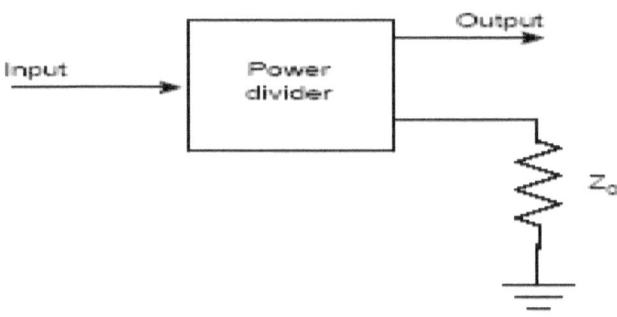

Figure P10.8

10.9) For a two-stage cascaded network with gain values of G_1 and G_2 and noise figures of F_1 and F_2 as shown in Figure P10.9, the input noise power is $N_i=kTB$. The output noise power is N_1 and N_2 at the output of the first and second stages. Are the following expressions correct:

a. $F_1=N_1/G_1N_i$

b. $F_2=N_2/G_2N_1$

c. $F_2=N_2/G_1G_2F_2N_i$

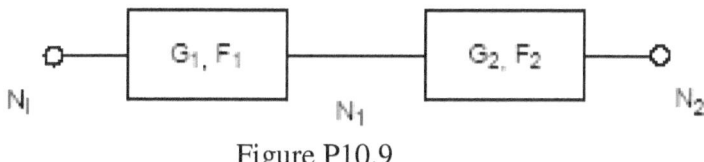

Figure P10.9

10.10) A receiver has the block diagram shown in Figure P10.10. Calculate:

a. The total gain (or loss) in dB,

b. The overall noise Figure in dB.

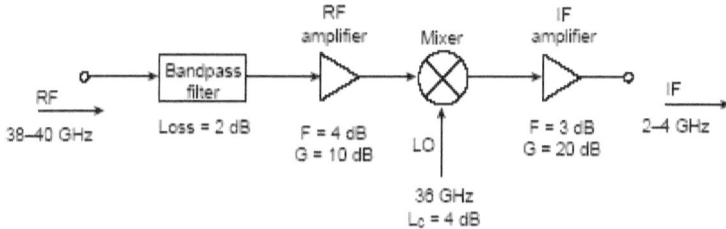

Figure P10.10

10.11) Two satellite receiver systems have the following specifications for
their components:

RF Amplifier: F=5 dB, G=10 dB

Mixer: L_c=5 dB

IF amplifier: F=2 dB, G=15 dB

Bandpass filter: IL=2 dB

Compare the two systems in terms of the overall gain and noise figure values (see Figure P10.11).

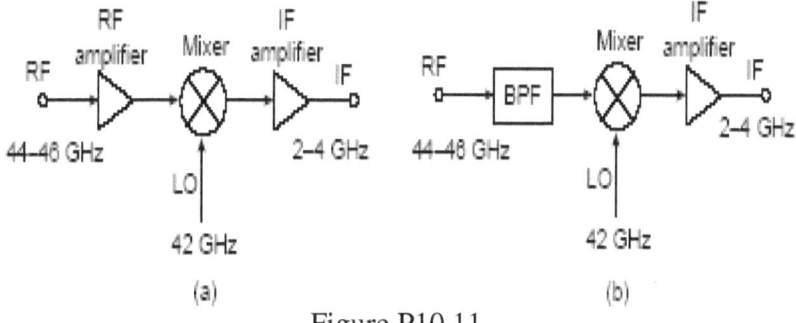

Figure P10.11

10.12) Calculate the overall noise Figure and gain for the receiver system shown in Figure P10.12.

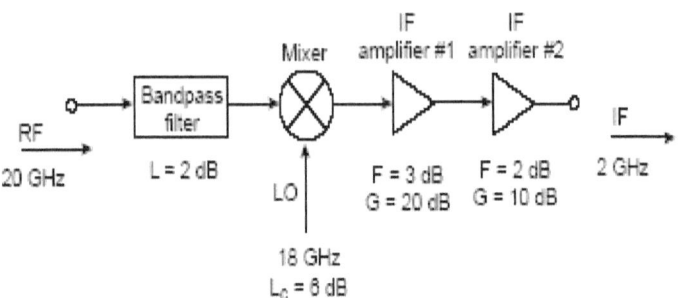

Figure P10.12

10.13) The S parameters and the noise parameters of a GaAs FET at 10 GHz are:

$$S = \begin{bmatrix} 0.6\angle -170° & 0.05\angle 16° \\ 2\angle 30° & 0.5\angle -95° \end{bmatrix}$$

F_{min}=2.5 dB
Γ_0=0.5∠145°
R_n =5 Ω

a. Is the transistor unconditionally stable?
b. Determine $G_{A,max}$
c. Determine the noise figure if the transistor is used in an amplifier designed for maximum available gain ($G_{A,max}$)

10.14) Consider the low noise block (LNB) shown in Figure P10.14. Calculate the total noise figure and the available gain of this LNB.

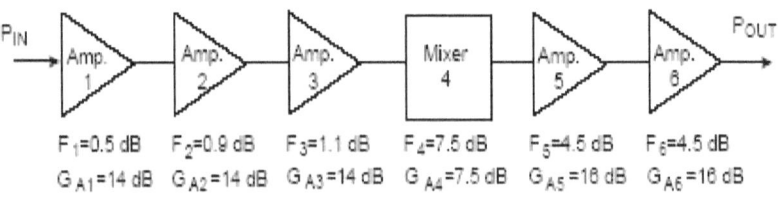

Figure P10.14

REFERENCES

[10.1] Ambrozy, A. *Electronic Noise.* New York: McGraw-Hill, 1982.

[10.2] Cappy, A. Noise Modeling and Measurement Techniques. *IEEE Transactions on Microwave Theory and Technique,* Vol. MTT-36, January 1988, pp. 1–10.

[10.3] Friis, H. T. Noise Figure of Radio Receivers. *Proceedings of IRE,* Vol. 32, July 1944, pp. 419 –22.

[10.4] Fukui, H. Available Power Gain, Noise Figure, and Noise Measure of Two-Ports and Their Graphical Representation. *IEEE Transactions on Circuit Theory,* June 1966, pp. 137–42.

[10.5] Gonzalez, G. *Microwave Transistor Amplifiers, Analysis and Design,* 2nd ed. Upper Saddle River: Prentice Hall, 1997.

[10.6] Haus, H. A. and R. B. Adler. *Circuit Theory of Linear Noisy Networks.* Cambridge: MIT Press, 1959 and New York: John Wiley & Sons, 1959.

[10.7] Pozar, D. M. *Microwave Engineering,* 2nd ed. New York: John Wiley & Sons, 1998.

[10.8] Radmanesh, M. M., and J. M. Cadwallader. Millimeter-Wave Noise Sources at V-Band (50 to 75 GHz), *Microwave Journal,* Vol. 36, No. 2, pp. 128–134, Sept. 1993.

[10.9] Radmanesh, M. M. and J. M. Cadwallader. Solid State Noise Sources at mm-Waves: Theory and Experiment. *Microwave Journal,* Vol. 34, No. 10, pp. 125–133, Oct. 1991.

[10.10] Radmanesh, M. M. *The Gateway to Understanding: Electrons to Waves and Beyond,* AuthorHouse, 2005.

[10.11] Radmanesh, M. M. *Cracking the Code of Our Physical Universe,* AuthorHouse, 2006.

[10.12] Vendelin, George D., Anthony M. Pavio, and Ulrich L. Rhode. *Microwave Circuit Design, Using Linear and Non-Linear Techniques.* New York: John Wiley & Sons, 1990.

PART III

APPENDICES

APPENDIX A

Physical Constants

Quantity	Symbol	Value
Angstrom unit	$A°$	$1\,A° = 10^{-4}\,\mu m = 10^{-10}\,m$
Avogadro constant	N_{AVO}	$6.02204 \times 10^{23}\,mol^{-1}$
Boltzmann constant	k	$1.38066 \times 10^{-23}\,J/K$
Charge of electron	q_e	$1.60218 \times 10^{-19}\,C$
Electron charge/mass ratio	q_e/m_e	$1.75880 \times 10^{11}\,C/kg$
Electron rest mass	m_e	$9.1095 \times 10^{-31}\,kg$
Electron volt	eV	$1\,eV = 1.60218 \times 10^{-19}\,J$
Intrinsic impedance (vacuum)	η_o	$120\pi\ (=377)\ \Omega$
Neutron rest mass	m_n	$1.67495 \times 10^{-27}\,kg$
Permeability (vacuum)	$\mu_o\ (=1/\varepsilon_o c^2)$	$4\pi \times 10^{-7} = 1.25663 \times 10^{-6}\,H/m$
Permittivity (vacuum)	$\varepsilon_o\ (=1/\mu_o c^2)$	$8.85418 \times 10^{-12}\,F/m$
Planck constant	h	$6.62617 \times 10^{-34}\,J\text{-}s$
Proton rest mass	M_p	$1.67264 \times 10^{-27}\,kg$
Speed of light (vacuum)	c	$2.99792 \times 10^8\,m/s$
Thermal voltage (at 293° K)	$V_T = kT/q$	$0.0252\,V \approx 25\,mV$

APPENDIX B

International System of Units (SI)

Quantity	Unit	Symbol	Dimension	Type
Capacitance	Farad	F	C/V	Derived unit
Charge	Coulomb	C	A-s	Derived unit
Conductance	Siemens	S	A/V	Derived unit
Current	Ampere	A	Basic dimension	Base unit
Energy	Joule	J	N-m	Derived unit
Force	Newton	N	$Kg\text{-}m/s^2$	Derived unit
Frequency	Hertz	Hz	1/s	Derived unit
Inductance	Henry	H	Wb/A	Derived unit
Length	Meter	m	Basic dimension	Base unit
Magnetic flux	Weber	Wb	V-s	Derived unit
Magnetic induction	Tesla	T	Wb/m^2	Derived unit
Mass	Kilogram	kg	Basic dimension	Base unit
Potential	Volt	V	J/C	Derived unit
Power	Watt	W	J/s	Derived unit
Pressure	Pascal	Pa	N/m^2	Derived unit
Resistance	Ohm	Ω	V/A	Derived unit
Temperature	Kelvin	K	Basic dimension	Base unit
Time	Second	s	Basic dimension	Base unit

APPENDIX C

Unit Prefixes & Conversions

Multiple	Prefix	Symbol
10^{18}	exa-	E
10^{15}	peta-	P
10^{12}	tera-	T
10^{9}	giga-	G
10^{6}	mega-	M
10^{3}	kilo-	k
10^{2}	hecta-	h
10	deka-	da
10^{-1}	deci-	d
10^{-2}	centi-	c
10^{-3}	milli-	m
10^{-6}	micro-	μ
10^{-9}	nano-	n
10^{-12}	pico-	p
10^{-15}	femto-	f
10^{-18}	atto-	a

Conversion Factors between Units

1 cm	393.7 mils
1 dB	0.115 nepers
1 foot	0.305 m
1 gauss	10^{-4} tesla
1 inch	2.54 cm = 25.4 mm
1 kg	2.2 lb = 1000 g
1 lb	453.6 g
1 micron	10^{-6} m = 10^{-8} cm
1 mil	10^{-3} inch = 2.54×10^{-3} cm
1 mile	1.61 km
1 neper	8.686 dB

APPENDIX D

Greek Alphabets

Letter	Lowercase	Uppercase
Alpha	α	A
Beta	β	B
Gamma	γ	Γ
Delta	δ	Δ
Epsilon	ε	E
Zeta	ζ	Z
Eta	η	H
Theta	θ	Θ
Iota	ι	I
Kappa	κ	K
Lambda	λ	Λ
Mu	μ	M
Nu	ν	N
Xi	ξ	Ξ
Omicron	o	O
Pi	π	Π
Rho	ρ	P
Sigma	σ	Σ
Tau	τ	T
Upsilon	υ	Y
Phi	φ	Φ
Chi	χ	X
Psi	ψ	Ψ
Omega	ω	Ω

APPENDIX E

Fragmented Energy Forms

I. POTENTIAL ENERGY

1) STATIC CHARGE (Q)

2) STATIC MAGNETIC POLE (M)

3) STATIC FORCE (ELECTRIC)

$$\overline{F} = Q\overline{E} \qquad (E.1)$$

4) STATIC FIELDS

    a) **Electric Field** $(\overline{E}, \overline{D})$

$$\overline{D}(x,y,z) = \varepsilon\overline{E}(x,y,z) \qquad (E.2)$$

    b) **Magnetic Field** $(\overline{H}, \overline{B})$

$$\overline{B}(x,y,z) = \mu\overline{H}(x,y,z) \qquad (E.3)$$

5) WORK

$$W = q\int_0^\ell \overline{E}\cdot\overline{d\ell} \qquad (E.4)$$

Where $\overline{E}$ is the electric field vector defined as the electrical force exerted on a unit of charge and $\overline{d\ell}$ is an infinitesimal displacement vector. Differentiating (E.4) with respect to "q" gives the differential form:

$$\frac{dW}{dq} = \int_0^\ell \overline{E}\cdot\overline{d\ell} \qquad (E.5)$$

Equation (E.5) in essence gives the equation for the work performed per unit charge between two points, which basically is equivalent to the concept of voltage.

If the particle moves in the same direction as the applied force, then equation for work simplifies into:

$$W=qE\ell,\qquad\qquad\text{(E.6)}$$

which simply states that work accomplished is the applied force (qE) multiplied by the distance (ℓ).

NOTE: *By close examination of the definition of "work", we can observe that it is the result of energy in action and represents "spent energy".*

6) **VOLTAGE (ELECTRICAL POTENTIAL DIFFERENCE)**

The work performed in moving a charge (q) from A to B is given by:

$$W_{BA} = \int_{A}^{B}(-q\overline{E})\cdot\overline{d\ell}\qquad\qquad\text{(E.7)}$$

The voltage (or potential difference) is defined in terms of the work performed and is mathematically expressed as:

$$V_{BA} = V_{B} - V_{A} = W_{BA}/q = -\int_{A}^{B}\overline{E}.\overline{d\ell}\qquad\text{(E.8a)}$$

Or in differential form:

$$V = \frac{dW}{dq}\qquad\qquad\text{(E.8b)}$$

NOTE: *Theoretically speaking, the reference point (or ground) at which the potential function (V) is assumed to be zero is at an infinite distance away from the point of measurement (i.e. at infinity); however, in practice and in actual circuit analysis the ground can be assumed to be any designated point in the circuit, purely by prior agreement.*

II. KINETIC ENERGY

7) MOMENTUM

Linear: $\bar{p} = m\bar{v}$ (E.9)

Angular: $\bar{L} = \bar{r} \times \bar{p}$ (E.10)

Where $\bar{r}$ and $\bar{v}$ are the lever arm and the velocity of a particle with mass m, respectively.

NOTE: *Mechanical Force* $(\bar{F})$ *on a body is a vector (having a magnitude and a direction) equal to the time rate of change of linear momentum* $(\bar{p} = m\bar{v})$, *i.e.,*

$$\bar{F} = \frac{d(\bar{p})}{dt}$$ (E.11)

Where m is the mass and $(\bar{v})$ is the velocity of the object.

8) TIME VARYING OR MOVING ELECTRIC CHARGE, Q(X,Y,Z,T), LEADING TO CURRENT (I) GIVEN BY:

I=dQ(x,y,z,t)/dt

9) DYNAMIC FORCES

a) Electric Force:

$\bar{F}_e(x,y,z,t) = Q\bar{E}(x,y,z,t)$ (E.12)

b) Magnetic Force:

$\bar{F}_m(x,y,z,t) = Q\bar{V} \times \bar{B}(x,y,z,t)$ (E.13)

10) DYNAMIC FIELDS

a) Electric Field $(\bar{E}, \bar{D})$

$\bar{D}(x,y,z,t) = \varepsilon\bar{E}(x,y,z,t)$ (E.14)

b) Magnetic Field $(\bar{H}, \bar{B})$

$\bar{B}(x,y,z,t) = \mu\bar{H}(x,y,z,t)$ (E.15)

11) POWER

$$P = \frac{dW}{dt} \qquad\qquad (E.16)$$

Or, in integral form:

$$W = \int_0^t Pdt \qquad\qquad (E.17)$$

Using the chain rule, Equation (E.16) can also be written as:

$$P = \frac{dW}{dt} = \frac{dW}{dq} \times \frac{dq}{dt} = VI \qquad\qquad (E.18)$$

Thus the total work performed or total energy transferred (or spent) is:

$$W = \int_0^t VIdt \qquad\qquad (E.19)$$

12) ELECTRONIC WAVES
   a) **The General Form:**
      A wave traveling in +x direction:
      $$U^+(x,y,z,t)=f^+(x-vt)=g^+(\omega t-\beta x) \qquad (E.20)$$

      A wave traveling in -x direction:
      $$U^-(x,y,z,t)=f^-(x+vt)=g^-(\omega t+\beta x) \qquad (E.21)$$
   b) **Sinusoidal Waves:**
      I. *Plane wave in ±x direction*
         Time Domain:
         $$u(x,y,z,t)=U_o\cos(\omega t \pm \beta x) \qquad (E.22a)$$
         Phasor Domain:
         $$U(x,y,z)= U_o e^{-(\pm j\beta x)} \qquad (E.22b)$$

APPENDIX F

Classical Laws of Electricity and Magnetism

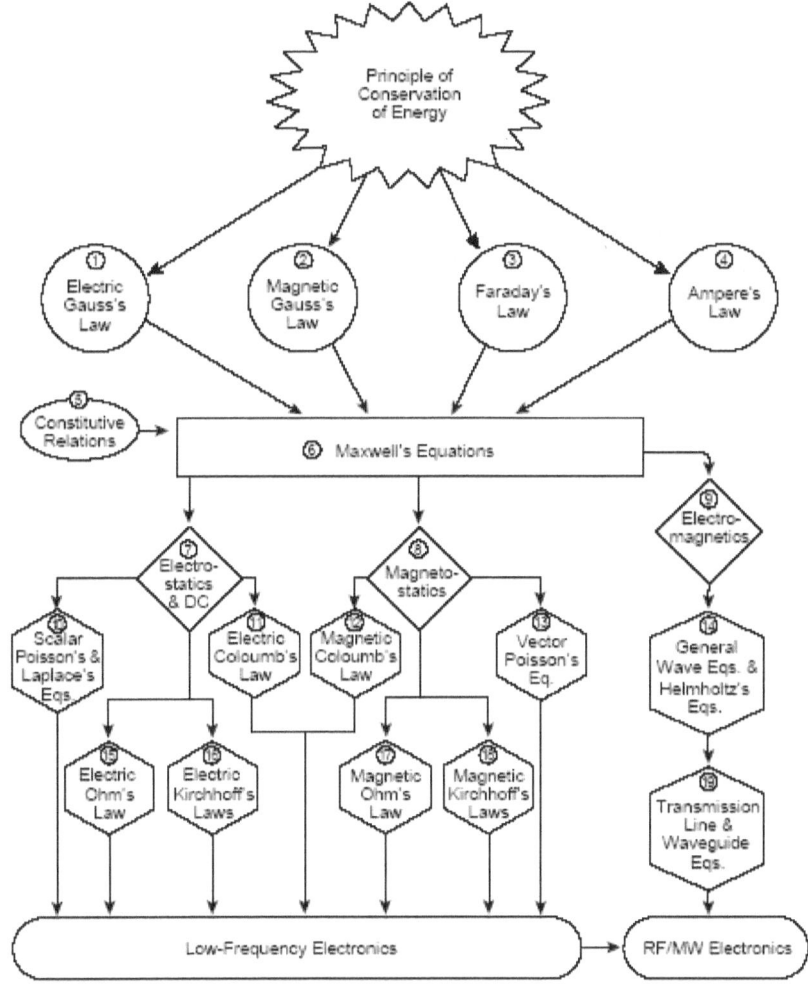

All Fundamental laws of electricity and magnetism.

SYMBOL LIST

$\vec{\mathbf{a}}_{R_{12}}$ – A unit vector in the direction of R_{12}.

$\overline{A}$ - Magnet potential vector.

$\overline{B}$ - Magnetic flux density vector.

C - Capacitance per unit length

$d\ell$ - An infinitesimal length.

$\overline{D}$ - Electric displacement vector.

emf – Electromotive force.

$\overline{E}$ - Electric field intensity vector.

$\overline{F}$ - Lorentz force vector.

$\overline{H}$ - Magnetic field intensity vector.

G - Conductance per unit length.

j – Unity imaginary number.

$\overline{\mathbf{J}}$ - Current density vector.

K – Wave propagation constant in unbounded space.

ℓ - Path length.

L - Inductance per unit length.

m - Number of branches.

mmf – Magnetomotive force.

M - Magnetic pole

Q – Electric charge.

R – Resistance per unit length.

R_{12} – Distance between point 1 and point 2.

t - Time

$\overline{\mathbf{v}}$ - velocity

v_P – Phase velocity.

V_{12} – Voltage difference between point 1 and point 2.

V – Voltage (or potential difference) function.

x – A length variable.

GREEKS

α - Attenuation constant.

β - phase constant.

ε - Permittivity.

ε_o – Free space permittivity.

γ - Propagation constant

ϕ - Potential function

Φ - Magnetic Flux.

λ - Wavelength.

μ - Permeability

μ_o - Free Space permeability.

ω - Frequency in radians.

ρ - Volume charge density.

$\mathfrak{R}$ - Magnetic reluctance.

σ - Electrical conductivity.

GENERAL LAWS AND EQUATIONS

1. Electric Gauss's Law:

$$\nabla \cdot \overline{D} = \rho$$

2. Magnetic Gauss's Law:

$$\nabla \cdot \overline{B} = 0$$

3. Faraday's Law:

$$\nabla \times \overline{E} = -\frac{\partial \overline{B}}{\partial t}$$

4. Ampere's Law:

$$\nabla \times \overline{H} = \frac{\partial \overline{D}}{\partial t} + \overline{J}$$

$$\overline{D} = \varepsilon \overline{E}$$

$$\overline{B} = \mu \overline{H}$$

$$\overline{J} = \sigma \overline{E}$$

5. Constitutive Relations:

6. Maxwell's Equations:

$$\nabla \times \overline{E} = -\frac{\partial \overline{B}}{\partial t}$$

$$\nabla \times \overline{H} = \frac{\partial \overline{D}}{\partial t} + \overline{J}$$

$$\nabla \cdot \overline{D} = \rho$$

$$\nabla \cdot \overline{B} = 0$$

7. Electrostatics:

$$\nabla \times \overline{E} = 0$$

$$\nabla \cdot \overline{D} = \rho$$

$$\overline{E} = -\nabla V$$

$$\overline{F} = q\overline{E}$$

8. Magnetostatics:

$$\nabla \times \overline{H} = \overline{J}$$

$$\nabla \cdot \overline{B} = 0$$

$$\overline{B} = \nabla \times \overline{A}$$

$$\overline{F} = Q(\overline{v} \times \overline{B})$$

($\overline{F}$ is the Lorentz Force, which can also be written as:
$\overline{F} = I\overline{dl} \times \overline{B}$)

9. Electromagnetics

$$\nabla \times \overline{E} = -\frac{\partial \overline{B}}{\partial t}$$

$$\nabla \times \overline{H} = \frac{\partial \overline{D}}{\partial t} + \overline{J}$$

$$\nabla \cdot \overline{D} = \rho$$

$$\nabla \cdot \overline{B} = 0$$

Under steady-state sinusoidal time dependence, the Maxwell's equations become:

$$\nabla \times \overline{E} = -j\omega\mu\overline{H}$$
$$\nabla \times \overline{H} = j\omega\overline{E} + \overline{J}$$
$$\nabla \cdot \overline{D} = \rho$$
$$\nabla \cdot \overline{B} = 0$$

where

$$\overline{J} = \sigma\overline{E}$$
$$\overline{D} = \varepsilon\overline{E}$$
$$\overline{B} = \mu\overline{H}$$

10. Scalar Poisson's & Laplace's Equations:
   a) Poisson's Equation:

$$\nabla^2\phi = -\frac{\rho}{\varepsilon}$$

   b) Laplace's Equation:

$$\nabla^2\phi = 0$$

11. Electric Coulomb's Law:

$$\overline{F}_2 = (\frac{Q_1 Q_2}{4\pi\varepsilon R_{12}^2})\hat{a}_{R_{12}}$$

Or, in terms of field quantity (*E*), we can write Electric Coulomb's Law as:

$$\overline{dE}_2 = \frac{dQ_1 \hat{a}_{R_{12}}}{4\pi\varepsilon R_{12}^2}$$

12. Magnetic Coulomb's Law:

$$\overline{F}_2 = (\frac{M_1 M_2}{4\pi\mu R_{12}^2})\hat{a}_{R_{12}}$$

Also Biot-Savart Law, considered by some to be the Coulomb's Law of Magnetostatics, is expressed in terms of field quantity (H) as:

$$\overline{dH}_2 = \frac{I_1 \overline{dl}_1 \times \hat{a}_{R_{12}}}{4\pi R_{12}^2}$$

13. Vector Poisson's Equation:

$$\nabla^2 \overline{A} = -\mu \overline{J}$$

14. General and Helmholtz's wave equations

a. General Wave Equations in a nonconducting medium ($\sigma=0$):

$$\nabla^2 \overline{E} - \mu\varepsilon \frac{\partial^2 \overline{E}}{\partial t^2} = 0$$

$$\nabla^2 \overline{H} - \mu\varepsilon \frac{\partial^2 \overline{H}}{\partial t^2} = 0$$

General Wave Equations in a lossy medium ($\sigma \neq 0$):

$$\nabla^2 \overline{E} - \mu\varepsilon \frac{\partial^2 \overline{E}}{\partial t^2} - \mu\sigma \frac{\partial \overline{E}}{\partial t} = 0$$

$$\nabla^2 \overline{H} - \mu\varepsilon \frac{\partial^2 \overline{H}}{\partial t^2} = 0$$

b. Helmholtz's Equations (Sinusoidal Waves):

Homogeneous Vector Eq.: $\quad \nabla^2 \overline{E} + k^2 \overline{E} = 0$,

$$\nabla^2 \overline{H} + k^2 \overline{H} = 0$$

Inhomogeneous Scalar Eq.: $\quad \nabla^2 \phi + k^2 \phi = -\dfrac{\rho}{\varepsilon}$

Inhomogeneous Vector Eq.: $\quad \nabla^2 \overline{A} + k^2 \overline{A} = -\mu \overline{J}$

Lorentz Condition: $\quad \nabla \cdot \overline{A} = -j\omega\mu\varepsilon\phi$

Wave Number:

a. Non-Conducting Media:
$$k = \omega\sqrt{\mu\varepsilon} = \omega/v_P = 2\pi/\lambda$$
b. Conducting Media:
$$k = \omega\sqrt{\mu_o\varepsilon}, \quad \varepsilon = \varepsilon_o(1 - j\frac{\sigma}{\omega\varepsilon_o})$$

15. Electric Ohm' Law:

$$V_{12} = RI$$

16. Electric Kirchhoff's Laws:

KVL: $\sum_{i=1}^{m}V_i = 0$

(or $\sum_{i=1}^{m_1} emf_i = \sum_{j=1}^{m_2}\Delta V_j$)

where $m=m_1+m_2$.

KCL: $\sum_{j=1}^{n}I_j = 0$

17. Magnetic Ohm' Law:
$$mmf_{12} = \Re\Phi$$

18. Magnetic Kirchhoff's Laws:

KVL: $\sum_{k=1}^{m_1}mmf_k = \sum_{i=1}^{m_2}H_i\ell_i$

KCL: $\sum_{j=1}^{n}\Phi_j = 0$

19. Under steady state sinusoidal conditions:
19a. TEM waves: Field equations; $E_z=0$, $H_z=0$

$$\nabla_t^2 \overline{E}_t + (k^2 - \beta^2)\overline{E}_t = 0$$

$$\nabla_t^2 \phi = 0$$

$$\overline{E} = -\nabla_t \phi e^{\mp jkz}$$

$$\overline{H} = \pm \frac{1}{\eta} \hat{z} \times \overline{E}$$

where

$$\eta = \sqrt{\frac{\mu}{\varepsilon}}$$

$$\beta = k = \omega\sqrt{\mu\varepsilon} = \omega/v_P = 2\pi/\lambda$$

19b. Transmission Line waves: Current and voltage Equations (Sinusoidal Waves) caused by TEM waves

$$\frac{d^2 V(x)}{dx^2} - \gamma^2 V(x) = 0$$

$$\frac{d^2 I(x)}{dx^2} - \gamma^2 I(x) = 0$$

Where

$$I(x) = (\frac{-1}{R + j\omega L}) \frac{dV(x)}{dx},$$

$$\gamma = \sqrt{(R + j\omega L)(G + j\omega C)} = \alpha + j\beta$$

For Lossless transmission lines (R, G=0). these Equations simplify into:

$$\frac{d^2V(x)}{dx^2} + \beta^2 V(x) = 0,$$

$$\frac{d^2I(x)}{dx^2} + \beta^2 I(x) = 0,$$

$$I(x) = (\frac{-1}{j\omega L})\frac{dV(x)}{dx}$$

$$\beta = \omega\sqrt{LC}$$

19c. TE waves: $E_z=0,\ H_z\neq0$

$$(\nabla_t^2 - \beta^2)(\overline{H}_t + \overline{H}_z) = 0$$

19d. TM waves: $E_z\neq0,\ H_z=0$

$$(\nabla_t^2 - \beta^2)(\overline{E}_t + \overline{E}_z) = 0$$

Where

$$\nabla = \nabla_t - j\beta\hat{z}$$

$$\nabla^2 = \nabla_t^2 - \beta^2$$

$$\overline{E} = \overline{E}_t + \overline{E}_z$$

$$\overline{H} = \overline{H}_t + \overline{H}_z$$

APPENDIX G

Materials Constants

TABLE G.1   Conductivity σ (S/m)

Material	σ
Aluminum	3.82×10^7
Bakelite	10^{-9}
Brass	1.50×10^7
Bronze	1.00×10^7
Clay	10^{-4}
Copper	5.80×10^7
Diamond	10^{-13}
Ferrite	10^{-2}
GaAs	2.5×10^{-7}
Germanium (Ge)	2.3
Glass	10^{-12}
Gold	4.10×10^7
Ground (wet)	$10^{-2} - 10^{-3}$
Iron	1.03×10^7
Marble	10^{-8}
Mica	10^{-14}
Nichrome	0.10×10^7
Nickel	1.45×10^7
Polystrene	10^{-16}
Porcelain	10^{-13}
Quartz	10^{-17}
Rubber (hard)	10^{-15}
Silicon	4.0×10^{-4}
Silver	6.17×10^7
Soil (sandy)	10^{-5}
Solder	0.70×10^7
Steel (stainless)	0.11×10^7
Tungsten	1.82×10^7
Water (distilled)	2×10^{-4}
Water (fresh)	10^{-3}
Water (sea)	$3-5$
Zinc	1.67×10^7

TABLE G.2   Relative Permittivity (ε_r) (Also Called Dielectric Constant)

Material	ε_r
Air	1
Alcohol	25
Bakelite	4.8
Gallium Arsenide(GaAs)	12.9
Germanium (Ge)	15.8
Glass	4–7
Ground (dry)	2–5
Ice	4.2
Indium Phosphide (InP)	14
Mica (ruby)	5.4
Nylon	4
Paper	2–4
Plexiglass	2.6–3.5
Polyethylene	2.25
Polystrene	2.55–6
Porcelain	6
Quartz (fused)	3.8
Rubber	2.5–4
Salt (NaCl)	5.9
Sand (dry)	4
Silica (fused)	3.8
Silicon	11.7
Snow	3.3
Soil (dry)	2.8
Styrofoam	1.03
Teflon	2.1
Water (Distilled)	80
Water (fresh)	80
Water (Sea)	20
Wood (dry)	1.5–4

TABLE G.3   Relative Permeability (μ_r)

Dielectric	μ_r
Aluminum	1.00000065
Beryllium	1.00000079
Bismuth	0.99999860
Cast Iron	60
Cobalt	60
Ferrite	1,000
Iron (pure)	4,000
Iron (transformer)	3,000
Machine Steel	300
Mumetal	20,000
Nickel	50
Parafin	0.99999942
Silicon Iron	4,000
Silver	0.99999981
Supermalloy	100,000
Wood	0.99999950

TABLE G.4   Semiconductor Substrate Material Constants

Property	Si	Si on Sapphire	GaAs	InP
Density (g/cm^3)	2.3	3.9	5.3	4.8
Dielectric constant (ε_r)	11.7	11.6	12.9	14
Mobility (cm^2/V–s)*	700	700	4300	3000
Operating Temperature (°C)	250	250	350	300
Radiation Hardness	Poor	Poor	Very good	Good
Resistivity (intrinsic) (Ω–cm)	10^3–10^5	$>10^{14}$	10^7–10^9	10^7
Saturation velocity (cm/s)	9×10^6	9×10^6	1.3×10^7	1.9×10^7
Semi-Insulating	No	Yes	Yes	Yes
Thermal Conductivity (W/cm–°C)	1.5	0.46	0.46	0.68

* Doping level = 10^{17} cm^{-3}

APPENDIX H

Conversion among the Y-Parameters

Let
  e =Common emitter (CE) configuration
  b =Common base (CB) configuration
  c =Common collector (CC) configuration
The conversions among the three configurations are given as follows:

1. CB & CC to CE

$$Y11,e = Y11,b + Y12,b + Y21,b + Y22,b = Y11,c$$

$$Y12,e = -(Y12,b + Y22,b) = -(Y11,c + Y12,c)$$

$$Y21,e = -(Y21,b + Y22,b) = ^-(Y11,c^+ Y21,c)$$

$$Y22,e = Y22,b = Y11,c + Y12,c + Y21,c + Y22,c$$

2. CE & CC to CB

$$Y11,b = Y11,e + Y12,e + Y21,e + Y22,e = Y22,c$$

$$Y12,b = ^-(^Y12,e^{+Y}22,e) = -(Y21,c + Y22,c)$$

$$Y21,b = ^-(^Y21,e^{+Y}22,e) = -(Y12,c + Y22,c)$$

$$Y22,b = Y22,e = Y11,c + Y12,c + Y21,c + Y22,c$$

3. CE & CB to CC

$$Y{11,}_c = Y{11,}_e = Y{11,}_b + Y{12,}_b + Y{21,}_b + Y{22,}_b$$

$$Y12,_c = - (Y11,_e{}^+ Y12,_e) = - (Y11,_b + Y21,_b)$$

$$Y21,_c = -(Y1 1,_e + Y21,_e) = - (Y11,_b + Y12,_b)$$

$$Y22,_c = Y{11,}_e + Y{12,}_e + Y{21,}_e + Y{22,}_e = Y{11,}_b$$

NOTE: *If other parameters other than Y-parameters (e.g., S-parameters, etc.) are desired, then a simple conversion (e.g., from Y- to S -parameters, etc.) can be used to accomplish such a task.*

APPENDIX I

MATHEMATICAL IDENTITIES

A) BINOMIAL FORMULAS

$(x \pm y)^2 = x^2 \pm 2xy + y^2$
$(x \pm y)^3 = x^3 \pm 3x^2 y + 3xy^2 \pm y^3$
$(x \pm y)^4 = x^4 \pm 4x^3 y + 6x^2 y^2 \pm 4xy^3 + y^4$
Or, in general:

$$(x + y)^n = x^n + nx^{n-1} + \frac{n(n-1)}{2!} x^{n-2} y^2 + \frac{n(n-1)(n-2)}{3!} x^{n-3} y^3 + \ldots + y^n$$

Where factorial n (n!) is defined by:
n!=1.2.3.......n
Note: Zero factorial is defined by: 0!=1

B) SPECIAL PRODUCTS

$x^2 - y^2 = (x-y)(x+y)$
$x^3 - y^3 = (x-y)(x^2 + xy + y^2)$
$x^3 + y^3 = (x+y)(x^2 - xy + y^2)$
$x^4 - y^4 = (x^2 - y^2)(x^2 + y^2) = (x-y)(x+y)(x^2 + y^2)$

C) TRIGONOMETRIC FUNCTION RELATIONS

$$\sin(-x) = -\sin x$$

$$\cos(-x) = \cos x$$

$$\tan x = \frac{\sin x}{\cos x}$$

$$\cot x = \frac{\cos x}{\sin x}$$

$$\sec x = \frac{1}{\cos x}$$

$$\csc x = \frac{1}{\sin x}$$

$$\sin^2 x + \cos^2 x = 1$$

$$\sin 2x = 2\sin x \cos x$$

$$\cos 2x = \cos^2 x - \sin^2 x = 1 - 2\sin^2 x = 2\cos^2 x - 1$$

$$\sin 3x = 3\sin x - 4\sin^3 x$$

$$\cos 3x = -3\cos x + 4\cos^3 x$$

$$\sin^2 x = \frac{1 - \cos 2x}{2}$$

$$\cos^2 x = \frac{1 + \cos 2x}{2}$$

$$\sin^3 x = \frac{3\sin x - \sin 3x}{4}$$

$$\cos^3 x = \frac{3\cos x + \cos 3x}{4}$$

$$\sin x \pm \sin y = 2\sin(\frac{x \pm y}{2})\cos(\frac{x \mp y}{2})$$

$$\cos x + \cos y = 2\cos(\frac{x + y}{2})\cos(\frac{x - y}{2})$$

$$\cos x - \cos y = -2\sin(\frac{x + y}{2})\cos(\frac{x - y}{2})$$

$$\sin x \sin y = \frac{1}{2}\left[\cos(x-y) - \cos(x+y)\right]$$

$$\cos x \cos y = \frac{1}{2}\left[\cos(x-y) + \cos(x+y)\right]$$

$$\sin x \cos y = \frac{1}{2}\left[\sin(x-y) + \sin(x+y)\right]$$

D) HYPERBOLIC FUNCTION RELATIONS

$$\sinh x = \frac{e^x - e^{-x}}{2}$$

$$\cosh x = \frac{e^x + e^{-x}}{2}$$

$$\tanh x = \frac{\sinh x}{\cosh x} = \frac{e^x - e^{-x}}{e^x + e^{-x}}$$

$$\coth x = \frac{1}{\tanh x}$$

$$\sec hx = \frac{1}{\cosh x}$$

$$\cosh x = \frac{1}{\sinh x}$$

$$\cosh^2 x - \sinh^2 x = 1$$

$$\sinh(-x) = -\sinh x$$

$$\cosh(-x) = \cosh x$$

$$\tanh(-x) = -\tanh x$$

$$\sinh(x \pm y) = \sinh x \cosh y \pm \cosh x \sinh y$$

$$\cosh(x \pm y) = \cosh x \cosh y \pm \sinh x \sinh y$$

$$\tanh(x \pm y) = \frac{\tanh x \pm \tanh y}{1 \pm \tanh x \tanh y}$$

$\sinh 2x = 2\sinh x \cosh x$

$\cosh 2x = \cosh^2 x + \sinh^2 x = 2\cosh^2 x - 1 = 1 + 2\sinh^2 x$

$\tanh 2x = \dfrac{2\tanh x}{1 + \tanh^2 x}$

$\sinh^2 x = \dfrac{1}{2}\left[\cosh 2x - 1\right]$

$\cosh^2 x = \dfrac{1}{2}\left[\cosh 2x + 1\right]$

$\sinh x \pm \sinh y = 2\sinh\dfrac{(x \pm y)}{2}\cosh\dfrac{(x \mp y)}{2}$

$\cosh x + \cosh y = 2\cosh\dfrac{(x + y)}{2}\cosh\dfrac{(x - y)}{2}$

$\cosh x - \cosh y = 2\sinh\dfrac{(x + y)}{2}\sinh\dfrac{(x - y)}{2}$

$\sinh x \sinh y = \dfrac{1}{2}\left[\cosh(x + y) - \cosh(x - y)\right]$

$\cosh x \cosh y = \dfrac{1}{2}\left[\cosh(x + y) + \cosh(x - y)\right]$

$\sinh x \cosh y = \dfrac{1}{2}\left[\sinh(x + y) + \sinh(x - y)\right]$

E) LOGARITHMIC RELATIONS

$\log_a xy = \log_a x + \log_a y$

$\log_a \dfrac{x}{y} = \log_a x - \log_a y$

$\log_a x^y = y\log_a x$

$\log_a x = \dfrac{\log_b x}{\log_b a}$

$\log_a a = 1$

F) COMPLEX NUMBERS

$x + jy = re^{j\theta}$ (conversion from rectangular to polar form)

and

$re^{j\theta} = r(\cos\theta + \sin\theta)$ (Euler's Identity)

Where,

$j = \sqrt{-1}$,

$r = \sqrt{x^2 + y^2}$

$\theta = \tan^{-1}(\frac{y}{x})$

$(re^{j\theta})^n = r^n e^{jn\theta}$

$(r_1 e^{j\theta_1})(r_2 e^{j\theta_2}) = r_1 r_2 e^{j(\theta_1 + \theta_2)}$

G) RELATIONSHIP BETWEEN EXPONENTIAL, TRIGONOMETRIC AND HYPERBOLIC FUNCTIONS

$e^{\pm j\pi} = -1$

$e^{\pm j\pi/2} = \pm j$

$e^{\pm j(x+2k\pi)} = e^{\pm jx}$

$e^{\pm j[x+(2k+1)\pi]} = -e^{\pm jx}$

$e^{\pm jx} = \cos x \pm j\sin x$ (Euler's identity)

$\sin x = \dfrac{e^{jx} - e^{-jx}}{2j}$

$\cos x = \dfrac{e^{jx} + e^{-jx}}{2}$

$\tan x = -j(\dfrac{e^{jx} - e^{-jx}}{e^{jx} + e^{-jx}})$

$\sin(jx) = j\sinh x$

$\cos(jx) = \cosh x$

$\tan(jx) = j\tanh x$

$\sinh(jx) = j\sin x$

$\cosh(jx) = \cos x$

$\tanh(jx) = j\tan x$

Where $j = \sqrt{-1}$ and k is an integer.

H) DERIVATIVES

$$\frac{d}{dx}(u^n) = nu^{n-1}\frac{du}{dx}$$

$$\frac{d}{dx}(uv) = u\frac{dv}{dx} + v\frac{du}{dx}$$

$$\frac{d}{dx}(\frac{u}{v}) = \frac{v(du/dx) - u(dv/dx)}{v^2}$$

$$\frac{d}{dx}\sin u = \cos u\frac{du}{dx}$$

$$\frac{d}{dx}\cos u = -\sin u\frac{du}{dx}$$

$$\frac{d}{dx}\tan u = \sec^2 u\frac{du}{dx}$$

$$\frac{d}{dx}\cot u = \csc^2 u\frac{du}{dx}$$

$$\frac{d}{dx}\log_a u = \frac{\log_a e}{u}\frac{du}{dx}$$

$$\frac{d}{dx}\log_e u = \frac{1}{u}\frac{du}{dx}$$

$$\frac{d}{dx}a^u = a^u\log_e a\frac{du}{dx}$$

$$\frac{d}{dx}e^u = e^u\frac{du}{dx}$$

$$\frac{d}{dx}\sinh u = \cosh u \frac{du}{dx}$$

$$\frac{d}{dx}\cosh u = \sinh u \frac{du}{dx}$$

$$\frac{d}{dx}\tanh u = \mathrm{sech}^2 u \frac{du}{dx}$$

$$\frac{d}{dx}\coth u = -\mathrm{csch}^2 u \frac{du}{dx}$$

I) INTEGRALS

$$\int u^n du = \frac{u^{n+1}}{n+1} + C$$

$$\int u\,dv = uv - \int v\,du$$

$$\int \frac{du}{u} = \log_e |u| + C$$

$$\int a^u du = \frac{a^u}{\log_e u} + C$$

$$\int e^u du = e^u + C$$

$$\int \log_e x = x \log_e x - x + C$$

$$\int \sin u\,du = -\cos u + C$$

$$\int \cos u\,du = \sin u + C$$

$$\int \tan u\,du = -\log_e \cos u + C$$

$$\int \cot u\,du = \log_e \sin u + C$$

$$\int \sinh u\,du = \cosh u + C$$

$$\int \cosh u\,du = \sinh u + C$$

$$\int \tanh u \, du = \log_e \cosh u + C$$

$$\int \coth u \, du = \log_e \sinh u + C$$

$$\int \sin^2 u \, du = \frac{u - \sin u \cos u}{2} + C$$

$$\int \cos^2 u \, du = \frac{u + \sin u \cos u}{2} + C$$

J) TAYLOR SERIES EXPANSION

$$f(x)|_{x=a} = f(a) + f'(a)(x-a) + \frac{f''(a)(x-a)^2}{2!} + \cdots + \frac{f^{(n-1)}(a)(x-a)^{n-1}}{(n-1)!} + \cdots$$

$$e^x|_{x=0} = 1 + x + \frac{x^2}{2!} + \frac{x^3}{3!} + \cdots$$

$$\sin x|_{x=0} = x - x^3/3! + x^5/5! - x^7/7! + \dots$$

$$\cos x|_{x=0} = 1 - x^2/2! + x^4/4! - x^6/6! + \dots$$

$$\ln(1+x)|_{x=0} = x - x^2/2! + x^3/3! - x^4/4! + \dots$$

K) EQUATION OF A CIRCLE

$$(x-a)^2 + (y-b)^2 = R^2$$

where (a,b) is the center of the circle having a radius R.

APPENDIX J

DC Bias Networks for an FET

There are five basic DC bias networks for an FET that uses one or two power supplies as described next[†].

[†] Vendelin, G. D. *Five Basic Bias Design for GaAs FET Amplifiers, Microwaves & RF,* February 1978. (Reproduced with the permission of Microwaves & RF.)

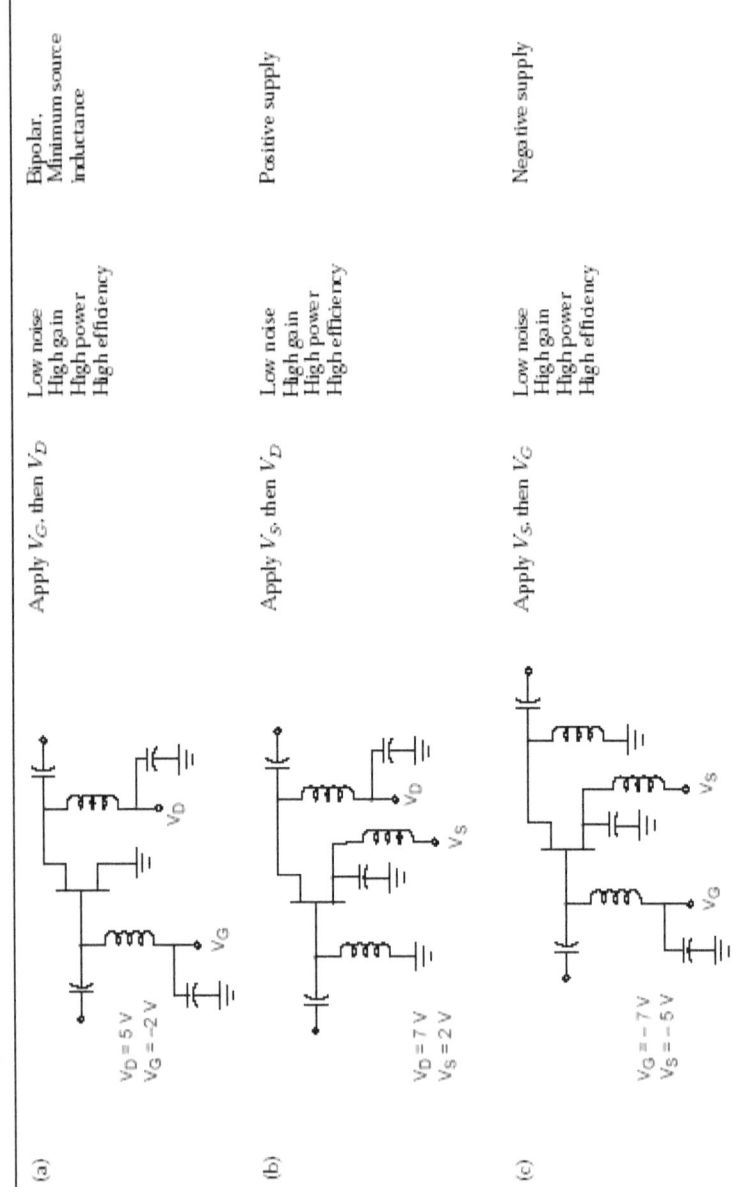

Figure	How	Amplifier Characteristics	Power Supply Used
(a) $V_D = 5$ V, $V_G = -2$ V	Apply V_G, then V_D	Low noise High gain High power High efficiency	Bipolar, Minimum source inductance
(b) $V_D = 7$ V, $V_S = 2$ V	Apply V_S, then V_D	Low noise High gain High power High efficiency	Positive supply
(c) $V_G = -7$ V, $V_S = -5$ V	Apply V_S, then V_G	Low noise High gain High power High efficiency	Negative supply

Figure	How	Amplifier Characteristics	Power Supply Used
(d) $V_D = 7V$ $V_S = 2V$ $= I_{DS}R_S$	Apply V_D	Low noise High gain High power Low efficiency Gain easily adjusted by varying R_S	Unipolar, incorporating R_S automatic transient protection
(e) $V_G = -7V$ $V_S = -5V$ $= -I_{DS}R_S$	Apply V_G	Low noise High gain High power Low efficiency Gain easily adjusted by varying R_S	Negative unipolar incorporating R_S

APPENDIX K

The Scientists Behind the Discoveries

There are several essential and auxiliary postulates in electrical engineering and physics. These postulates deal with quantities, which need to be measured precisely. The units used for the measurement of electrical quantities in physics or electrical engineering are usually named after a major contributor or inventor who has advanced this field of study materially and substantially[†]. The scientists behind the electric, magnetic and other important units in chronological order are described next.

[†]Radmanesh, M. M. *The Gateway to Understanding: Electrons to Waves and Beyond,* AuthorHouse, 2005.

Quantity	Unit name	Scientist	Contribution
1) Magnetomotive Force (cgs)	Gilbert (Gi)	William Gilbert (1544-1603), English royal Physician	Earth as a giant Magnet & Magnetism (1st Monumental Discovery)
2) Force	Newton (N)	Sir Isaac Newton (1642-1727), English mathematician and philosopher	Laws of gravity & motion
3) Temperature	Fahrenheit (°F)	Gabriel Daniel Fahrenheit (1686-1736), German physicist.	Mercury thermometer
4) Temperature	Celsius (°C)	Anders Celsius (1701-1744), Swedish astronomer.	Centigrade thermometer
5) Electric Charge	Coulomb (C)	Charles Augustin Coulomb (1736-1806), French engineer & physicist	Law of charges (1st Monumental Discovery)
6) Power	Watt (W)	James Watt (1736-1819), Scottish engineer/inventor	Birth of steam power
7) Electric Potential	Volt (V)	Alessandro Volta (1745-1827), Italian physicist	Electric Batteries (2nd Monumental Discovery)
8) Electric current	Ampere (A)	Andre Marie Ampere (1775-1836), French physicist & mathematician	Circulation of current (4th Monumental Discovery)

9) Magnetic Field Intensity (cgs)	Oersted (Oe)	Hans Christian Oersted (1777-1851), Danish physicist	Connection between electricity and magnetism (**3rd Monumental Discovery**)
10) Magnetic Field Strength (cgs)	Gauss	Karl Friedrich Gauss (1777-1855), German mathematician & astronomer	Terrestrial magnetics (**1st supplemental Discovery**)
11) Resistance	Ohm (Ω)	George Simon Ohm (1787-1854) German physicist & astronomer	Resistor law (**2nd & 3rd supplemental Discovery**)
12) Capacitance	Farad (F)	Michael Faraday (1791-1867), English scientist	Induction & laws of electrolysis (**5th Monumental Discovery**) (**5th & 6th supplemental Discovery**)
13) Inductance	Henry (H)	Joseph Henry (1797-1878), U.S. physicist	A glimpse of electromagnetic waves (**4th supplemental Discovery**)
14) Magnetic Flux	Weber (Wb)	Wilhelm Eduard Weber (1804-1891), German physicist	Terrestrial magnetic field
15) Magnetic Flux (cgs system)	Maxwell (Mx)	James Clerk Maxwell (1831-1879), Scottish physicist	Celebrated equations (**6th Monumental Discovery**)
16) Conductance	Siemens (S)	Werner (1816-1892, German) & William (1823-1883, British) Siemens (brothers), engineers	World telegraphy (Werner), practical dynamo (William)

17) Work (or Energy)	Joule (J)	James Prescott Joule (1818-1889), English physicist	Thermodynamics & Electricity
18) Temperature	Kelvin (K)	William Thomson Kelvin (1824-1907), British Physicist & mathematician	Heat & Electricity
19) None		Gustav R. Kirchhoff (1824-1887), German Professor of Physics	Current & voltage laws **(7th & 8th supplemental Discovery)**
20) Magnetic Flux Density	Tesla (T)	Nikola Tesla (1856-1943), U.S. inventor (born in Croatia)	AC power system, Many great inventions
21) Frequency	Hertz (Hz)	Heinrich Rudolph Hertz (1857-1894), German physicist	Birth of radio, Experimental proof of Maxwell's Equations.

Note: cgs is Centimeter-Gram-Second system of units.

APPENDIX L

CD ROM Download

A. INSTRUCTIONS

1. To Download the CD ROM, please type in the following link exactly:

http://www.csun.edu/~matt/ADVANCED-RFMW.zip

2. Once a pop up window shows up, click on "open" to unzip the file. Make sure you have the "WinZip" software to unzip the files properly.

3. After unzipping the files, create a folder called "E-book" in the C: drive.

4. Save all of the files in this folder.

5. To start the software, double click on the Microsoft® file called "StartMenu" file.

B. MAIN FEATURES

A CD containing software in the form of an electronic book (E-book), which contains all numerical examples from the text, is bound into the back of each textbook. The solutions are programmed using Visual Basic software, which is built into the "Microsoft Excel®" application software. The main features of this CD are as follows:

1. It is a powerful interactive tool for learning the textbook content and also for solving the numerical problems.

2. The software includes 90 solved problems based on the numerical

examples in the book.

3. The big advantage of the interactive software tool is its use of live math. Every number and formula is interactive. The reader can change the starting parameters of a problem and watch as the final results change before his or her eyes. This feature allows the reader to experiment with every number, formula, etc.

4. Each solved problem becomes a worksheet that the reader can modify to solve dozens of related problems.

5. The electronic book takes advantage of the powerful Microsoft Excel® environment to perform many tedious and complicated RF and microwave design calculations (usually using complex numbers), allowing the student to focus on the essential concepts.

6. This is an excellent tool for students, engineers, and educators, to:
   a.   Understand the fundamentals and practical concepts of RF and Microwaves and,
   b.   Encourage applications and new RF/Microwave circuit designs using the concepts presented in the book.

C. How to Start the Program

The following steps need to be carried out before the software is ready to use:

1. Either one of the following two methods may be used to utilize the contents of the CD-ROM:
   a.   Read all the files directly from the CD-ROM or,
   b.   Create a folder called "E-book" in the C: Drive. Copy the entire content of the CD-ROM into the folder entitled "E-book."

2. Open Microsoft Excel 2000 software (or Excel 97 with SR-1 or SR-2 revision) and open the "E-book" folder.

3. You may begin the program by double clicking on the "Start Menu" file.

NOTE: *You may double click on the* "Start Menu" *file directly as a shortcut without opening the Excel software.*

4. Once the "Start Menu" file opens up, click on "Analysis Hub" and then open up the "about the CD" file at the top left corner by clicking on it.

5. Carefully read all the information in the "about the CD" file and close it by either

    a. holding the ALT + TAB keys and choosing EXCEL or,

    b. clicking "return to Start Menu" ARROW to return to the E-book Start Menu.

6. Turn on both *Analysis Toolpack* and *Analysis Toolpack— VBA* as discussed below.

7. Proceed to the desired example by clicking on it and selecting "Enable Macros."

D. HOW TO USE THE E-BOOK SOFTWARE

Before proceeding to the worked-out examples, we need to select from the toolbar menu "Tools," "Add-ins," and, from the dialog box, select both "Analysis Tool-pack" and "Analysis Toolpack—VBA " in order to set up the software properly.

When a particular example is selected and clicked for interactive use, the user will encounter a dialogue box where "Enable Macros" must be selected. When the example is opened, the user will observe that each numerical example consists of several sections, which can be briefly summarized as:

1.   Problem statement: Word for word text taken from the book that describes the nature of the problem.

2.   Input data: This section provides all the manipulatable data, which the reader may have at his or her own disposal to vary interactively and experiment with, in order to examine different scenarios and obtain answers to "what if" questions. *Inside the input data box, the user may change the values of the parameters only, and not any of the units. The user should type the new value in the appropriate box and press enter/return to observe the desired change. This is the only place where the user is allowed to make any changes to the software.*

3.   Output data: This section contains a step-by-step solution of the problem as well as easy explanations provided for quick assimilation of the results. Most of the complicated calculations are done in complex numbers using the Visual Basic programming technique, which is part of the Microsoft Excel® software.

4. Problem format and color codes: All problems are formatted and color-coded in the same manner throughout the software. This is done for the user's easy recognition and reference, and is delineated as follows:

Example xx.xx	Cyan
Problem text	Tan
Solution:	Navy Blue
Input data (interactive part)	
Heading	Red
Content	Turquoise/Brown
Output data	
Heading	Sky Blue
Content	Yellow
Interactive Answers	Green
Caution	Red
Note/Conclusion	Pink
Reference	Violet

E. SOFTWARE KNOWLEDGE REQUIRED
A rudimentary knowledge of Microsoft Excel® is required to operate the software successfully. The user does *not* need to know Visual Basic programming techniques to work with the examples' solutions interactively.

F. MINIMUM SOFTWARE/HARDWARE REQUIREMENTS
The user needs to have the following:
1. **Hardware requirements:** A personal computer (PC) with a Pentium chip, preferably.
2. **Software requirements:** Windows 95/97/XP/NT operating system and Microsoft Excel 2000/2002 (or Excel 97 with: SR-1 or SR-2 revisions).

NOTE 1: *To obtain the Service Release 1 or 2 (SR-1 or SR-2), the user should download the required software from the following Website: http://www.microsoft. com/*

Go to the search link and find the SR-1 or SR-2 revision, which is suitable to the version of Microsoft Office that you own. Download and install the SR-1 or SR-2 upgrade to repair all known bugs in Microsoft Excel 97. Without this correction, Excel 97 gives incomplete values for the worked-out examples in the textbook CD.

NOTE **2:** *If you own Microsoft Excel 2000/2002, please ignore "Note 1." Install the textbook CD directly without any changes to the Microsoft Excel software using the procedure outlined in the previous section. If you experience any problems, you need to download and install the SR-1A revision for Microsoft Excel 2000/2002 from the site mentioned previously.*

G. TROUBLESHOOTING PROBLEMS

If the following problems occur, you may correct them as follows:

1.   If "###### " appears in place of a numerical answer, it means that the cell is too small and you have to resize that cell in order to display the final numerical result correctly. To resize the cell, go to the Excel toolbar menu and select **Format, Column,** and **Autofit Selection.**

2.   If "#VALUE!" appears, it means that any of the following conditions may have occurred:

a.   Divide by zero.

b.   Negative number under a square root.

c.   The number is out of range.

d.   Excel 97 software is not used with SR-1 or SR-2 revision.

e.   The *Analysis Toolpack* and *Analysis Toolpack— VBA* are not turned on.*

* You need to select from the toolbar menu "Tools," "Add-ins," and from the dialog box select *Analysis Toolpack* and *Analysis Toolpack—VBA* in order to set up the software properly.

APPENDIX M

Essential Graphs & Charts

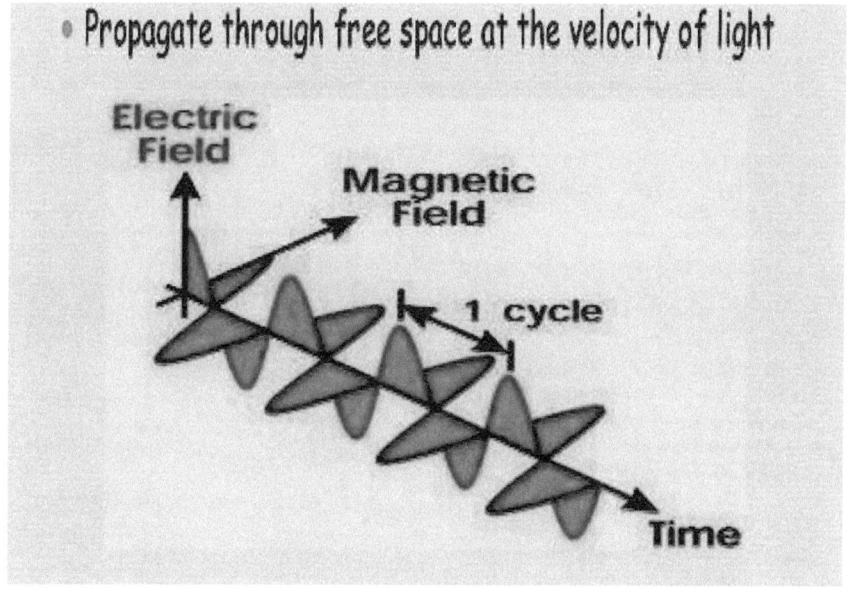

An "Electromagnetic Wave" consists of
oscillating electric and magnetic fields,
both perpendicular to each other and the
direction of propagation.

Electromagnetic Spectrum

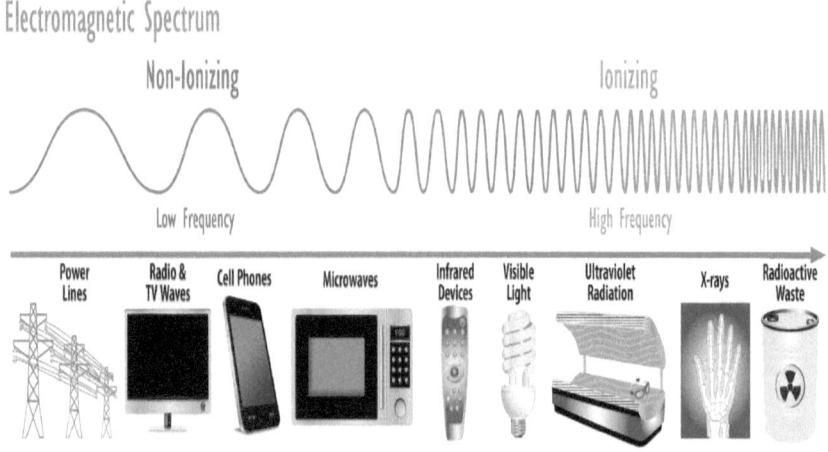

THE ELECTROMAGNETIC SPECTRUM

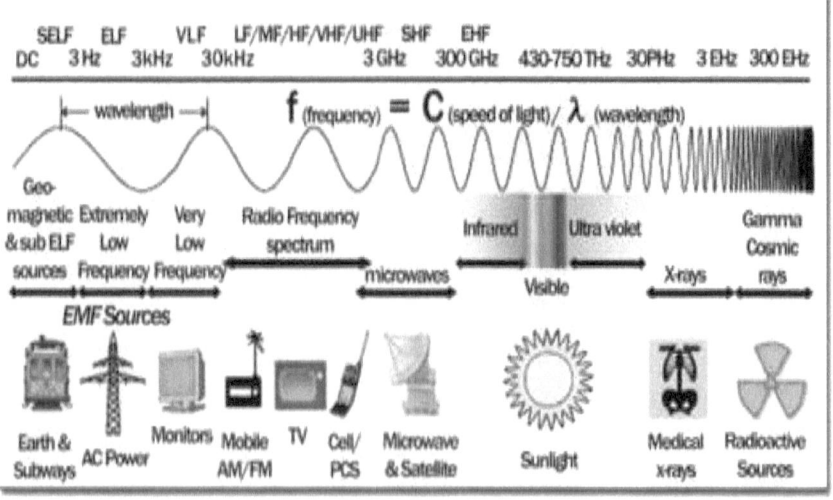

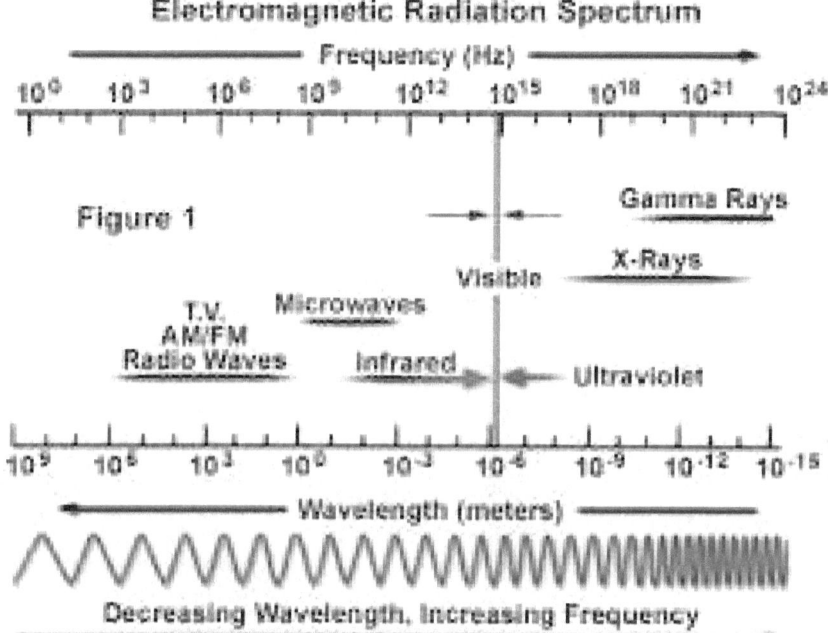

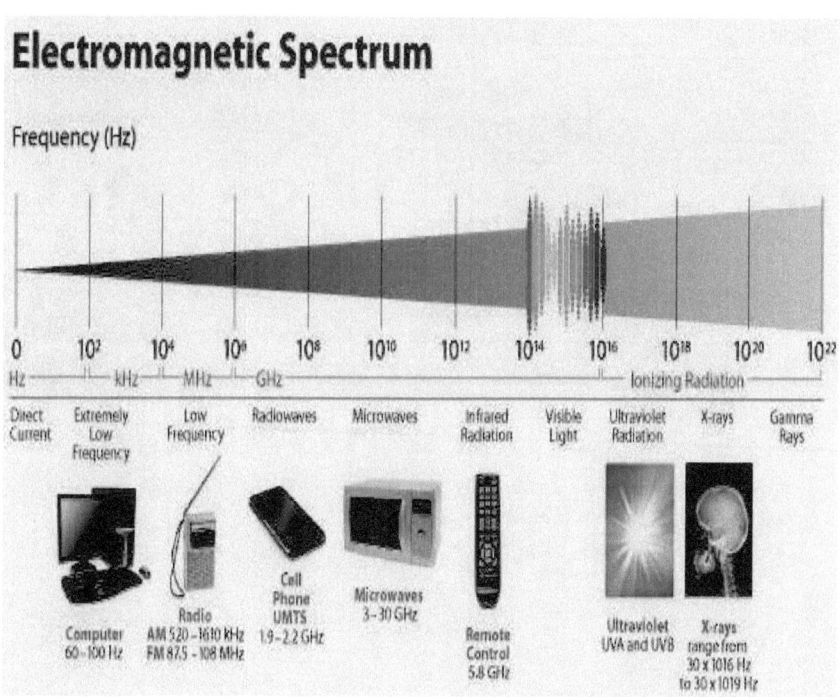

THE ELECTROMAGNETIC SPECTRUM

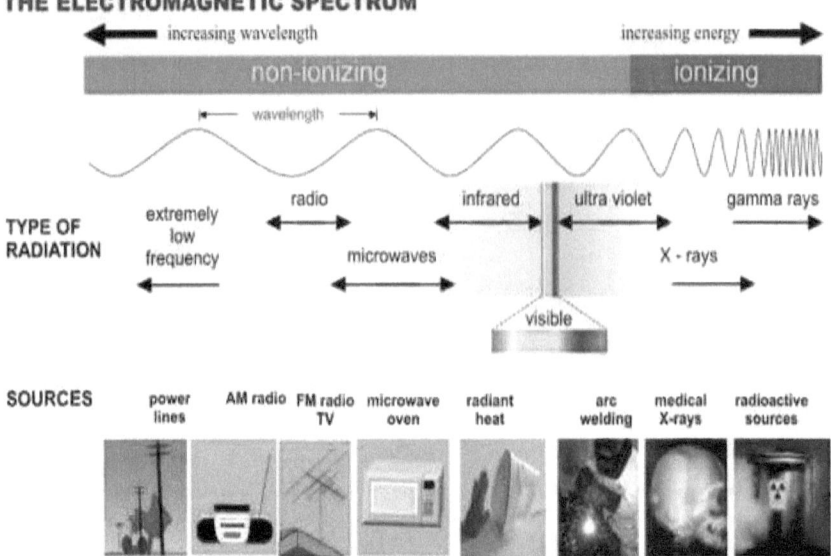

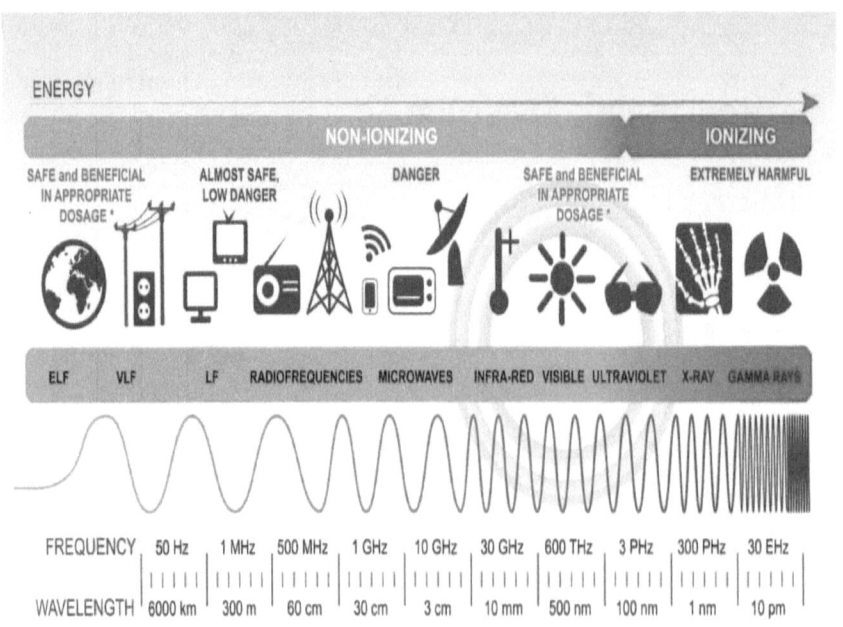

Our exposure to radiation on the electromagnetic spectrum

NON-IONIZING RADIATION
Traditionally perceived as harmless due to its lack of potency

IONIZING RADIATION
Can cause cellular and/or DNA damage with prolonged exposure

Frequency (Hz)

0	10^2	10^4	10^5	10^8	10^{10}	10^{12}	10^{14}	10^{16}	10^{18}	10^{20}	10^{22}

Computer	Radio	Cellphone	Microwave	Remote control	Visible light	Ultraviolet	X-rays	Gamma rays
80-100 Hz	520 KHz-108 MHz	1.9 GHz-2.2 GHz	3 GHz-30GHz	5.8 GHz		UVA and UVB		

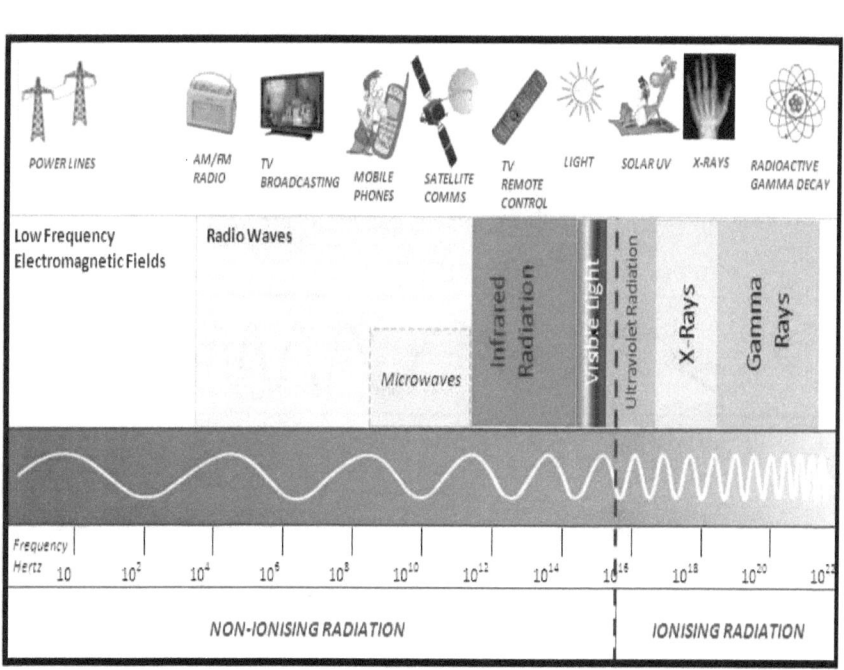

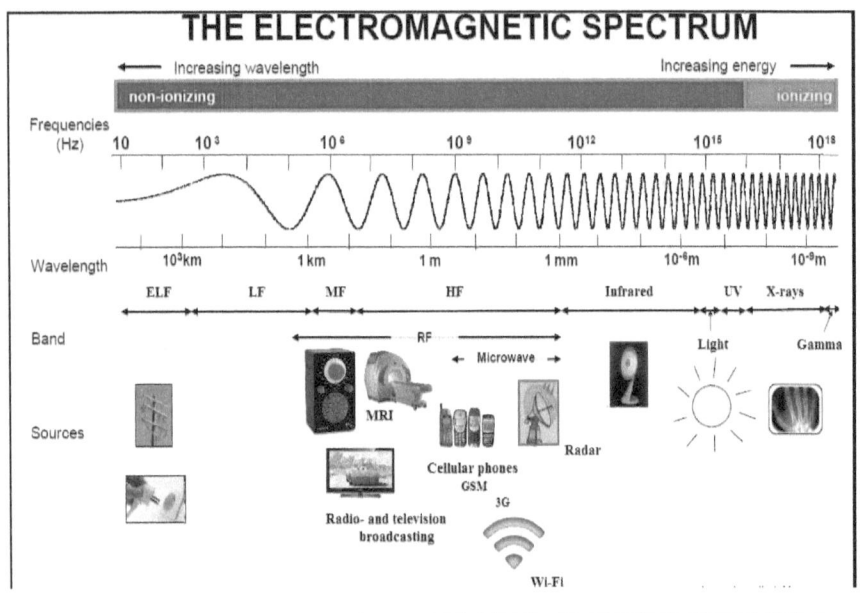

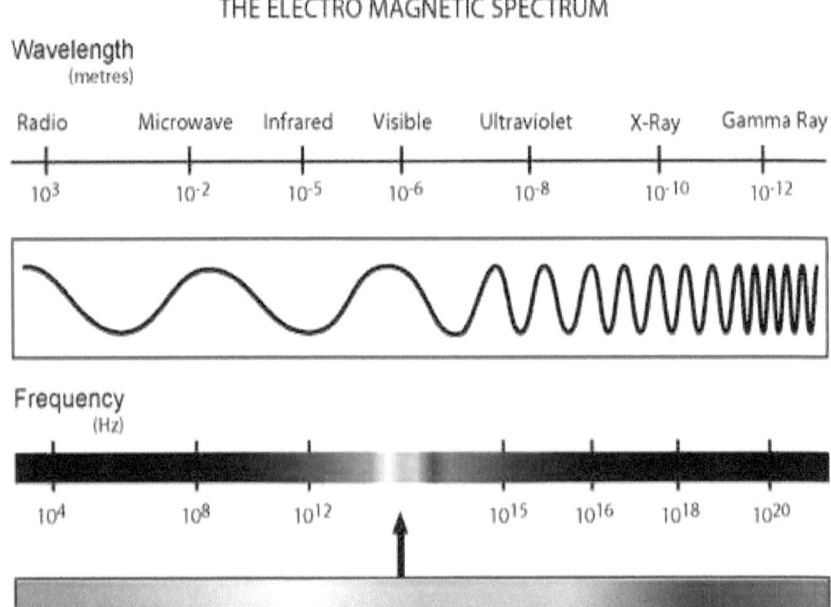

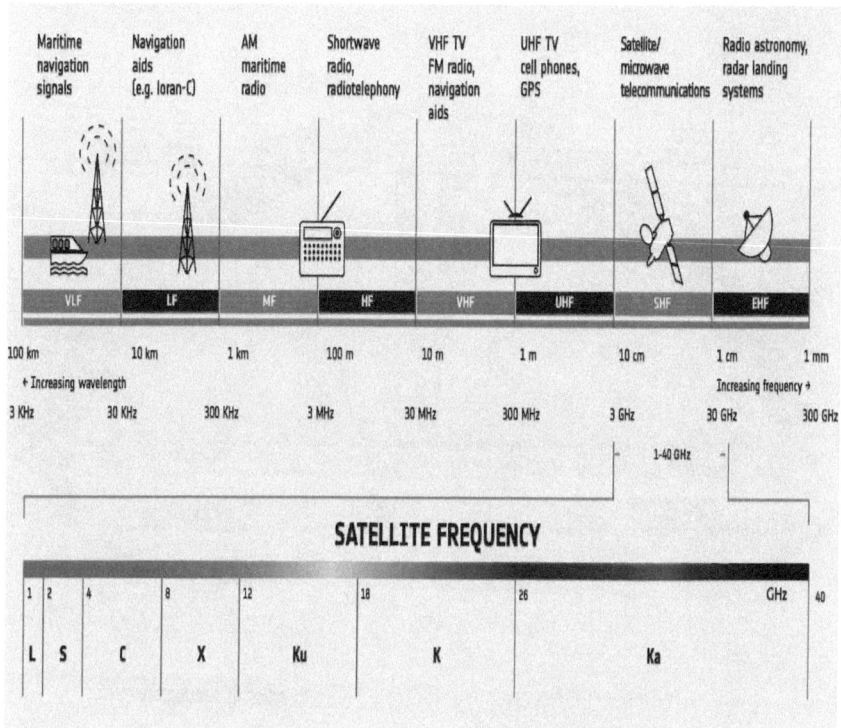

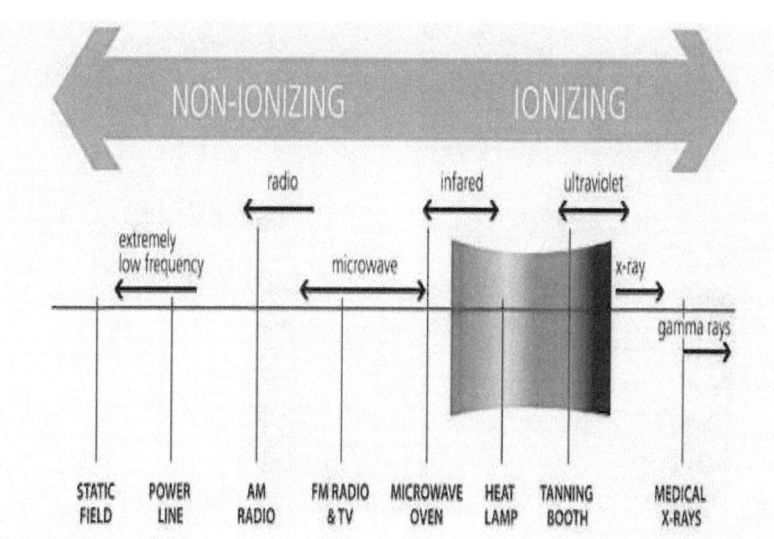

The Spectrum of Electromagnetic Waves

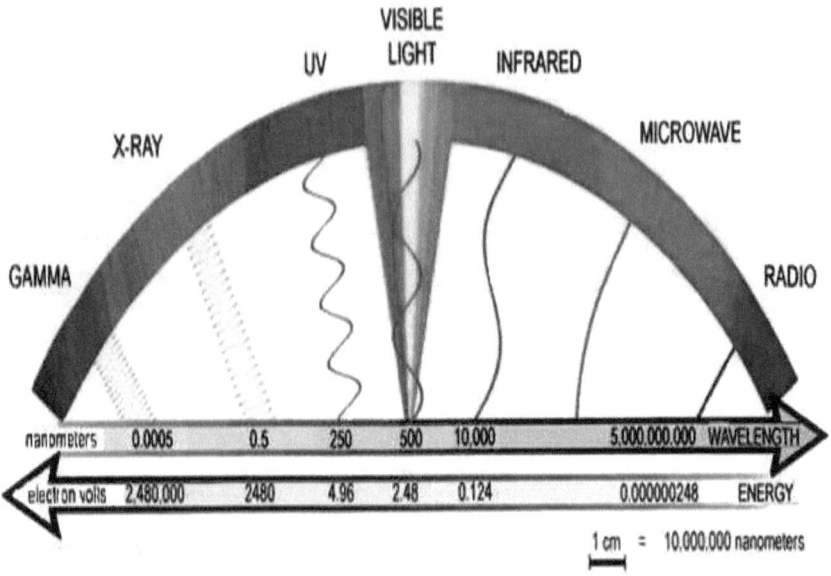

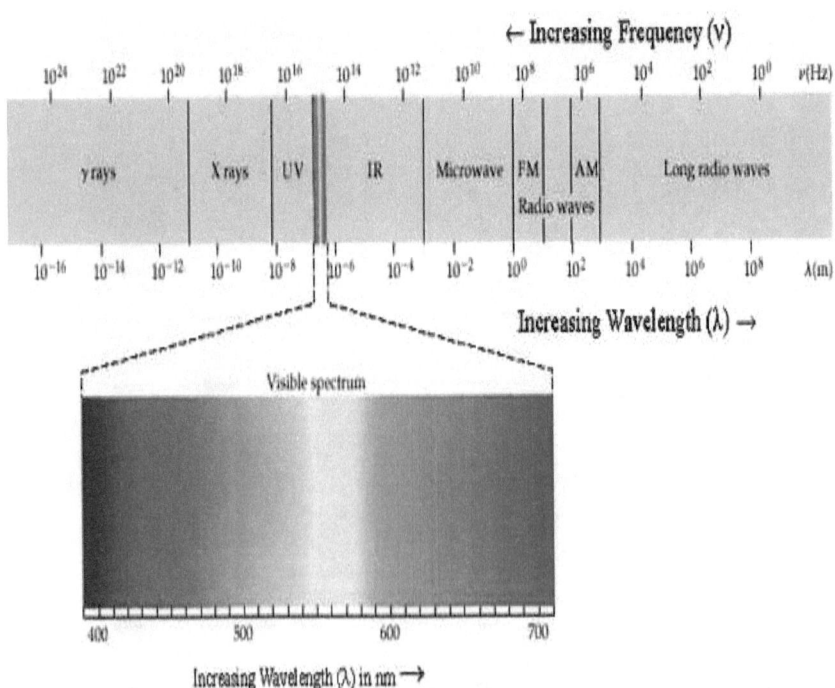

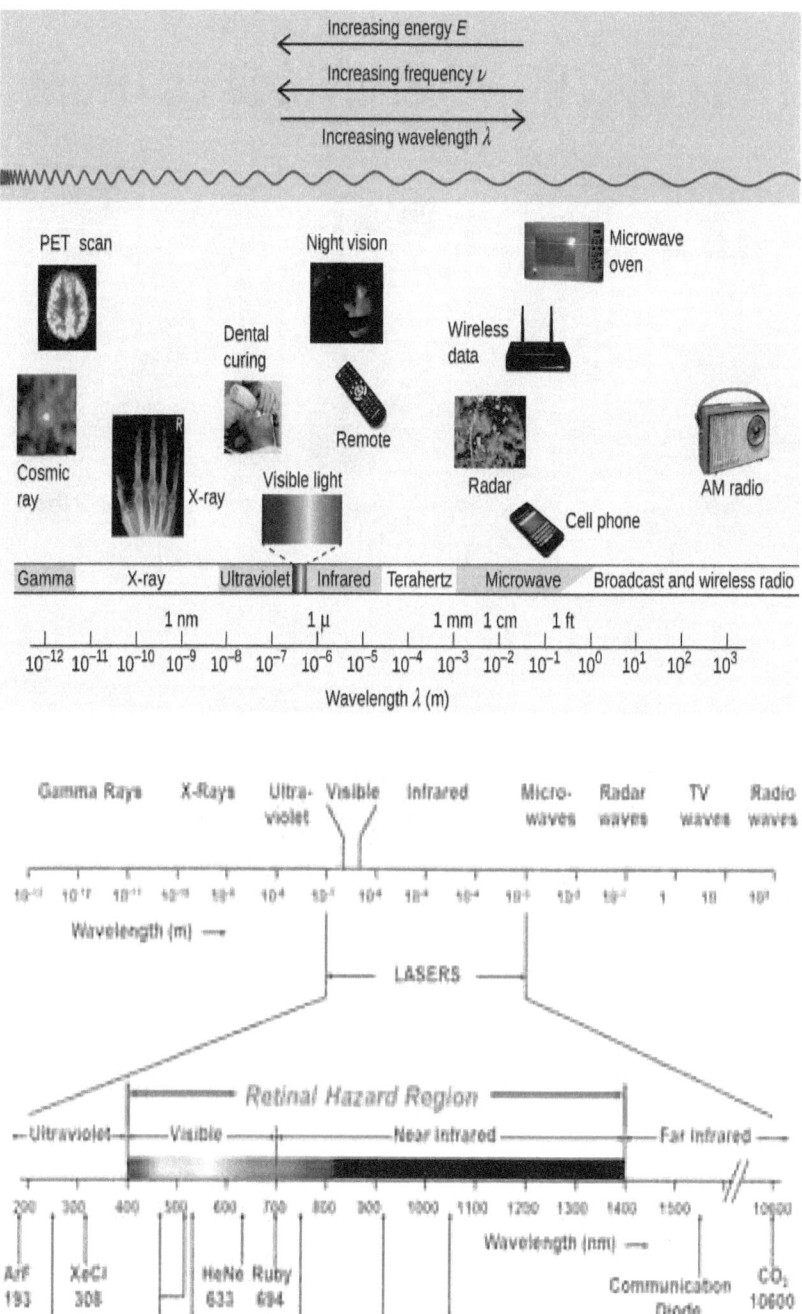

THE ELECTROMAGNETIC SPECTRUM

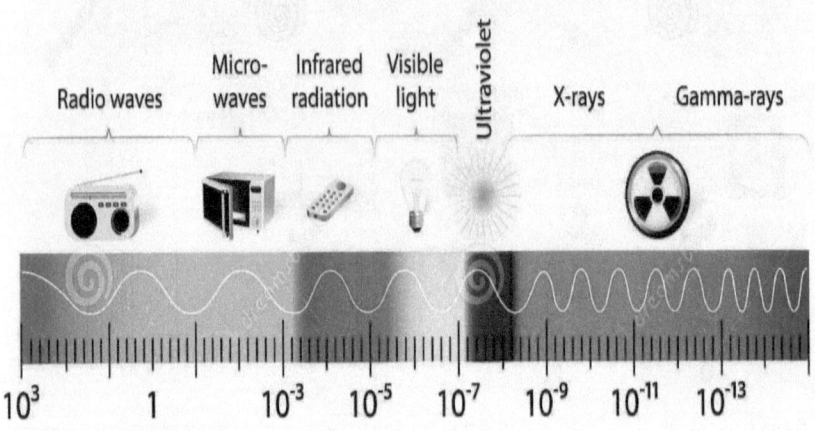

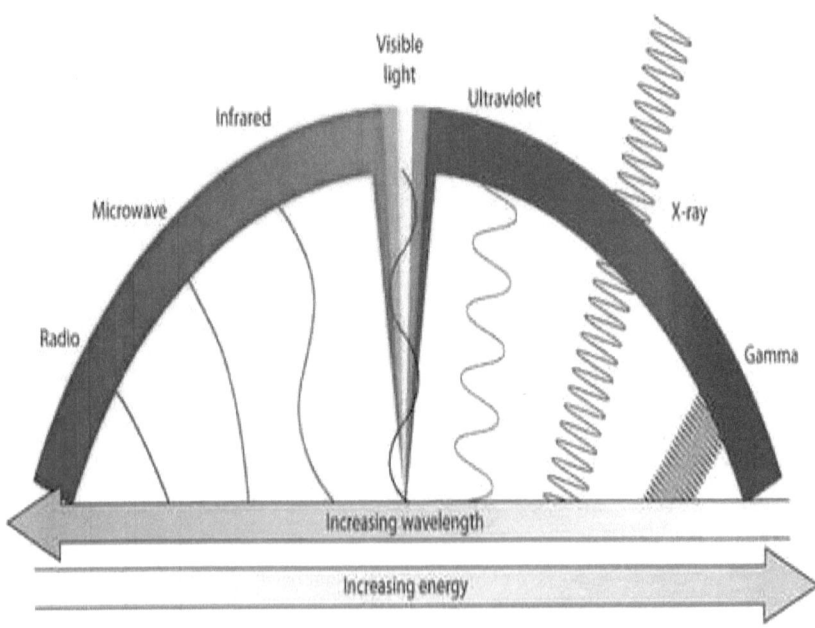

THE ELECTRO MAGNETIC SPECTRUM

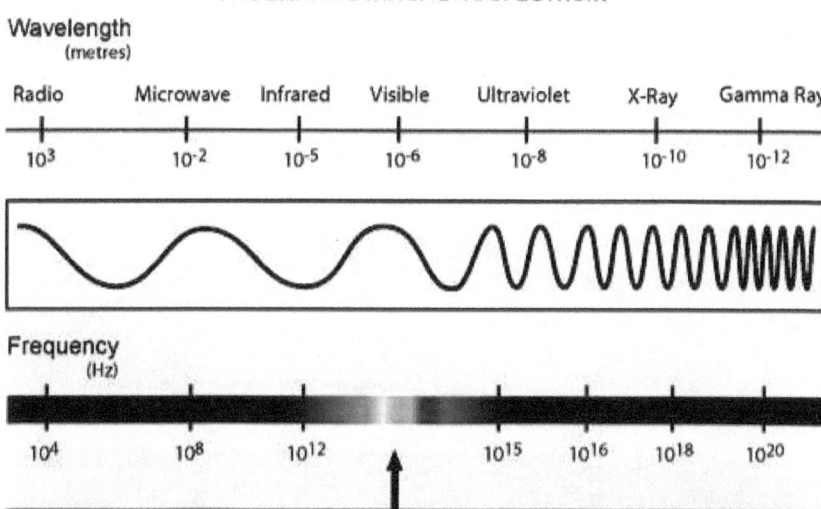

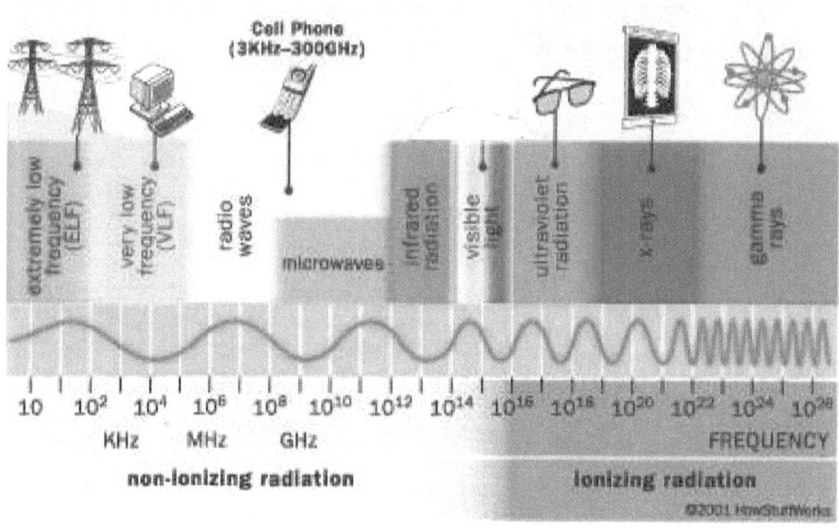

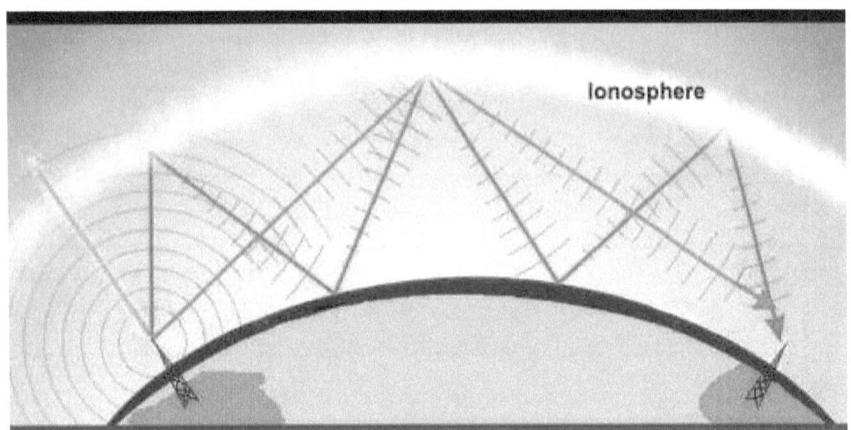

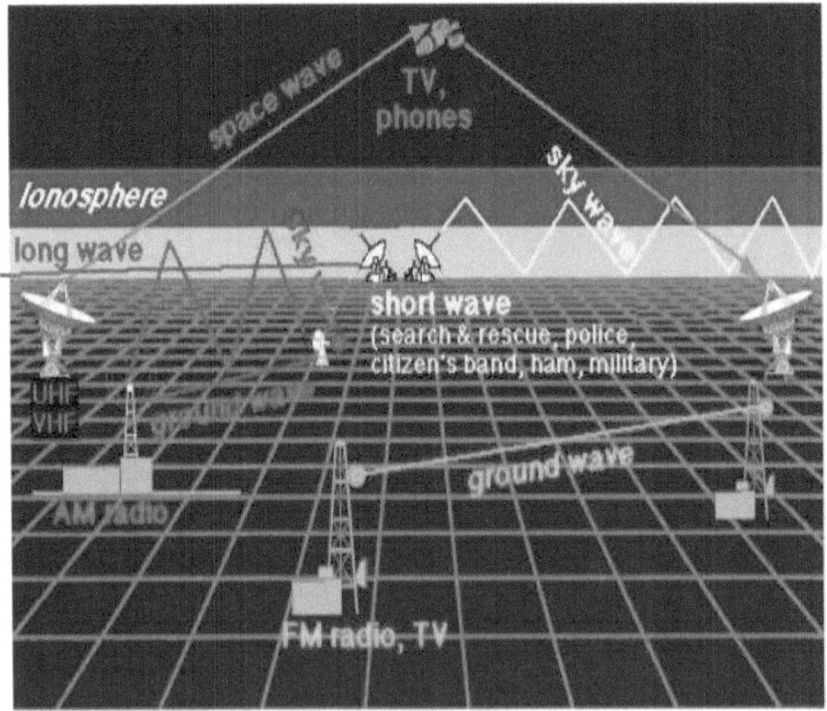

Radio Wave Propagation

Glossary of Technical Terms

The following glossary supplements the presented materials in the text, but does not replace the use of an unabridged technical dictionary, which is a must for mastery of sciences.

Absolute

a) That which is without reference to anything else and thus not comparative or dependent upon external conditions for its existence (opposed to relative), **b)** That which is free from any limitations or restrictions and is thus unconditionally true at all times.

Absolute Temperature Scale

A scale with which temperatures are measured relative to absolute zero (the temperature of –273.15 °C or –459.67 °F or 0 K). The absolute temperature scale leads to the absolute temperatures, which are: a) The temperature in Celsius degrees, relative to –273.15 °C (giving rise to the Kelvin scale), and b) The temperature in Fahrenheit degrees, relative to –459.67 °F (giving rise to the Rankine scale). At absolute zero temperature, molecular motion theoretically vanishes and a body would have no heat energy. The absolute zero temperature is approachable but never attainable. *See also* **Temperature**.

Active Device

An electronic component such as a transistor that can be used to produce amplification (or gain) in a circuit.

Admittance

The measure of ease of AC current flow in a circuit, the reciprocal of impedance expressed in Siemens (symbol Y or y).

Ampere (A)
The unit of electric current defined as the flow of one Coulomb of charge per second. Alternately, it can also be defined as the constant current that would produce a force of 2×10^{-7} Newton per meter of length in two straight parallel conductors of infinite length, and of negligible cross section, placed one meter apart in a vacuum.

Ampere's Law
Current (either conduction or displacement) flowing in a wire or in space generates a magnetic flux that encircles the wire in a clockwise direction when the current is moving away from the observer. The direction of the magnetic field follows the right-hand rule (This law may also be referred to as the law of magnetic field generation).

Differential form: $\mathrm{Curl}\overline{H} = \overline{J} + \dfrac{\partial \overline{D}}{\partial t}$,

Integral form : $\oint_{C} \overline{H} \cdot \overline{d\ell} = I + \int_{S} \dfrac{\partial \overline{D}}{\partial t} \cdot \overline{dS}$

Amplitude
The extent to which an alternating current or pulsating current or voltage swings from zero or a mean value.

Analog
Pertaining to the general class of devices or circuits in which the output varies as a continuous function of the input.

Anode
The positive electrode of a device (such as a diode, etc.) toward which the electrons move during current flow.

Application Mass
All of the related masses that are connected and/or obtained as a result of the application of a science. This includes all physical devices, machines, experimental setups, and other physical materials that are directly or indirectly derived from and are a result of the application. In this book when we say application mass, we really mean "technical application mass." See also **Generalized application mass, Technical application mass and personalized application mass**.

Attenuation
The decrease in amplitude of a signal during its transmission from one point to another.

Attenuation Constant
The real component of the propagation constant.

Attenuator
A resistive network that provides a reduction of the amplitude of an electrical signal without introducing appreciable phase or frequency distortion.

Average Power
The power averaged over one cycle.

Axiom
A self-evident truth accepted without proof as a basis for a science.

B

Bias
The steady and constant current or voltage applied to an electrical device to establish an operating point for proper operation of the device.

Bidirectional
Responsive in both directions.

Bilateral
Having a voltage-current characteristic curve that is symmetrical with respect to the origin. If a positive voltage produces a positive current magnitude, then an equal negative voltage produces a negative current of the same magnitude.

Brewster Angle
The angle of incidence of light reflected from a dielectric surface at which the reflection coefficient becomes zero when the light's electrical field vector lies in the plane of incidence (parallel polarization). In other words, if a parallel polarized wave is incident at a dielectric surface at the Brewster angle, all of the wave will be transmitted through and there will be no reflection. Generally

speaking, the concept of Brewster angle applies to any electronic wave of any frequency, not just light waves [$\theta_B = \tan^{-1}(\varepsilon_2/\varepsilon_1)^{1/2}$].

C

Capacitance
The property that permits the storage of electrically separated charges when a potential difference exists between two conductors. The capacitance of a capacitor is defined as the ratio between the electric charge of one electrode, and the difference in potential between the electrodes.

Capacitor
A device consisting essentially of two conducting surfaces separated by an insulating material (or a dielectric) such as air, paper, mica, etc., that can store electric charge.

Cathode
The portion or element of a two-terminal device that is the primary source of electrons during operation.

Cavity (Also Called a Cavity Resonator)
A metallic enclosure inside which resonant fields at microwave frequencies are excited in such a way that it becomes a source of electromagnetic oscillations frequencies.

Cell
A single and basic unit for producing electricity by electrochemical or biochemical action. For example, a battery consists of a series of connected cells.

Celsius (°C)
$1/100^{th}$ of the temperature difference between the freezing point of water (0°C) and the boiling point of water (100°C) on the Celsius temperature scale given by:

$$T(°C) = T(K) - 273.15 = \frac{5}{9}\{T(°F) - 32\}.$$

Characteristic Impedance
The driving-point impedance of a transmission line if it were of infinite length. This can also be defined as the ratio of the voltage to current at every point along a transmission line on which there are no standing waves. It is given in general by:

$$Z_0 = \sqrt{(R + j\omega L)/(G + j\omega C)}$$

Charge
A basic property of elementary particles of matter (electrons, protons, etc.) that is capable of creating a force field in its vicinity. The built-in force field is a result of stored electric energy.

Chip
A single substrate upon which all the active and passive circuit elements are fabricated using one or all of the semiconductor techniques of diffusion, passivation, masking, photoresist, epitaxial growth, etc.

Circuit
The interconnection of a number of devices in one or more closed paths to perform a desired electrical or electronic function.

Classical Mechanics (Also Called Classical Physics, Non-Quantized Physics or Continuum Physics)
Is the branch of physics based on concepts established before quantum physics, and includes materials in conformity with Newton's mechanics and Maxwell's electromagnetic theory.

Coaxial Transmission Line (Also Called Coaxial Cable)
A concentric transmission line in which one conductor completely surrounds the other, the two being separated by a continuous solid dielectric or by dielectric spacers. Such a line has no external field and is not susceptible to external fields.

Coulomb (C)
The unit of electric charge defined as the charge transported across a surface in one second by an electric current of one ampere. An electron has a charge of 1.602×10^{-19} Coulomb.

Coulomb's Laws
The laws that state that the force (F) of attraction or repulsion between two electric charges (or magnetic poles) is directly

proportional to the product of the magnitude of charges, Q (or magnetic pole strengths, M), and is inversely proportional to the square of distance (d) between them; that is,

Electric: $F = \dfrac{Q_1 Q_2}{4\pi\varepsilon d^2}$,

Magnetic: $F = \dfrac{M_1 M_2}{4\pi\mu d^2}$

The force between unlike charges, Q_1 and Q_2 (or poles, M_1 and M_2) is an *attraction*, and between like charges (or poles) is a *repulsion*.

Communication Principle (Also Called Universal Communication Principle)
A fundamental concept in life and livingness that is intertwined throughout the entire field of sciences, and states that for a communication to take place between two or more entities, three elements must be present: a source point, a receipt point, and an imposed space or distance between the two.

Complex Power
Power calculated based on the reactance of a component.

Component
A packaged functional unit consisting of one or more circuits made up of devices, which in turn may be part of an operating system or subsystem.

Conductivity
The ratio of the current density (J) to the electric field (E) in a material. It represents the ability to conduct or transmit electricity.

Conductor
a) A material that conducts electricity with ease, such as metals, electrolytes, and ionized gases; b) An individual metal wire in a cable, insulated or un-insulated.

Curl Operation
Curl is an operation on a vector field, which creates another vector whose magnitude measures the maximum net circulation per unit area of the vector field at any given point and has a direction perpendicular to the area, as the area size tends toward zero. The

cause of the curl of a vector field is a vortex source. For example electric current (conduction or displacement) is the vortex source for the magnetic field.

Current
Net transfer of electrical charges across a surface per unit time, usually represented by (I) and measured in Ampere (A). Current density (J) is current per unit area.

D

DC (Also Called Direct Current)
A current that always flows in one direction (e.g., a current delivered by a battery).

Decibel (dB)
The logarithmic ratio of two powers or intensities or the logarithmic ratio of a power to a reference power, multiplied by 10. It is one-tenth of an international unit known as *Bel*: $N(dB)=10\log_{10}(P_2/P_1)$.

Device
A single discrete conventional electronic part such as a resistor, a transistor, etc.

Diamagnetics
are materials (such as glass, wood, lead, sulfur and others), which avoid magnetic lines of force.

Dichotomy
Two things or concepts that are sharply or distinguishably opposite to each other.

Die (Also Called Chip)
A single substrate on which all the active and passive elements of an electronic circuit have been fabricated. This is one portion taken from a wafer bearing many chips, but it is not ready for use until it is packaged and provided with terminals for connection to the outside world.

Dielectric
A material that is a non-conductor of electricity. It is characterized by a parameter called *dielectric constant* or *relative permittivity* (ε_r).

Dielectric Constant
The property of a dielectric defined as the ratio of the capacitance of a capacitor (filled with the given dielectric) to the capacitance filled with air as the dielectric, but otherwise identical in geometry.

Diffraction
Is the redistribution of the intensity of waves in space, which results from the presence of an object (such as a grating, consisting of narrow slits or grooves) in the path of the beam of light waves. This shall split up the beam into many rays, causing interference and thus producing patterns of dark and light bands downstream (i.e., regions with variations of wave amplitude and phase).

Digital
Circuitry in which data-carrying signals are restricted to either of two voltage levels.

Discovery
The gaining of knowledge about something previously unknown.

Discrete Device
An individual electrical component such as a resistor, capacitor, or transistor as opposed to an integrated circuit that consists of several discrete components.

Distributed Element
An element whose property is spread out over an electrically significant length or area of a circuit instead of being concentrated at one location or within a specific component.

Divergence
a) The emanation of many flows from a single point, or reversely, the convergence of many flows to one point; b) (of a vector field, F) The net outflux per unit volume at any given point in a vector field, as the volume size shrinks to zero (symbolized by divF). The cause of the divergence of a vector field is called a flow source. For example, a positive electric charge is the flow source for the electric field and creates a net outflux of electric field per unit volume at any given point.

Dual
Two concepts, energy forms or physical things that are of comparable magnitudes but of opposite nature, thus becoming counterpart of each other.

Duality Theorem
States that when a theorem is true, it will remain true if each quantity and operation is replaced by its dual quantity and operation. In circuit theory, the dual quantities are "voltage and current" and "impedance and admittance." The dual operations are "series and parallel" and "meshes and nodes."

E

Electric Charge (or Charge)
(**Microscopic**) A basic property of elementary particles of matter (e.g., electron, protons, etc.) that is capable of creating a force field in its vicinity. This built-in force field is a result of stored electric energy. (**Macroscopic**) The charge of an object is the algebraic sum of the charges of its constituents (such as electrons, protons, etc.), and may be zero, a positive or a negative number.

Electric Current (or Current)
The net transfer of electric charges (Q) across a surface per unit time.

Electric Field
The region about a charged body capable of exerting force. The intensity of the electric field at any point is defined to be the force that would be exerted on a unit positive charge at that point.

Electric Field Intensity
The electric force on a stationary positive unit charge at a point in an electric field (also called *electric field strength*, *electric field vector*, and *electric vector*).

Electrical Noise (or Noise)
Any unwanted electrical disturbance or spurious signal. These unwanted signals are random in nature, and are generated either

internally in the electronic components or externally through impinging electromagnetic radiation.

Electricity
Is a form of energy, which can be subdivided into two major categories: a) Electrostatics, and b) Electrokinetics.

Electrodynamics
Is a scientific field of study dealing with the various phenomena of electricity in motion, including the interactions between current-carrying wires as well as the forces on current wires in an independent magnetic field.

Electrokinetics
Is that broad and general field of study dealing with electric charges in motion. It studies moving electric charges (such as electrons) in electric circuits and electrified particles (such as ions, etc.) in electric fields.

Electrolysis
The action whereby a current passing through a conductive solution (called an *electrolyte*) produces a chemical change in the solution and the electrodes.

Electrolyte
A substance that ionizes when dissolved in a solution. Electrolytes conduct electricity, and in batteries they are instrumental in producing electricity by chemical action.

Electrolytic Cell
In general, a cell containing an electrolyte and at least two electrodes. Examples include voltaic cells, electrolytic capacitors, and electrolytic resistors.

Electromagnetic (EM) Wave
A radiant energy flow produced by the oscillation of an electric charge as the source of radiation. In free space and away from the source, EM rays of waves consist of vibrating electric and magnetic fields that move at the speed of light (in vacuum), and are at right angles to each other and to the direction of motion. EM waves propagate with no actual transport of matter, and grow weaker in amplitude as they travel farther in space. EM waves include radio,

microwaves, infrared, visible/ultraviolet light waves, X-ray, gamma rays, and cosmic rays.

Electromagnetics
The branch of physics that deals with the theory and application of electromagnetism.

Electromagnetism
a) Magnetism resulting from kinetic electricity; b) Electromagnetics.

Electron
A stable elementary particle of matter, which carries a negative electric charge of one electronic unit equal to q= $-1.602x10^{-19}$ C and has a mass of about $9.11x10^{-31}$ kg and a spin of ½.

Electronics
The study, control, and application of the conduction of electricity through different media (e.g., semiconductors, conductors, gases, vacuum, etc.).

Electroplating
Depositing one metal on the surface of another by electrolytic action.

Electrostatics
The branch of physics concerned with static charges and charged objects at rest.

Elementary Particle
A particle, which can not be described as a compound of other particles and is thus one of the fundamental constituents of all matter (e.g. electron, proton, etc.).

Energy
The capacity or ability of a body to perform work. The energy of a body is either potential motion (called *potential energy*) or due to its actual motion (called *kinetic energy*).

F

Fahrenheit (°F)
$1/180^{th}$ of the temperature difference between the freezing point of water (32°F) and the boiling point of water (212°F) on the Fahrenheit temperature scale.

$$T(°F) = T(°R) - 459.67 = \frac{9}{5} T(°C) + 32$$

Where °R and °C are symbols for degrees Rankin and Celsius, respectively.

Farad (F)
The unit of capacitance in the MKSA system of units equal to the capacitance of a capacitor that has a charge of one Coulomb when a potential difference of one volt is applied.

Faraday's Law (also called the law of electromagnetic induction)
When a magnetic field cuts a conductor, or when a conductor cuts a magnetic field, an electrical current will flow through the conductor if a closed path is provided over which the current can circulate; i.e.,

Differential form: $\text{Curl}\overline{E} = \dfrac{-\partial\overline{B}}{\partial t}$,

Integral form: $\oint_c \overline{E} \cdot \overline{d\ell} = -\int_s \dfrac{\partial\overline{B}}{\partial t} \cdot \overline{dS} = -\dfrac{d\Phi}{dt}$

Ferrimagnetics
Ferrimagnetics are materials with the relative permeability (μ_r) much greater than that of vacuum having $\mu_r=1$. Ferrimagnetic materials are materials made of iron oxides (chemical formula: XFe_2O_3, where X is a metal ion), where their internal magnetic moments are not all aligned in one direction, that is to say, some are aligned antiparallel but with smaller magnitudes, so that the net magnetic field output is still much higher than a paramagnetic material. Examples of ferrimagnetics include materials such as manganese-zinc ferrite, barium ferrite, and a whole class of materials, having a high

electrical resistance, called ferrites. Ferrimagnetic materials exhibit hysteresis, which is a type of material behavior characterized by an inability to retrace exactly the input-output curve when the magnetizing force is reversed. This nonlinear behavior is caused by the fact that the material will retain some of the magnetic effects internally (called the remnant magnetism) even when the external magnetizing force is completely removed.

Ferromagnetics
ferromagnetics are materials with the relative permeability (μ_r) much greater than that of vacuum ($\mu_r=1$), the amount depending on the magnetizing force. Ferromagnetic materials are a group of materials whose internal magnetic moments align in a common direction such as iron, nickel, cobalt, and their alloys. Ferromagnetic materials exhibit hysteresis, which is a type of material behavior characterized by an inability to retrace exactly the input-output curve when the magnetizing force is reversed. This nonlinear behavior is caused by the fact that the material will retain some of the magnetic effects internally (called the remnant magnetism) even when the external magnetizing force is completely removed.

Field
An entity that acts as an intermediary agent in interactions between particles, is distributed over a region of space, and whose properties are a function of space and time, in general.

Field Theory
The concept that, within a space in the vicinity of a particle, there exists a field containing energy and momentum, and that this field interacts with neighboring particles and their fields.

Flow
The passage of particles (e.g., electrons, etc.) between two points. Example: electrons moving from one terminal of a battery to the other terminal through a conductor. The direction of flows is from higher to lower potential energy levels.

Force
That form of energy that puts an unmoving object into motion, or alters the motion of a moving object (i.e., its speed, direction or both). Furthermore, it is the agency that accomplishes work.

Frequency

The number of complete cycles in one second of a repeating quantity, such as an alternating current, voltage, electromagnetic waves, etc.

G

Gain

The ratio that identifies the increase in signal or amplification that occurs when the signal passes through a circuit.

Gauss

The unit of magnetic induction (also called *magnetic flux density*) in the cgs system of units equal to one line per square centimeter, which is the magnetic flux density of one Maxwell per square centimeter, or 10^{-4} Tesla.

Gauss's Law (electric)

The summation of the normal component of the electrical displacement vector over any closed surface is equal to the electric charges within the surface, which means that the source of the electric flux lines is the electric charge; i.e.,

Differential form: $\mathrm{Div}\overline{\mathrm{D}} = \rho$

Integral form: $\oint_{S} \overline{\mathrm{D}} \cdot \overline{\mathrm{dS}} = \int_{V} \rho \mathrm{dv} = \mathrm{Q}$

Gauss's Law (magnetic)

The summation of the normal component of the magnetic flux density vector over any closed surface is equal to zero, which in essence means that the magnetic flux lines have no source or magnetic charge; i.e.,

Differential form: $\mathrm{Div}\overline{\mathrm{B}} = 0$,

Integral form: $\oint_{S} \overline{\mathrm{B}} \cdot \overline{\mathrm{dS}} = 0$

Generalized Application Mass (G.A.M.)

In general, is any created space, which contains created energies and created matter of any form, shape or size existing as a function of time. In simple terms, generalized application mass is any matter and energy, condensed and packaged into an object form, which exists in a time-stream (from its inception to now). The generalized concept of application mass includes the entire mechanical space containing all energies and matter such as electrons, atoms, molecules and all the existing gigantic masses of planets, stars, galaxies, which are not the direct byproduct of Man's sciences.

Generalized Ohm's Law

When dealing with linear circuits under the influence of time-harmonic signals, Ohm's law can be restated under the steady-state condition in the phasor domain as $V=ZI$, where Z is a complex number called impedance and V and I are voltage and current phasors, respectively.

Gilbert (Gi)

The unit of magnetomotive force in the cgs system of units, equal to the magnetomotive force of a closed loop of one turn in which there is a current of $10/4\pi$ amperes. One Gilbert equals $10/4\pi$ Ampere-turn.

Gradient (of a scalar function)

Gradient (of a scalar function) is a vector, which lies in the direction of maximum rate of increase of the function at any given point and therefore is normal to the constant-value surfaces. Mathematically, it is a vector obtained from a real function $f(x,y,z)$, whose components are the partial derivatives of $f(x,y,z)$, e.g., in the Cartesian coordinate system we can write: $\mathrm{grad} f=(\partial f/\partial x, \partial f/\partial y, \partial f/\partial z)$.

Ground

(a) A metallic connection with the earth to establish zero potential (used for protection against short circuit); (b) The voltage reference point in a circuit. There may or may not be an actual connection to earth but it is understood that a point in the circuit said to be at ground potential could be connected to earth without disturbing the operation of the circuit in any way.

H

Henry (H)
The unit of self and mutual inductance in the MKSA system of units equal to the inductance of a closed loop that gives rise to a magnetic flux of one Weber for each ampere of current that flows through.

Hertz (Hz)
The unit of frequency equal to the number of cycles of a periodic function that occur in one second.

Hole
A vacant electron energy state near the top of the valence band in a semiconductor material. It behaves like a positively charged particle having a certain mass and mobility. It is the dual of electron, unlike a proton which is the dichotomy of an electron.

Hypothesis
An unproven theory or proposition tentatively accepted to explain certain facts or to provide a basis for further investigation.

I

Impedance
The total opposition that a circuit presents to an AC signal, and is a complex number equal to the ratio of the voltage phasor (V) to the current phasor (I).

Incident Wave
A wave that encounters a discontinuity in a medium, or encounters a medium having different propagation characteristics.

Inductance (L)
The inertial property of an element (caused by an induced reverse voltage), which opposes the flow of current when a voltage is applied; it opposes a change in current that has been established.

Inductor
A conductor used to introduce inductance into an electric circuit, normally configured as a coil to maximize the inductance value.

Input
The current, voltage, power, or other driving force applied to a circuit or device.

Insulator
A material in which the outer electrons are tightly bound to the atom and are not free to move. Thus, there is negligible current through the material when a voltage is applied.

Integrated Circuit (IC)
An electrical network composed of two or more circuit elements on a single semiconductor substrate.

Isolation
Electrical separation between two points.

J

Joule (J)
The unit of energy or work in the MKSA system of units, which is equal to the work performed as the point of application of a force of one Newton moves the object through a distance of one meter in the direction of the force.

Junction
A joining of two different semiconductors or of semiconductor and metal.

Junction Capacitance
The capacitance associated with a junction such as the capacitance of a region of transition between p- and n-type semiconductor materials.

K

Kelvin (K)
The unit of measurement of temperature in the absolute scale (based on Celsius temperature scale), in which the absolute zero is at − 273.15 °C. It is precisely equal to a value of 1/273.15 of the absolute temperature of the triple point of water, being a particular pressure and temperature point, 273.15 K, at which three different phases of water (i.e., vapor, liquid, and ice) can coexist at equilibrium. *See also* **temperature**.

Kinetic
(*Adjective*) Pertaining to motion or change. (*Noun*) Something which is moving or changing constantly such as a piece of matter.

Kinetic Energy (K.E.)
The energy of a particle in motion. The motion of the particle is caused by a force on the particle.

Kirchhoff's Current Law (KCL)
The law of conservation of charge that states that the total current flowing to a given point in a circuit is equal to the total current flowing away from that point.

Kirchhoff's Voltage Law (KVL)
An electrical version of the law of conservation of energy that states that the algebraic sum of the voltage drops in any closed path in a circuit is equal to the algebraic sum of the electromotive forces in that path.

Knowledge
Is a body of facts, principles, data, and conclusions (aligned or unaligned) on a subject, accumulated through years of research and investigation, that provides answers and solutions in that subject.

L

Law
An exact formulation of the operating principle in nature observed to occur with unvarying uniformity under the same conditions.

Law of Conservation of Energy (Excluding All Metaphysical Sources of Energy)
This fundamental law simply states that any form of energy in the physical universe can neither be created nor destroyed, but only converted into another form of energy (also known as the principle of conservation of energy).

Leyden Jar
The first electric capacitor (or condenser) capable of storing charge; it consists of a glass jar with a coat of tin foil outside and inside and a metallic rod passing through the lid and connecting with the inner tin lining. It is named after the city of Leyden (also written as Leiden) in Holland, where it was invented.

Light Waves
Electromagnetic waves in the visible frequency range, which ranges from 400 nm to 770 nm in wavelength.

Linear Network
A network in which the parameters of resistance, inductance, and capacitance of the lumped elements are constant with respect to current or voltage, and in which the voltage or current sources are independent of or directly proportional to other voltages and currents or their derivatives, in the network.

Load
The impedance to which energy is being supplied.

Lossless
A theoretically perfect component that has no loss and hence, transmits all of the energy fed to it.

Lumped Element
A self-contained and localized element that offers one particular electrical property throughout the frequency range of interest.

M

Magnet
A piece of ferromagnetic or ferromagnetic material whose internal domains are sufficiently aligned so that it produces a considerable net magnetic field outside of itself and can experience a net torque when placed in an external magnetic field.

Magnetic Field
The space surrounding a magnetic pole, a current-carrying conductor, or a magnetized body that is permeated by magnetic energy and is capable of exerting a magnetic force. This space can be characterized by magnetic lines of force.

Magnetic Field Intensity (H)
The force that a magnetic field would exert on a unit magnetic pole placed at a point of interest, which expresses the free space strength of the magnetic field at that point (also called *magnetic field strength*, *magnetic intensity*, *magnetic field*, *magnetic force*, and *magnetizing force*).

Magnetostatics
The study of magnetic fields that are neither moving nor changing direction.

Man
Homo sapiens (literally, the knowing or intelligent man); mankind.

Mathematics
Mathematics is a short-hand methods of stating, analyzing, or resolving real or abstract problems and expressing their solutions by symbolizing data, decisions, conclusions, and assumptions.

Matter
Matter particles are a condensation of energy particles into a very small volume.

Maxwell (Mx)
The unit for magnetic flux in the cgs system of units, equal to 10^{-8} Weber.

Maxwell's Equations

A series of four advanced classical equations developed by James Clerk Maxwell between 1864 and 1873, which describe the behavior of electromagnetic fields and waves in all practical situations. They relate the vector quantities for electric and magnetic fields as well as electric charges existing (at any point or in a volume), and set forth stringent requirements that the fields must satisfy. These celebrated equations are given as follows:

Differential form   Integral form

1) Ampere's Law: $\text{Curl}\overline{H} = \overline{J} + \dfrac{\partial \overline{D}}{\partial t}$, $\oint_C \overline{H} \cdot \overline{d\ell} = I + \int_S \dfrac{\partial \overline{D}}{\partial t} \cdot \overline{dS}$

2) Faraday's Law: $\text{Curl}\overline{E} = \dfrac{-\partial \overline{B}}{\partial t}$, $\oint_C \overline{E} \cdot \overline{d\ell} = -\int_S \dfrac{\partial \overline{B}}{\partial t} \cdot \overline{dS} = -\dfrac{d\Phi}{dt}$

3) Gauss's Law (electric): $\text{Div}\overline{D} = \rho$, $\oint_S \overline{D} \cdot \overline{dS} = \int_V \rho dv = Q$

4) Gauss's Law (magnetic): $\text{Div}\overline{B} = 0$, $\oint_S \overline{B} \cdot \overline{dS} = 0$

From these equations, Maxwell predicted the existence of electromagnetic waves whose later discovery made radio possible. He showed that where a varying electric field exists, it is accompanied by a varying magnetic field induced at right angles, and vice versa, and the two form an electromagnetic field pair that could propagate as a transverse wave. He calculated that in a vacuum, the speed of the wave was given by $1/\sqrt{(\varepsilon_0\mu_0)}$, where ε_0 and μ_0 are the permittivity and permeability of vacuum. The calculated value for this speed was in remarkable agreement with the measured speed of light, and Maxwell concluded that light is propagated as electromagnetic waves.

Mechanics

The totality of the three categories of application mass: a) Generalized application Mass; b) Technical application mass, and c) Personalized application mass. See also classical mechanics and quantum mechanics.

Microelectronics
The body of electronics that is associated with or applied to the realization of electronic systems from extremely small electronic parts.

Microstrip Line
A microwave transmission line that is composed of a single conductor supported above a ground plane by a dielectric.

Microwave Integrated Circuit (MIC)
A circuit that consists of an assembly of different circuit functions that are connected by Microstrip transmission lines. These different circuits all incorporate planar semiconductor devices, passive lumped elements, and distributed elements.

Microwaves
Waves in the frequency range of 1 GHz to 300 GHz.

Millimeter Waves
Electromagnetic radiations in the frequency range of 30 to 300 GHz, corresponding to wavelengths ranging from 10 mm to 1 mm.

Model
A physical (e.g., a small working replica), abstract (e.g., a procedure) or a mathematical representation (e.g., a formula) of a process, a device, a circuit, or a system and is employed to facilitate their analysis.

Monolithic Integrated Circuit
An integrated circuit that is formed in a single block or wafer of semiconductor materials. The term is derived from Greek, "monolithos", which means "made of one stone."

Monolithic Circuits
Are integrated circuits entirely on a single chip of semiconductor.

Monolithic Microwave Integrated Circuit (MMIC)
A microwave circuit obtained through a multilevel process approach comprising of all active and passive circuit elements as well as interconnecting transmission lines, which are formed into the bulk or onto the surface of a semi-insulating semiconductor substrate by some deposition scheme such as epitaxy, ion implantation, sputtering, evaporation, diffusion, etc.

Monumental Discovery

Any of the six major un-ravelings or breakthroughs of knowledge about a significant phenomenon in the field of electricity, which shifted the subject in a substantial way and expanded all of the hitherto knowledge amply.

Natural Laws

A body of workable principles considered as derived solely from reason and study of nature.

Neper (Np)

A unit of attenuation used for expressing the ratio of two currents, voltages, or fields by taking the natural logarithm (logarithm to base e) of this ratio. If voltage V_1 is attenuated to V_2 so that $V_2/V_1 = e^{-N}$, then N is attenuation in Nepers (always a positive number) and is defined by: N (Np)$=\log_e(V_1/V_2)=\ln(V_1/V_2)$, where $V_1 > V_2$.

Neutron

One of uncharged stable elementary particles of an atom having the same mass as a proton. A free neutron decomposes into a proton, an electron, and a neutrino. A neutrino is a neutral uncharged particle but is an unstable particle since it has a mass that approaches zero very rapidly (a half-life of about 13 minutes).

Network

A collection of electric devices and elements (such as resistors, capacitors, etc.) connected together to form several interrelated circuits.

Newton (N)

The unit of force in MKSA system of units equal to the force that imparts an acceleration of one m/s^2 to a mass of one kilogram.

Noise

Random unwanted electrical signals that cause unwanted and false output signals in a circuit.

Nomenclature
The set of names used in a specific activity or branch of learning; terminology.

Nonlinear
Having an output that does not rise and fall in direct proportion to the input.

Nucleus
The core of an atom composed of protons and neutrons, having a positive charge equal to the charge of the number of protons that it contains. The nucleus contains most of the mass of the atom, pretty much like the sun containing most of the mass of the solar system.

O

Occam's (or Ockham's) Razor Doctrine
A principle that assumptions introduced to explain a thing must not be multiplied beyond necessity. In simple terms, it is a principle stating that the simplest explanation of a phenomenon, which relates all of the facts, is the most valid one. Thus by using the Occam's razor doctrine a complicated problem can be solved through the use of simple explanations, much like a razor cutting away all undue complexities (after William of Occam, an English philosopher, 1300-1349, who made a great effort to simplify scholasticism).

Oersted (Oe)
The unit of magnetic field in the cgs system of units equal to the field strength at the center of a plane circular coil of one turn and 1-cm radius when there is a current of $10/2\pi$ ampere in the coil.

Ohm (Ω)
The unit of resistance in the MKSA system of units equal to the resistance between two points on a conductor through which a current of one ampere flows as a result of a potential difference of one volt applied between the two points.

Ohm's Law
The potential difference V across the resistor terminals is directly proportional to the electrical current flowing through the resistor. The proportionality constant is called resistance (R); i.e., V=RI. Ohm's Law can also be interpreted as the conversion of potential energy (V) into kinetic energy (I), which is a simple statement expressing the principle of conservation of energy.

Original Postulates
A series of exact postulate (space, energy, change) that have gone into the construction of the physical universe. See primary postulates.

Oscillator
An electronic device that generates alternating-current power at a frequency determined by constants in its circuits.

Output
The current, voltage, power, or driving force delivered by a circuit or device.

P

Paramagnetics
are materials (such as aluminum, beryllium, etc.), which accept magnetism.

Particle
Any tiny piece of matter, so small as to be considered theoretically without a magnitude (i.e., zero size), though having mass, inertia, and the force of attraction. Knowing zero size is an absolute and thus impossible in the physical universe, practical particles range in diameter from a fraction of angstrom (as with electrons, atoms, and molecules) to a few millimeters (as with large raindrops).

Passive
A component that may control but does not create or amplify electrical energy.

Perfect Conductor
Is a conductor having infinite conductivity or zero resistivity.

Personalized Application Mass (P.A.M.)

Is the category of application mass, which has been created and is based solely upon the viewpoint's own postulates and considerations. Examples of this category include such things as one's own customized possessions, any piece of artwork or music, one's own body characteristics (such as hairdo, clothing, shape, etc.), a book's layout or cover design, so on and so forth. see also **application mass, Technical application mass, and Generalized application mass.**

Phase

The angular relationship of a wave to some time reference or other wave.

Phase Constant

The imaginary component of the propagation constant for a traveling wave at a given frequency.

Phasor

A result of a mathematical transformation of a sinusoidal waveform (voltage, current, or EM wave) from the time domain into the complex number domain (or frequency domain) whereby the magnitude and phase angle information of the sinusoid is retained.

Physical Universe (Also Called Material Universe; The Universe)

Is a universe based upon three postulates, called original postulates (space, energy and change) and has four main components (matter, energy, space and time).

Plane Wave

A wave whose wavefronts are plane surfaces and normal to the direction of propagation.

Plating

See electroplating.

PN Junction

An abrupt transition between p-type and n-type semiconductor regions within a crystal lattice. Such a junction possesses specific electrical properties such as the ability to conduct in only one direction, and is used as the basis for semiconductor devices, such as diodes, transistors, etc.

Port
Access point to a system or circuit.

Postulate
a) (NOUN) is an assumption or assertion set forth and assumed to be true unconditionally and for all times without requiring proof; especially as a basis for reasoning or future scientific development; **b) (VERB)** To put forth or assume a datum as true or exist without proof.

Potential Difference (or Voltage)
The electrical pressure or force between any two points caused by accumulation of charges at one point relative to another, which has the capability of creating a current between the two points.

Potential Energy (P.E.)
Any form of stored energy that has the capability of performing work when released. This energy is due to the position of particles relative to each other.

Power
The rate at which work is performed; i.e., the rate at which energy is being either generated or absorbed.

Primary Postulates
A series of four postulates derived from original postulates. These postulates are responsible for the four basic components of the physical universe: matter, energy, created space, and mechanical time. See original postulates.

Principle
A rule or law illustrating a natural phenomenon, operation of a machine, the working of a system, etc.

Processing
The act of converting material from one form into another more desired form, such as in integrated circuit fabrication where one starts with a wafer and through many steps ends up with a functional circuit on a chip.

Propagation
The travel of electromagnetic waves through a medium.

Propagation Constant
A number showing the effect (such as losses, wave velocity, etc.) a transmission line has on a wave as it propagates along the line. It is a complex term having a real term called the *attenuation factor* and an imaginary term called the *phase constant.*

Proton
An elementary particle, which is one of the three basic subatomic particles, with a positive charge equivalent to the charge of an electron (q= +1.602x10^{-19} C) and has a mass of about 1.67x10^{-27} kg with a spin of ½. Proton together with neutron is the building block of all atomic nuclei.

Pulse
A variation of a quantity, which is characterized by a rise to a certain level (amplitude), a finite duration, and a decay back to the normal level.

Pyramid of Knowledge
Workable knowledge forms a pyramid, where from a handful of common denominators efficiently expressed by a series of basic postulates, axioms and natural laws, which form the foundation of a science, an almost innumerable number of devices, circuits and systems can be thought up and developed. The plethora of the mass of devices, circuits and systems generated is known as the "application mass", which practically approaches infinity in sheer number.

Q

Quantum Mechanics (Also Called Quantum Physics or Quantum Theory)
Is the study of atomic structure which states that an atom or molecule does not radiate or absorb energy continuously. Rather, it does so in a *series of steps, each step being the emission or absorption of an amount of energy packet (E) called a quantum.* Quantum physics is the modern theory of matter, electromagnetic radiation and their interaction with each other. It differs from

classical physics in that it generalizes and supersedes it, mainly in the realm of atomic and subatomic phenomena.

Quark
A hypothetical basic particle having a fraction of charge of an electron (such as 1/3 or 2/3) from which many of the elementary particles (such as electrons, protons, neutrons, mesons, etc.) may be built up theoretically. No experimental evidence for the actual existence of free quarks has been found.

R

Radio Frequency (RF)
Any wave in the frequency range of a few kHz to 300 MHz, at which coherent electromagnetic radiation of energy is possible.

Rankine (°R)
The unit of measurement of temperature in the absolute scale (based on Fahrenheit temperature scale), in which the absolute zero is at -459.67 °F. *See also* **temperature**.

Reactance
Is a parameter that is the measure of the opposition to the flow of alternating current (Symbolized by X).

Reactive Element
Is an element, which impedes the flow of current in a wire. An inductor or a capacitor are reactive elements. A purely reactive element does not dissipate energy as does a resistor, but stores it in the associated electric and/or magnetic fields.

Rectifier
Is a device having an asymmetrical conduction characteristic such that current can flow in only one direction through the device.

Reflected Waves
The waves reflected from a discontinuity back into the original medium, in which they are traveling.

Reflection Coefficient
The ratio of the reflected wave phasor to the incident wave phasor.

Resistance
A property of a resistive material that determines the amount of current flow when a voltage is applied across it. The resistor value is dependent upon geometrical dimensions, material, and temperature.

Resistor
A lumped bilateral and linear element that impedes the flow of current, i(t), through it when a potential difference, V (t), is imposed between its two terminals. The resistor's value is found by: R=V(t)/i(t).

Resonant Frequency
The frequency at which a given system or circuit will respond with maximum amplitude when driven by an external sinusoidal force.

Right-Hand Rule
For a current-carrying wire, the rule that if the fingers of the right hand are placed around the wire so that the thumb points in the direction of the current flow, the finger curling around the wire will be pointing in the direction of the magnetic field produced by the wire.

S

Science
A branch of study concerned with establishing, systematizing, and aligning laws, facts, principles, and methods that are derived from hypothesis, observation, study and experiments.

Semiconductor
A material having a resistance between that of conductors and insulators, and usually having a negative temperature coefficient of resistance.

Signal
An electrical quantity (such as a current or voltage) that can be used to convey information for communication, control, etc.

Silicon (Si)
A semiconductor material element in column IV of the periodic table used as in device fabrication.

Sinusoidal
Varying in proportion to the sine or cosine of an angle or time function. For example, the ordinary AC signal is a sinusoidal.

Small Signal
A low-amplitude signal that covers such a small part of the operating characteristic curve of a device that operation is nearly always linear.

Solid-State Device
Any element that can control current without moving parts, heated filaments, or vacuum gaps. All semiconductors are solid-state devices, although not all solid-state devices (such as transformers, ferrite circulators, etc.) are semiconductors.

Space (Also Called Created Space)
The continuous three-dimensional expanse extending in all directions, within which all things under consideration exist.

Standing Wave
A standing, apparent motionless-ness, of particles causing an apparent no out-flow, no in-flow. A standing wave is caused by two energy flows, impinging against one another, with comparable magnitudes to cause a suspension of energy particles in space, enduring with a duration longer than the duration of the flows themselves.

Standing Wave Ratio (SWR)
The ratio of current or voltage on a transmission line that results from two waves having the same frequency and traveling in opposite directions meeting and creating a standing wave.

Static
(**Adjective**) Pertaining to no-motion or no-change. (**Noun**) Something which is without motion or change such as truth (an abstract concept). In physics, one may consider a very distant star (a physical universe object) a static on a short term basis, but it is not totally correct because the distant star is moving over a long period of time, thus is not truly a static but only an approximation, or a physical analogue of a true static.

Subjective Time
Is the consideration of time in one's mind, which can be a nonlinear or linear quantity depending on one's viewpoint.

Substrate
A single body of material on or in which one or more electronic circuit elements or integrated circuits are fabricated.

Superposition Theorem
This theorem states that in a linear network, the voltage or current in any element resulting from several sources acting together is the sum of the voltages or currents resulting from each source acting alone, while all other independent sources are set to zero; i.e.,
$f(v_1+v_2+\ldots\ldots\ldots+v_n)=f(v_1)+f(v_2)+\ldots\ldots\ldots+f(v_n)$

Supplemental Discovery
Any of the eight subordinate discoveries (along with their magnetic duals), which fill in the gaps left behind by the six monumental discoveries of electricity.

Switch
A mechanical or electrical device that completes or breaks the path of the current or sends it over a different path.

Switching
Is the making, breaking, or changing of connections in an electronic or electric circuit.

Symbiont
An organism living in a state of association and interdependence with another kind of organism, especially where such association is of mutual advantage, such as a pet. Such a state of mutual interdependence is called "symbiosis."

T

Technical Application Mass (T.A.M.)
Is the category of man-made application mass that is produced directly as a result of application of a science using its scientific postulates, axioms, laws and other technical data. Examples include such things as a television set, a computer, an automobile, a power

generator, a telephone system, a rocket, etc. See also **Application mass, Personalized application mass, and Generalized application mass.**

Technology
The application of a science for practical ends.

Temperature
The degree of hotness or coldness measured with respect to an arbitrary zero or an absolute zero, and expressed on a degree scale. Examples of arbitrary-zero degree scales are Celsius scale (°C) and Fahrenheit scale (°F); and examples of absolute-zero degree scales are Kelvin degree scale (based on Celsius degree scale) and Rankine degree scale (based on Fahrenheit degree scale).

Tesla (T)
The unit of magnetic field in the MKSA system of units equal to one Weber per square meter.

TEM (Transverse Electro-Magnetic) Wave
Waves having the electric and magnetic fields perpendicular to each other and to the direction of propagation. These waves have no field components in the direction of propagation.

Theorem
A proposition that is not self-evident but can be proven from accepted premises and therefore, is established as a principle.

Theory
An explanation based on observation and reasoning, which explains the operation and mechanics of a certain phenomenon. It is a generalization reached by inference from observed particulars and implies a larger body of tested evidence and thus a greater degree of probability. It uses a hypothesis as a basis or guide for its observation and further development.

Thermal Noise (Johnson Noise or Nyquist Noise)
The most basic type of noise that is caused by thermal vibration of bound charges and thermal agitation of electrons in a conductive material. This is common to all passive or active devices.

Time (Also Called Mechanical Time or Objective Time)

That characteristic of the physical universe at a given location that orders the sequence of events on a microscopic or macroscopic level. It proceeds from the interaction of matter and energy and is merely an "index of change," used to keep track of a particle's location. The fundamental unit of time measurement is supplied by the earth's rotation on its axis while orbiting around the sun. It can also alternately be defined as the co-motion and co-action of moving particles relative to one another in space. See also subjective time.

Torque

A force that tends to produce rotation or twisting.

Transformer

An electrical device that, by electromagnetic induction, transforms electric energy from one (or more) circuit(s) to one (or more) other circuit(s) at the same frequency, but usually at a different voltage and current value.

Transmission Line (T.L.)

Any system of conductors suitable for conducting electric or electromagnetic energy efficiently between two or more terminals.

Transmitted wave

That portion of an incident wave that is not reflected at the interface, but actually travels from one medium to another.

Two-port network

A network that has only two access ports, one for input or excitation, and one for output or response.

U

Unidirectional

Flowing in only one direction (e.g., direct current).

Unilateral

Flowing or acting in one direction only causing a non-reciprocal characteristic.

Universal Communication Principle (Also Called Communication Principle)

A fundamental concept in life and livingness that is intertwined throughout the entire field of sciences and states that for communication to take place between two or more entities, three elements must be present: a source point, a receipt point, and an imposed space or distance between the two.

Universe (Derived From Latin Meaning "Turned Into One", "A Whole)

Is the totality or the set of all things that exist in an area under consideration, at any one time. In simple terms, it is an area consisting of things (such as ideas, masses, symbols, etc.) that can be classified under one heading and be regarded as one whole thing.

V

Viewpoint

Is a point on a mental plane from which one creates (called postulating viewpoint) or observes (called observing viewpoint) an idea, an intended subject or a physical object.

Volt (V)

The unit of potential difference (or electromotive force) in the MKSA system of units equal to the potential difference between two points for which one Coulomb of charge will do one joule of work in going from one point to the other.

Voltage

Voltage (or potential difference) between two points is defined to be the amount of work done against an electric field in order to move a unit charge from one point to the other.

Voltage Source

The device or generator connected to the input of a network or circuit.

Voltage Standing Wave Ratio (VSWR)

The ratio of maximum voltage to the minimum voltage on a transmission. The standing wave on a line results from two voltage (or current) waves having the same frequency, and traveling in opposite directions.

W

Wafer
A thin semiconductor slice of silicon or germanium on which matrices of microcircuits or individual semiconductors can be formed using manufacturing processes. After processing, the wafer is separated into chips (or *die*) containing individual circuits.

Watt (W)
The unit of power in MKSA system of units defined as the work of one joule done in one second.

Wave
A disturbance that propagates from one point in a medium to other points without giving the medium as a whole any permanent displacement.

Wave Propagation
The travel of waves (e.g., electromagnetic waves) through a medium.

Waveguide
A transmission medium comprised of a hollow conducting tube within which electromagnetic waves are propagated.

Wavelength
The physical distance between two points having the same phase in two consecutive cycles of a periodic wave along a line in the direction of propagation.

Weber (Wb)
The unit of magnetic flux in the MKSA system of units equal to the magnetic flux, which linking a circuit of one turn, produces an electromotive force of one volt when the flux is reduced to zero at a uniform rate in one second.

Work
The advancement of the point of application of a force on a particle.

Index

About the Author

Matthew M. Radmanesh received his BSEE degree from Pahlavi University in electrical engineering in 1978, his MSEE and Ph.D. degrees from the University of Michigan, Ann Arbor, in Microwave Electronics and Electro-Optics in 1980 and 1984, respectively.

He has worked in academia for Kettering University (formerly GMI Engineering & Management) and in industry for Hughes Aircraft Co., Maury Microwave Corp., and Boeing Aircraft Co. He is currently a faculty member in the Electrical and Computer Engineering (ECE) department at California State University, Northridge, CA.

Dr. Radmanesh is a member of Eta Kappa Nu, and Tau Beta Pi Honor societies and a past president (three years) of the SFV Chapter of the IEEE Microwave Theory and Technique (MTT) society. His many years of experience in both microwave industry and academia have led to over 40 technical papers in national and international journals and several design handbooks in microwave engineering, in solid state devices and integrated circuit engineering. His current research interests include the design of RF and Microwave devices and circuits, millimeter-wave circuit applications, photonic engineering as well as engineering and business education in higher principles of investment. He received the distinguished lecturer award at the 1994 IEEE international Microwave Symposium and was awarded twice by IEEE LA council for his contributions to the MTT society (1994, 1995). He also received two awards for commitment and dedication to education from IEEE in 2002, 2003, and 2018.

Dr. Radmanesh won the MPD divisional award while at Hughes Aircraft Co. for his pioneering work in the development and design of solid state millimeter-wave noise sources in Ka-band as well as V-band, and a similar award for his outstanding contributions to the HERF project from Boeing Aircraft Co. He holds two patents for his pioneering work and Novel designs of two millimeter-wave noise sources.

Dr. Radmanesh has authored several popular books including "The Essentials of Lifelong Investing" in 2019, "The Modern Philosophy and Science of Investment" in 2017, "The Gateway to Prosperity System" in 2015, "Advanced Principles of Success and Prosperity" in 2012, "The Ultimate Keys to Success in Business & Science," in 2008, "Cracking the Code of Our Physical Universe," in 2006, and another "The Gateway to Understanding," accompanied by a comprehensive WORKBOOK in 2005, all published by AuthorHouse; as well as another textbook in electronics entitled "Radio Frequency and Microwave Electronics Illustrated" published by Prentice Hall in 2001, with its Chinese edition (ISBN 7-5053-7628-4) published in 2002, and the Korean language translation (ISBN 89-7283-264-2) in 2005.

He has also created and produced fourteen multi-CD audiobooks on education, engineering, sciences and business management, most notable amongst them are: "*The Unique science of Investment,*" "*The crowning Philosophy of Investment,*" "*The Ten Supreme Laws to Power,*" "*The Thirty Monumental Principles to Affluence,*" "*The ten Ultimate Keys to Abundance,*" "*The Twenty Golden Rules to Eminence,*" "*The Million Dollar Concepts in Business,*" "*Cracking the Code Series,*" "*The Superior Foundation for Engineering & Sciences,*" and "*The Essence of Being a Genius.*" His hobbies include chess, philosophy, soccer, and tennis.

Dr. Radmanesh has created a wealth of technical information in electronics, engineering, sciences, education and the business world using the scientific methodology as the main tool. In the scientific communities as well as the business circles around the Globe, he intends to bring about a higher level of understanding about the "basic principles of life and livingness" where the knowledge about specific subjects, such as RF & microwave engineering, sciences, business, etc., forms merely a subset of a larger arena.

Other Books By The Author

Radio frequency and Microwave Electronics
Illustrated, Prentice Hall, 2001.

The Gateway to Understanding: Electrons to
Waves and Beyond, AuthorHouse, 2005.

The Gateway to Understanding: Electrons to
Waves and Beyond WORKBOOK,
Author House, 2005.

Cracking The Code of Our Physical
Universe Waves and Beyond, AuthorHouse,
2006.

**RF & Microwave Design Essentials
AuthorHouse, 2007.**

**Advanced RF & Microwave Circuit
Design, AuthorHouse, 2009.**

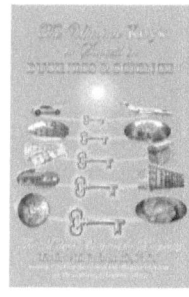

**The Ultimate Keys to Success in Business
& Science, AuthorHouse, 2008.**

For more information or to order any of the books, please visit:
www.KRCbooks.com

The Complete Smith Chart (ZY)

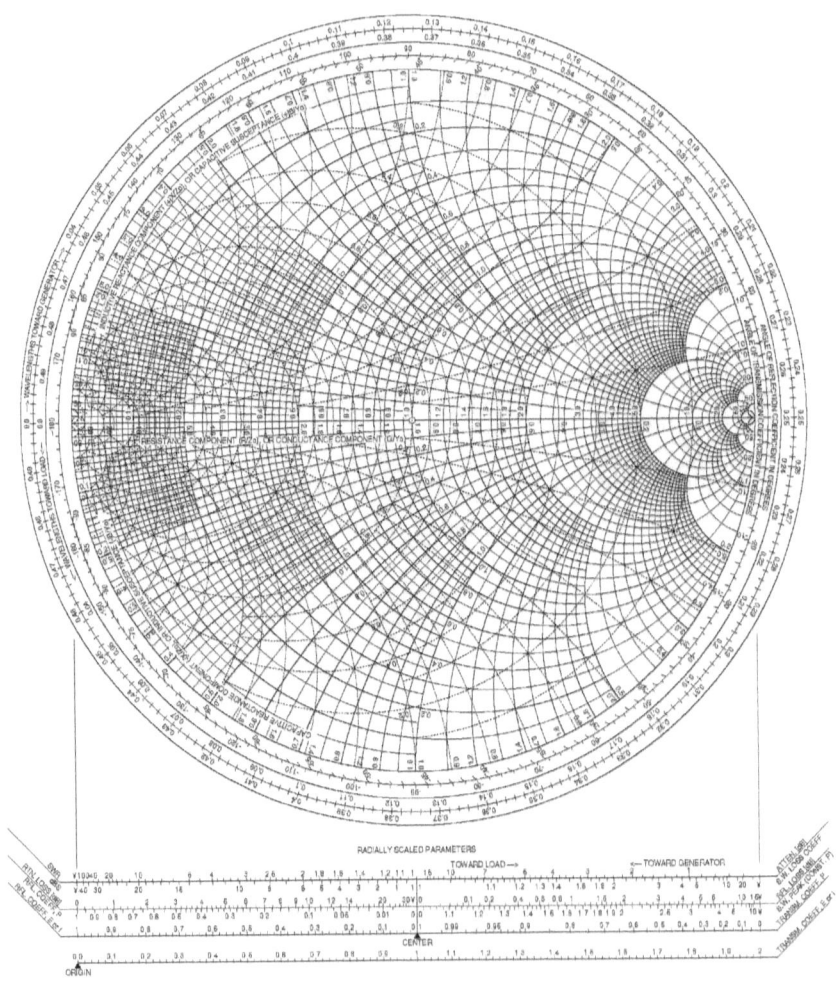

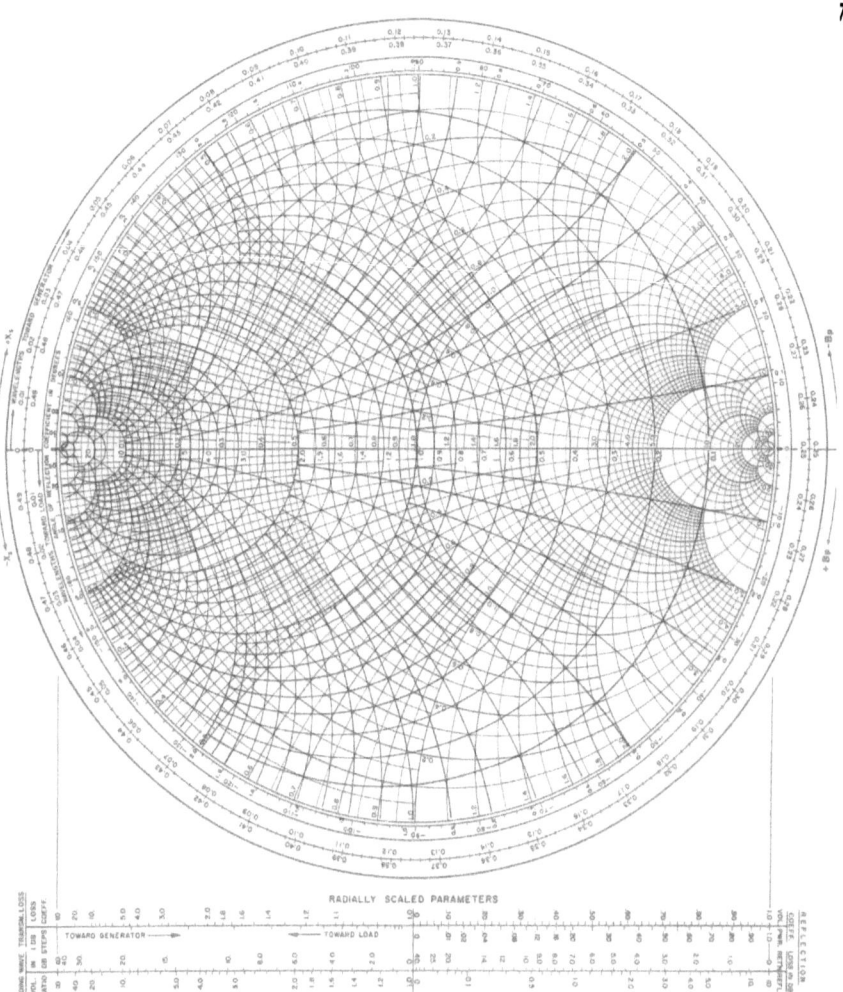

RADIALLY SCALED PARAMETERS

TOWARD GENERATOR ⟶   ⟵ TOWARD LOAD

CENTER

www.ingramcontent.com/pod-product-compliance
Lightning Source LLC
Chambersburg PA
CBHW021347210526
45463CB00001B/6